Climate Change and Sub-Saharan Africa

The Vulnerability and Adaptation of Food Supply Chain Actors

Edited by

John K. M. Kuwornu
School of Environment, Resources and Development,
Asian Institute of Technology, Thailand

Series on Climate Change and Society
Vernon Press

www.vernonpress.com

In the Americas:
Vernon Press
1000 N West Street,
Suite 1200, Wilmington,
Delaware 19801
United States

In the rest of the world:
Vernon Press
C/Sancti Espiritu 17,
Malaga, 29006
Spain

Series on Climate Change and Society

Library of Congress Control Number: 2018961774

ISBN: 978-1-62273-265-4

TABLE OF CONTENTS

INTRODUCTION

The adverse effects of climate change and variability have become some of the biggest environmental and socio-economic challenges for society as a whole and food supply chain actors in particular. Climate change serves as a serious inhibitor to the attainment of food security (i.e., the availability, accessibility, stability and utilization of nutritious food and quality drinking water). Climate change has attracted the attention of the academic community, and governmental and non-governmental organizations. Numerous studies have examined the effects of climate change on food production and the livelihoods of farming households dependent on semi-subsistence agriculture for their continued existence.

Nevertheless, entire food supply chains are believed to be adversely affected by climate change and variability. The key questions are: How vulnerable are food supply chain actors (i.e., input suppliers, farmers, wholesalers, processors, middlemen, exporters, retailers) to climate change and climatic variability? What adaptation strategies are they adopting? How is the resilience of food supply chains being supported? By what means are they adopting the adaptation strategies? Are they being financed and/or supported by international organizations to cope with climate change? And what governmental support are they receiving to help cope with climate change? These and many related questions are addressed in this book.

This book empirically examines these issues to shed light on the effects of climate change across entire food supply chains with special reference to smallholder farmers, and to provide an exposition on the policy environment of climate change adaptation. This is a "must read book" and an essential resource for students, lecturers, researchers, agribusinesses, marketing firms, agricultural institutions, climate change adaptation institutions, policymakers and many others with an interest in agricultural development and the global food industry.

Editor
John K. M. Kuwornu

LIST OF FIGURES

LIST OF TABLES

LIST OF ABBREVIATIONS

Adaptation Policy Framework	APF
African Development Bank	AfDB
Agricultural Training Programme	ATP
Clean Development Mechanism	CDM
Climate Investment Funds	CIF
Climate Vulnerability Index	CVI
Community-based Health Planning and Services	CHPS
Demographic Health Survey	DHS
Divisional Delegate of Agriculture and Rural Development	DDARD
Environmental Protection Agency	EPA
Farmer Based Organization	FBO
Food and Agricultural Organization	FAO
Ghana Statistical Service	GSS
Global Climate Financing Mechanism	GCFM
Global Environmental Facility	GEF
Green Care Association	GCA
Household Vulnerability Index	HVI
Information Communication Technology	ICT
Institute of Statistical, Social and Economic Research	ISSER
Intergovernmental Panel on Climate Change	IPCC
International Air Passenger Adaptation Levy	IAPAL
International Development Research Centre	IDRC
International Institute for Environment and Development	IIED
International Institute for Sustainable Development	IISD
International Institute of Tropical Agriculture	IITA

International Maritime Emission Reduction Scheme	IMERS
Least Developed Countries	LDC
Livelihood Vulnerability Index	LVI
Ministry of Energy, Science, Technology and Innovation	MESTI
Ministry of Food and Agriculture	MoFA
National Association of Counties	NACO
National Adaptation Programme of Action	NAPA
National Oceanic and Atmospheric Administration	NOAA
Non-governmental Organization	NGO
Official Development Assistant	ODA
Organization of Economic Co-operation and Development	OECD
Pilot Program for Climate Resilience	PPCR
Propensity Score Matching	PSM
Roll Black Malaria	RBM
Shisong, Strategic Humanitarian Services	SHUMAS
Special Climate Change Fund	SCCF
Standard International Occupational Prestige Scale	SIOPS
Statistics, Research and Information Directorate	SRID
Strategic Priority for Adaptation	SPA
Third Assessment Report	TAR
United Nation Development Program	UNDP
United Nations Employment Programme	UNEP
United Nations Framework Convention on Climate Change	UNFCCC
Upper Nun Valley Development Authority	UNVDA
West and Central African Council for Agricultural Research and Development	WECARD
World Health Organization	WHO

LIST OF CONTRIBUTORS

John K. M. Kuwornu holds a PhD in Agricultural Economics and Marketing from Wageningen University, Netherlands. He is currently an Associate Professor in Agribusiness Management Field of Study at the School of Environment, Resources and Development, Asian Institute of Technology, Thailand. His areas of teaching and research include: Agribusiness Management, Agricultural Marketing and Trade, Agricultural Development, Agricultural Policy, Marketing Channels, Consumer Behaviour, Supply Chain Management, Commodity Futures Markets, Econometrics, Quantitative Methods, Climate Change, Food Security, Agrobiodiversity and Stock Markets.

Al-Hassan M. Ramatu has a BSc (Hons) Agriculture, MA Agricultural Economics and PhD Agricultural Economics. She is currently an Associate Professor at the University of Ghana, Legon. Her research area is in smallholder agricultural development in Ghana, including analysis of productivity, food security, poverty, and market access. Her areas of teaching are Research Methods, Production Economics, Farm Management, and Agriculture in Development Policy.

Derick T. Adu holds Master of Philosophy degree in Agricultural Economics from University of Ghana, Legon. He works as a Part-Time Lecturer in Faculty of Agriculture and Social Sciences at Anglican University College of Technology (Ghana), and a Project Officer at Freshmacs Ghana Limited. His teaching and research interests include Applied Econometrics, Managerial Economics, Macroeconomics, Environmental Economics, Climate Change Economics, Education Economics, Microeconomics, and Development Studies. He reviews for a number of reputable international journals. He has published in international peer-reviewed journals including Cogent Economics and Finance, Springerplus, Kasetsart Journal of Social Sciences, Cogent Food and Agriculture, and International Journal of Education Economics and Development.

Suhiyini Issah Alhassan is a Ghanaian by birth and Tamale is his home town. He holds Master of Philosophy degree in Agribusiness (University of Ghana), Bachelor of Arts degree in Development Studies (University for Development Studies, Ghana), Diploma in Education (University of Education, Winneba, Ghana) and a certificate in Information Technology (Technonet Ghana Ltd). Suhiyini is an educationist and works with The Ghana Education Service, Ministry of Education Ghana. He is also a researcher and engages in consultancy services for individuals, private enterprises and other Non-governmental Organisations. His research interest and expertise include but not limited to agribusiness, food security, climate change, supply and value chain management, land use management, poverty and small and medium scale enterprises development. He remains focused on developing himself as a distinguished researcher and welcomes new ideas in his field of research. He has an understanding of both quantitative and qualitative analytical tools. Hitherto, he had served as the Saboba district director of the ADPL Nationwide Farming Scheme and also the Team Leader of Enterprise Life Assurance Company Limited, Tamale. Suhiyini Issah Alhassan is currently a PhD student in the Department of Agricultural and Resource Economics, Faculty of Agribusiness and Communication Sciences, University of Development Studies, Tamale, Ghana.

Yaw B. Osei-Asare holds a PhD Development Economics from the University of Bonn, Germany. He is currently a Senior Lecturer in the Department of Agricultural Economics and Agribusiness, University of Ghana, Legon. He teaches Microeconomics and Macroeconomics, Quantitative Methods and Operations Research, and Environmental Economics. His areas of research interest include Environmental, Economic and Social Impact Assessments and Baseline Studies, Resource and environmental economics, ecosystem valuations, trade policy, poverty reduction and general economic policy issues.

Suiven John Paul Tume (Cameroon) has a Bachelor's Degree in Geography (2004), a Postgraduate Diploma in Education (2006) and Master's Degree in Geography (2008) all from the University of Buea. He is currently a PhD research fellow working on indigenous adaptations to climate change in the agriculture and water sectors. He volunteers at Green Care Association, Shisong—Cameroon as a researcher on indigenous climate change adaptation. He is also a consultant and volunteer researcher at Green Future Consulting where he works on climate change vulnerability and adaptation. He is a part-time instructor at the Department of Geography and Environmental Studies of the Catholic University of

Cameroon (CATUC), Bamenda and at the Department of Geography and Planning at the University of Bamenda. He has several publications in diverse subjects on climate change and the environment.

Mbu Dora Nyukighan (Cameroon) holds Bachelor's Degree in English Language from the University of Buea (1996), Post Graduate Diploma in Education from the National Teachers' Institute Kaduna, Nigeria (2009) and a master's degree in African Literature from the Ahmadou Bello University in Zaria, Nigeria (2012). She is a PhD research fellow on 'Gender Representations in Cameroon Drama of English Expression' at the University of Yaounde I. She is a lecturer in the Department of English and Literary Studies, Federal University Lokoja, Nigeria. She has carried out academic research on gender-related environmental issues in Anglophone Cameroon and papers articles on female power and empowerment, traditional and modern education of the Cameroonian child. She is interested vulnerability and adaptation of rural women to climate change in the Northwest Region of Cameroon.

Moye Eric Kongnso (Cameroon) holds a Bachelor's Degree in Geography and Environmental Management (2007) and a Masters Degree in Geography (Applied Agro-climatology, 2011), both from the Dschang. He is currently a PhD fellow. He is a member of a number of professional and research groups in environmental issues. He has co-published articles and book chapters on climate variability and change in the agriculture and water sectors.

Bankui Andrew Dzeaye (Cameroon) holds a Bachelors degree in International Business Management at the Furtwangen University of Applied Sciences, Germany and a Master's degree in International Management at the Fulda University of Applied Sciences, Germany. He has been Senior Project Manager in SHUMAS - Cameroon (Strategic Humanitarian Services - Cameroon) in partnership with the German Technical Cooperation (GIZ) for expatriates. He is an assistant lecturer at the Department of Agricultural Economics of the Catholic University of Cameroon (CATUC), Bamenda. He is currently working on his PhD thesis on economic and social impacts of adopting soil fertility management technology: the case of western highlands of Cameroon at the Faculty of Agronomy and Agricultural Sciences, University of Dschang.

Mairong Frederick Nsaikii (Cameroon) holds a higher teachers' diploma in biology, a bachelor's degree in biology and a master's degree in plant biology. He has several years of teaching experience in secondary and high schools. He is currently enrolled for a PhD at the Faculty of Agronomy and

Agricultural Sciences, University of Dschang. His research area is on climate change and food security in the Northwest region of Cameroon.

Njodzeka Gilbert Njodzeka (Cameroon) has a Diploma in Tree Nursery and Soil Conservation (1997) at Rural Training Centre (RTC), Mfonta. He is the Coordinating Manager and Head of Project Development at Green Care Association, Shisong—Cameroon. He is a crusader in combating environmental degradation through tree planting, apiculture, water catchment protection, developing renewable energy and animal husbandry through non-conventional livestock. He has won several international awards such as Bui Family Union, New Jersey Chapter–Appreciation of Service, 2010; United Kingdom-Commonwealth Mid-Career Fellowship (2010) and United Kingdom-World Environment Radio-First Prize, Best Feedback (2008).

Ali Essossinam is a lecturer. Essossinam teaches International Negotiation of Agricultural Products, Agricultural Finance, Microeconomics and Macroeconomics in Faculty of Economics and Management Sciences at University of Kara (TOGO) and Production Economics and Agricultural Policy in School of Agriculture at University of Lomé (TOGO). He has his Ph.D. in Applied Agricultural Economics and Policy from University of Ghana, Legon, Accra, (GHANA) and MSc in Applied Macroeconomics from University of Abomey Calavi-UAC, (BENIN) under the New Graduate Interuniversity Program (NPTCI). He got his MA and B.Sc degree in Economics from University of Kara (TOGO). The areas of research interests of Dr. ALI include Climate Change and Poverty, the Analysis of demand of Weather Index-based Insurance in Developing countries, the Economic Valuation of Non-market Commodities, Food Security Issues, Economic Impact Valuation Research, Agricultural Finance, and Environmental Issues.

Sampson Osei is a PhD candidate in Development Studies at the University of the Western Cape, South Africa. He holds Masters Degrees in Development Management and Development Studies from the Ruhr University, Germany and the University of the Western Cape, respectively.

Abdulrazak Karriem is a Senior Lecturer in the Institute for Social Development at the University of the Western Cape in Cape Town, South Africa. He received his PhD in City and Regional Planning from Cornell University. Prior to entering academia, he worked for the Department of Land Affairs in the Nelson Mandela government.

Abugri A. Stephen is an MPhil Graduate of the Agricultural Economics Programme in the University for Development Studies (UDS), Tamale, Ghana. He has a strong background in action and development research.

Dr. Joseph Amikuzuno holds a PhD Degrees in Agricultural Economics and is an expert in Price Transmission and Climate Change Impact analyses. He is currently a Senior Lecturer and the Head of the Climate Change and Food Security Department of the UDS.

Dr. Osman Tahidu Damba holds a PhD in Agricultural Economics from Ataturk University, Turkey, an MPhil in Agribusiness from the University of Ghana and BSc. Agricultural Technology from the University for Development Studies, Tamale, Ghana. Dr. Damba is currently a Lecturer at the Department of Climate Change and Food Security, Faculty of Agribusiness and Communication Sciences (FACS) at the University for Development Studies. He is focused on food security and climate change dynamics, Agricultural Markets, time series forecasting for volatility, value chains, FBO Development and agribusiness start-ups in developing countries. Dr. Damba has firm and extensive understanding in applied research and hands-on field practical techniques. He has extensive experience in the not-for-profit sector and has had stints with local and international organizations through roles at the Association of Church Development Projects (ACDEP) as a Market Access Coordinator, Farmers Organization Network in Ghana (FONG) as Agribusiness Cluster Officer, Pan African Organization for Sustainable Development (POSDEV) as a Project Officer as well as the International Fertilizer Development Center (IFDC) as a Business Development Advisor. Dr. Damba has further worked closely with several local farmer associations in capacity development and also facilitated linkages for Agricultural Commodity companies in Northern Ghana. Dr. Damba currently teaches Social and Environmental Impacts of Climate Change, Food Economics as well as Agricultural Finance and Farm Management at the graduate and postgraduate levels.

Armah Ralph Nii Armah holds an MPhil in Agricultural Economics from the University of Ghana. He is currently a Doctoral Candidate at Kansas State University, USA.

Mohammed Tiyumtaba Shaibu is a Senior Technologist working for Ghana's Council for Scientific and Industrial Research-Animal Research Institute. He had his education in Ghana and holds a BSc in Agricultural Economics and Extension and an MPhil in Agribusiness from the University for Development Studies and the University of Ghana respectively. He has one journal publication and 9 conference papers to his credit. He has attended several workshops/trainings/short courses both locally and internationally. Shaibu's interest is in socioeconomics of livestock production including Crop-Livestock Systems, Agricultural Marketing, Agricultural Commodity Value Chains and Climate Change Adaptation of Farmers. He is currently implementing a project on climate change and variability implications for fodder production and goat value chains in Northern Ghana.

Dr. Franklin Kodzo Avornyo is a Principal Research Scientist who works at the Council for Scientific and Industrial Research – Animal Research Institute, Nyankpala Station near Tamale in Ghana. He holds an MSc. in Sustainable Agriculture and a PhD. His specialty is in Animal Science and has published works in the areas of improving livestock productivity, livestock value chains, integration of crop-livestock systems and climate change and livestock production interface. He has 25 refereed journal publications, 25 edited conference papers, one book and other publications to his credit. His interests include addressing challenges in the livestock sector.

Dr. Adelina Maria Mensah, a Senior Research Fellow at the Institute for Environment and Sanitation Studies (IESS), University of Ghana (UG), has an interdisciplinary background in environmental sciences with a focus on aquatic sciences and the human dimensions that impact on its management and sustainability. Dr. Mensah has worked extensively with numerous national and international institutions and stakeholders on a wide range of environmental and sustainable development issues, including aquatic ecosystem functioning, natural resource management, and climate adaptation. Her current interests are research and innovation in science and technology for promoting sustainable futures.

Dr. Elaine Tweneboah Lawson has over 14 years' experience in teaching, research and academic programme coordination at the University of Ghana. Her teaching and research have focused on coupled human-environment systems at a variety of scales. Her research interests include climate change adaptation, coastal ecosystems, dimensions of poverty, environmental policy processes, gender, and social dimensions of zoonotic diseases. Dr. Lawson has a doctorate degree in Environmental Resource management from the Brandenburg University of Technolgy, Cottbus, Germany.

Prof. Christopher Gordon holds Bachelor of Science and Master of Science degrees in Zoology from the University of Ghana, Legon and a UNESCO Post-Graduate Certificate in Limnology awarded by the Austrian

Academy of Sciences. He has a Doctorate degree in Human Environmental Sciences from King's College, University of London, United Kingdom. He has a distinguished academic career spanning over thirty years at the University of Ghana in the fields of research, teaching as well as advisory and fundraising services at both local and international levels. He has successfully supervised over 100 graduate students, - two of whom have won the Silver Medal of the Ghana Academy of Arts and Sciences for Best Post-graduate thesis in Ghana.

Dr. Avishek Datta is an Associate Professor of Agricultural Systems and Engineering Programme at the Asian Institute of Technology, Thailand. He obtained his PhD in Agronomy from the University of New England, Armidale, New South Wales, Australia and conducted his Postdoctoral Research at the University of Nebraska-Lincoln, Nebraska, USA. He has more than ten years of experience working in multidisciplinary agricultural research and development activities. His main areas of expertise include crop production under conventional and organic farming systems in relation to food security and climate change.

FOREWORD
BY SURESH BABU

Achieving UN Sustainable Development Goals will greatly depend on how developing countries tackle the deleterious consequences of the climate change on the welfare indicators. While the governments of the developing countries are trying to address the issue of climate change and develop measures to mitigate and adapt to the changing environments, they are often faced with limited information on the indicators and causal factors that affect the food system in general and the food supply chains in particular.

In this book, Dr. John Kuwornu compiles a set of studies that deal with several key policy and programmatic issues related to adaptation of the key actors and players of the food supply chains. The chapters of this book address the nature of the adaptive measures the actors of the food supply chains take to manage the effects of climate change. The chapters also address the challenges related to building resilience of the food supply system.

The case studies compiled in this book are also useful as cross-community learning as the context of the chapters take us through various scenarios under which the vulnerable nature of the food systems is managed through adaptive measures. The book will be an essential reference for those involved in analyzing the effects of climate change on the socioeconomic conditions of actors in food supply chains in Africa and rest of the World.

Dr. Suresh Babu
Head, Capacity Strengthening,
International Food Policy Research Institute,
Washington D.C.

FOREWORD BY VENKATACHALAM ANBUMOZHI

One aspect of the climate change that has been studied with fewer details so far is related to its economic impacts along the food value chain. This is due, perhaps, to the intrinsic complexity among the different actors that characterizes long-term social predictions and short-term adaptation measures. The truth is that some communities in the food value chain will suffer severe vulnerabilities from the new climate conditions.

We should worry more about the food value chains and the communities associated with that, which have never enjoyed technological advances but adjusted with ecosystem-based adaptation measures, but now see their economic welfare is threatened by accelerated climate risks. For instance, farming communities from sub-Saharan Africa, whose lands are increasingly becoming less productive and dry, providing testimony of what might happen in other parts of the world such as South East Asia, if the average temperature rises more than 2^0 C. The vulnerability of food value chains to climate change in terms of geography and timescale have diverse difficulties to understand the consequences for the major crops and small-scale farming communities.

This book explores the impacts of climate change on a wide variety of value chains and places in Sub-Saharan Africa, their vulnerabilities, and mitigative and adaptive capacities. In this book, we find a very useful exploration of agronomic and economic approximations connected to these consequences as well as innovative ideas of micro and macro adaptation along the value chain. For sure, the new assessment tools and adaptation measures developed in the chapters of this book will be extremely helpful for building an accurate diagnosis and to achieve a greater resilience of small-scale farming communities, all of which face the long-term consequences of climate change.

Dr. Venkatachalam Anbumozhi
Senior Economist
Economic Research Institute for ASEAN and East Asia (ERIA), Jakarta, Indonesia

FOREWORD BY JOYASHREE ROY

Several recent expert opinion-seeking surveys show that hunger is one of the 17 SDGs with the greatest potential and chance of being solved in the next decade. However, this will need multi-dimensional strategies. IPCC's 'Special report on Global Warming of 1.5°C above pre-industrial levels and the likely impacts and pathways in the context of strengthening the global response to the threat of climate change, sustainable development and efforts to eradicate poverty' reports high confidence in a relatively reduced impact but it still reports an increase in climate-related risks to food security at 1.5°C compared to 2°C and the disproportionate effect on vulnerable populations that are dependent on agriculture and coastal livelihoods in dryland regions and less developed countries. Overlapping risks across livestock, water, energy, and food exacerbate current hazards, exposure and vulnerabilities. Regional examples are limited by the availability of local scale studies. Also, the assessment is confident in the role of redistributive policies to ameliorate adverse impacts on already disadvantaged populations of multiple mitigation options when accelerated and scaled up in the short window of opportunity within the next decade.

This present book, based on mostly field level first-hand primary evidence from within Africa, can be considered as a major contribution to the bridging of this knowledge gap through methodological rigor. The selected articles for this volume bring up one point very clearly that in Africa there is diversity in vulnerability and exposure. Hence, the resultant risks to households, farmer groups, region-specific crop varieties, and subnational administrative units highlight the need for a wide variety of attention and interventions towards adaptive capacity enhancement and risk reduction. The vulnerability of particular crop varieties, communities, indigenous populations, farmer categories by gender and responsible climate parameters are well covered. Offering regional diversity through studies of Ghana, Cameroon, and Togo, high-quality sets of data are presented that could inform global assessments. The book not only focuses on the food production system and producers' vulnerability but also on the vulnerability of the supply chain is also assessed in relation to Uganda. Policy recommendation provides a starting point for a more extensive research agenda to understand what works better under what circumstances.

In this book, the conclusions based on empirical evidence from various studies support the portfolio approach by taking into consideration synergies and tradeoff among multiple SDGs, which are also mentioned in the assessment of the IPCC special report on the global warming of 1.5°C.

The new granular insights brought up by the various articles in the book are very useful for informing the larger global debate. From an intellectual point of view, the editor's compilation of articles that study various regions of Africa and the arrangement of each article in a similar format helps in the understanding of the vulnerability and adaptation of farmers and other food supply chain actors to climate change.

The articles in this book will be of considerable interest and use to the authors of the IPCC Sixth Assessment Report and the development community including academia and decision makers like national governments, bilateral and multilateral funding agencies, and civil societies engaged in developmental actions in the region. This book will generate interest for further studies in Africa and in other countries in order to advance deeper regional and global debate, struggle and understanding.

Joyashree Roy
Bangabandhu Chair Professor
Asian Institute of Technology, Thailand
Professor of Economics (On lien) and Founder Advisor to Global Change Programme & SYLFF-JU Programme of Jadavpur University, India.

PREFACE

This book provides an exposition on the effects of climate change on food supply chains. It seeks to explore the effects of climate change at the farm level, the effects on intermediaries in the food supply chains, and the role of policymakers and international institutions regarding adaptation to climate change.

Chapter 1 by Derick T. Adu and John K. M. Kuwornu provides an overview of the policy environment of climate change adaptation.

Chapter 2 by Suhiyini Issah Alhassan, Yaw Bonsu, Osei-Asare, and John K.M. Kuwornu examines the vulnerability of women rice farmers to climate variability in the Northern Region of Ghana using the Livelihood Vulnerability Index Approach.

Chapter 3 by Suiven John Paul Tume, Mbu Dora Nyuykighan, Moye Eric Kongnso, Bankui Andrew Dzeaye, Mairong Frederick Nsaikii, and Njodzeka Gilbert Njodzeka assesses food crop vulnerability to due climate variability and change at the household level in Bui Division in the northwest of Cameroon.

Chapter 4 by Ali Essossinam examines the farm households' perception of climate change, traditional beliefs and determinants of the adaptation decisions in northern Togo.

Chapter 5 by Sampson Osei and Abdulrazak Karriem examines social capital and climate change adaptation among smallholder farmers in the Central Region of Ghana.

Chapter 6 by Suhiyini Issah Alhassan, Yaw Bonsu Osei-Asare, and John K.M., Kuwornu examines the factors influencing women rice farmers' vulnerability to climate change in the Northern Region of Ghana.

Chapter 7 by Armah Ralph Nii Armah, Al-Hassan M. Ramatu, John K. M. Kuwornu examines the impact of participation in climate change projects on maize farmers' resilience to climatic shocks, yields and income.

Chapter 8 by Mustapha Abubakar Sadiq, Al-Hassan M. Ramatu, John K. M. Kuwornu examines the vulnerability of smallholder maize farming households to climate variability in the Eastern Region of Ghana.

Chapter 9 by Suhiyini Issah Alhassan, John K.M. Kuwornu, and Yaw Bonsu Osei-Asare examines the climate change adaptation strategies by women rice farmers in the Northern Region of Ghana.

Chapter 10 by Mohammed Tiyumtaba Shaibu, Suhiyini Issah Alhassan, Franklin Kodzo Avornyo, Elaine Tweneboah Lawson, Adelina Mensah and Christopher Gordon explores the perceptions and determinants of adoption of indigenous strategies for adaptation to climate change by smallholder livestock farmers in the North-Western Region of Ghana.

Chapter 11 by Joseph Amikuzuno, John K. M. Kuwornu and Damba, T. Osman examines gender-based climate change impacts and adaptation strategies among smallholder farmers in northern Ghana.

Chapter 12 by Derick T. Adu, John K. M. Kuwornu and Avishek Datta explores smallholder maize farmers' constraints to climate change adaptation strategies in the Brong-Ahafo Region of Ghana.

Chapter 13 by John K. M. Kuwornu and Derick T. Adu provides a general discussion on the vulnerability and adaptation of food supply chain actors to climate change.

Finally, chapter 14 by John K. M. Kuwornu provides conclusions and recommendations regarding the vulnerability and adaptation of food supply chain actors to climate change.

ACKNOWLEDGEMENTS

I would like to thank all the authors for their invaluable contributions this book. My special appreciations also go to the reviewers of the individual chapters and the entire book. I am really grateful to Mr. Derick T. Adu for formatting the text in the book. My special gratitude also goes to Vernon Press for publishing this book. Thanks also to Dr. John Meadows for proofreading the final draft of the manuscript.

CHAPTER 1

The Policy environment of climate change adaptation

Derick T. Adu and John K.M. Kuwornu

1.1 Introduction

Natural and man-made systems, infrastructure and economies are likely to be adversely affected by climate change. Uncertainties surrounding climate change necessitate comprehensive responses and policy strategies at the local, regional, national and global levels. Two basic response strategies have been emphasized by the United Nations Framework Convention on Climate Change (UNFCCC): mitigation and adaptation. According to Fussel and Klein (2002), while mitigation strives to limit climate change by decreasing greenhouse gas emissions and enriching 'sink' opportunities, adaptation aims to alleviate the negative effects through system-specific actions. In the quest to tackle the climate change challenge and create an effective and all-encompassing international climate change regime, mitigation and adaptation measures must be put in place. More interest has been drawn to mitigation in the past (i.e., around 1990), both in scientific and policy debates. This could be attributed to the fact that climate change is a problem linked to lasting interruptions to global geo-biochemical cycles and related impacts on the climate system (Cohen et al., 1998).

In recent years, attention has been devoted to climate change adaptation following the Intergovernmental Panel on Climate Change's (IPCC) Third Assessment Report (TAR). Parry, Hammill and Drexhage (2005) suggested that adaptation has currently surfaced as a crucial policy priority, stimulating action both within and outside the climate change negotiations. Adger, Khan and Brooks (2003) cited in Dalei (2016, p. 5) also noted that "adaptation encompasses natural resource management, strengthening food security, development of social and human capital, and reinforcement of institutional systems". The IPCC (2001) argued that regardless of the robust nature of mitigation measures, a certain level of climate change is unavoidable owing to historical emissions in addition to the inertia of the climate system and that the manifestation of mitigation measures takes a long time whiles adaptation actions take effect almost immediately. There are quite a number of other adaptation definitions. Smithers and Smit (1997) define it as "changes in a system in response to some force or perturbation, in our case related to climate" (p.135), while Pielke (1998) suggest it is an "adjustment in individual, group and institutional behaviour in order to reduce society's vulnerabilities to climate" (p. 159).

The IPCC (2001) also defines adaptation as "adjustments in ecological, social or economic systems in response to actual or expected stimuli and their effects or impacts. This term refers to changes in processes, practices and structures to moderate potential damages or to benefit from opportunities associated with climate change" (p. 89). Adaptation can be classified into reactive or anticipatory, private or public, and planned or autonomous contingent on the goal, timing and reasons for implementation. It can also be categorized as short- or long-term, and localized or widespread (IPCC, 2001). Adaptation is a "process by which strategies to moderate, cope with, and take advantage of the consequences of climatic events are enhanced, developed and implemented" (UNDP, 2004, p. 248). Adaptation is the "process or outcome of a process that leads to a reduction in harm or risk of harm or realization of benefits associated with climate variability and climate change" (UK Climate Impact Programme" (UKCIP), 2005, p. 11).

1.2 Policies

Numerous policies have been developed, for instance, the Marrakesh Funds[1] at the CoP7 (Conference of Paris), the Delhi Declaration at CoP8 which reaffirmed economic and social development and poverty eradication, and the Buenos Aires Programme at Work on Adaptation and

---

[1] The LDC Fund (Least Developed Country) and SCCF (Special Climate Change Fund) to be managed by the Global Environmental Facility (GEF) and Adaptation Fund under the Kyoto Protocol.

Response Measures at CoP10[2]. A five-year Work Programme on Adaptation was agreed upon at the CoP11 in Montreal in 2005 organized by the United Nations Department of Economic and Social Affairs Division for Sustainable Development). Several other climate change adaptation programmes have been instituted. These are the UK's Climate Impacts Programme, Canada's Climate Change Impacts and Adaptation Programme, and the Caribbean Community Climate Change Centre. Smit and Pilifosova (2001) argued that for an adaptation project to be effective, it must decrease the vulnerability of the system and build in the potential to anticipate and act on future climatic changes. It must also be consistent with indigenous environmental conditions and satisfy the needs of the local dwellers. Importantly, every response and measure of the project must be 'mainstreamed' into development and poverty alleviation processes.

Adaptation strategies are crucial for susceptible people. This is because according to Downing, Ringius, Hulme and Waughray (1997), a failure to adapt to the adverse impacts of climate change results in "considerable deprivation, social disturbance and population displacement, and even morbidity and mortality" (p. 30). One challenge in choosing an adaptation strategy is identifying whether it will be beneficial to the target group. For example, strategies such as large-scale agriculture, irrigation and hydroelectric development may benefit large groups, or national interests, but may harm local, poor, Indigenous populations. This suggests that adaptation may not produce the same benefit for everyone and "win-win" circumstances are very unlikely when it comes to climate change adaptation. Thus, some segments of society will lose, and others will benefit. It is therefore imperative to consider the secondary effects of adaptation, as well as the losses suffered by groups avoided or relegated (Kates, 2000). Table 1.1 presents a global typology of climate change adaptation measures.

---

[2] The decision at CoP10 adds to earlier work on adaptation undertaken by the SBI (Subsidiary Body for Implementation) and SBSTA (Subsidiary Body for Scientific and Technical Advice). It initiates a process for the development of a new 5-year work programme within the SBSTA.

Table 1. 1: Typology of climate change adaptation action

Adaptation category	Description	Examples of actions in the category	Similar classification in the literature
Capacity Building	Developing human resources, institutions and communities, and equipping them with the capability to adapt to climate change.	Training/workshops for knowledge/skills development, public outreach and education, dissemination of information to decisionmakers/stakeholders, identification of best practices, and development of training materials.	Educational/informational (Smit and Skinner, 2002; Wilbanks and Kates, 1999; Huq et al., 2004; Smit et al., 2000; Carter et al., 1994; Tompkins et al., 2010)
Management and Planning	Incorporating understanding of climate science, impacts, vulnerability and risk into government and institutional planning and management.	Developing an adaptation plan, livelihood diversification, drought planning, coastal planning, ecosystem-based planning, and changing natural resource management.	Administrative/institutional/ organizational (Smit and Skinner, 2002; Wilbanks and Kates, 1999; Huq et al., 2004; Smit et al., 2000; Carter et al., 1994; Tompkins et al., 2010)
Practice and Behavior	Revisions or expansion of practices and on-the-ground behaviors that are directly related to building resilience.	Soil/land management techniques, climate-resilient crops or livestock practices, post-harvest storage, rainwater collection, and expanding integrated pest management.	Behavioral (Smit and Skinner, 2002; Wilbanks and Kates, 1999; Huq et al., 2004)
Policy	The creation of new policies or policy and regulatory reviews to allow flexibility to adapt to a changing climate.	Mainstreaming adaptation into development policies, land-use specific policies, improvement of water resource governance, revised design parameters, and ensuring compliance with existing regulations.	Legislative/Legal (Smit et al., 2000; Carter et al., 1994)

Information	Systems for communicating climate information to help build resilience towards climate impacts (other than communication for early warning systems).	Decision support tools, communication tools, data acquisition efforts, digital databases, and remote communication technologies.	Infrastructural/structural (Smit et al., 2000; Carter et al., 1994) Educational/informational (Smit and Skinner, 2002; Wilbanks and Kates, 1999; Huq et al., 2004; Smit et al., 2000; Carter et al., 1994)
Physical infrastructure	Any new or improved hard physical infrastructure aimed at providing direct or indirect protection from climate hazards.	Climate-resilient buildings, reservoirs for water storage, irrigation systems, canal infrastructure, and sea walls.	Infrastructural/structural (Smit et al., 2000; Carter et al., 1994)
Warning or observing systems	Implementation of new or enhanced tools and technologies for communicating weather and climate risks, and for monitoring changes in the climate system.	Developing, testing and deploying monitoring systems, and upgrading weather or hydromet services.	Research and development (Smit et al., 2000; Carter et al., 1994)
"Green" infrastructure	Any new or improved natural infrastructure aimed at providing direct or indirect protection from climate hazards.	Revegetation, afforestation, woodland management, and increased landscape-scale vegetation cover.	Infrastructural/structural (Smit et al., 2000; Carter et al., 1994)

Financing	New financing or insurance strategies to prepare for future climate disturbances.	Insurance schemes, microfinance, and contingency funds for disaster management.	Financial (Smit and Skinner, 2002; Wilbanks and Kates, 1999; Huq et al., 2004; Smit et al., 2000; Carter et al., 1994) Market mechanisms (Smit et al., 2000; Carter et al., 1994)
Technology	Develop or expand climate-resilient technologies.	Technologies to improve water use or water access, solar energy capacity, biogas, water purification, and solar salt production.	Technological (Smit and Skinner, 2002; Wilbanks and Kates, 1999; Huq et al., 2004; Smit et al., 2000; Carter et al., 1994)

Source: Biagini et al. (2014)

In developing any adaptation strategy, policy and measure to build resilience to climate change and climatic variability, a structured approach known as an Adaptation Policy Framework (APF) has been proposed by the Global Environmental Facility (GEF). The APF is defined as "a structural process for developing adaptation strategies, policies and measures to enhance and ensure human development in the face of climate change, including climate variability. The APF is designed to link climate change adaptation to sustainable development and other global environmental issues. It consists of five basic components: scoping and designing an adaptation project, assessing current vulnerability, characterizing future climate risks, developing an adaptation strategy, and continuing the adaptation process" (UNDP, 2004, p. 248).

The APF establishes a foundation by which economies can appraise and amend prevailing planning processes and practices to address climate change impacts (Ebi and Semenza, 2008). This method connects adaptation to sustainable development as well as global environmental issues (UNDP/GEF, 2003). The framework identifies the following four principles on which actions aimed at developing climate change and variability adaptation can be based.

- "adaptation to short-term climate variability and extreme events for reducing vulnerability to long-term climate change";
- "adaptation policies and measures assessed in a developmental context";
- "adaptation occurs at different levels in society, including the local level"; and
- "the adaptation strategy and the process by which it is implemented are equally important" (UNDP/GEF, 2003, p. 4).

The Adaptation Policy Framework is presented in Figure 1.1.

Figure 1. 1: Adaptation Policy Framework

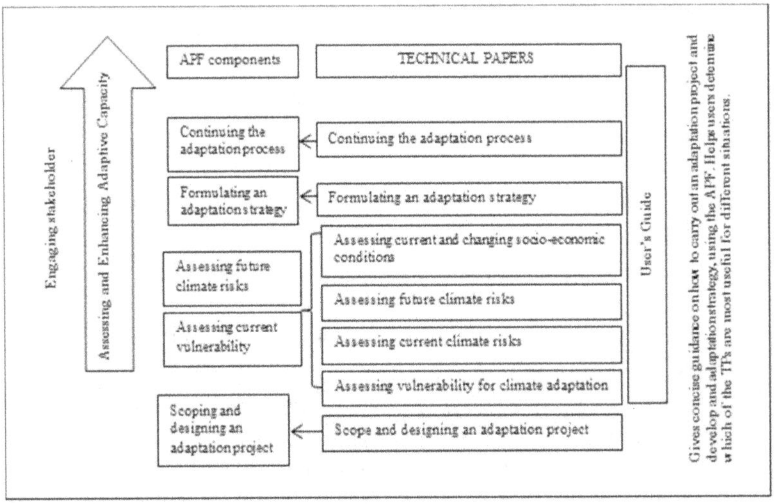

Source:<http://www.undp.org/gef/undp-gef_publications/publications/apf%20sectionI_ugb.pdf>

Usage of the APF is dependent on the expected yield. It can be used to achieve the following outputs. The APF can be expanded to identify policy options to lessen adverse effects of climate change, either to improve society's flexibility or to expand the variety of coping strategies (Lim et al., 2004). This policy option may be aimed at certain aspects of a national development strategy, at specific geographic areas, or at important sectors of the national economy (e.g., agriculture, forestry, water resources, transportation, coastal zone management, public health, ecosystem management and risk management). Adaptation within one sector frequently has effects on another. For example, lessening the effects of drought can result in better nutrition levels and health among the general public. Hence, the APF has been designed to facilitate a process of integrated assessment, including a consultation process in which links between sectors can be identified and assessed. Such assessments can also offer valuable input to national communications under the UNFCCC. The APF process can be used for formulating adaptation projects, or for exploring the potential to add adaptation considerations to other types of projects. These projects can focus on any population scale, from the local to the national level.

The APF has number of components. Lim et al. (2004) summarized each component of the APF as follows:

- *Scoping and designing an adaptation project* includes ensuring that the project is well designed and can be integrated into the national policy process.
- *Assessing current vulnerability* involves an assessment of the present situation. It addresses the questions: "Where does society stand today with respect to vulnerability to climate risks?" "What factors determine its current vulnerability?" And "How successful are its efforts to adapt to current climate risks?".
- *Assessing future climate risks* involves developing scenarios of future climate, vulnerability, and socioeconomic and environmental trends as a basis for considering future climate risks.
- *Formulating an adaptation strategy* involves the creation of a set of adaptation policy options and measures in response to current vulnerability and future climate risks.
- *Continuing the adaptation process* involves implementing, monitoring, evaluating and sustaining the initiatives of the adaptation project.
- *Engaging stakeholders in the adaptation process* is crucial to the successful implementation of adaptation. This cross-cutting component involves creating and sustaining an active dialogue among affected individuals and groups.
- *Assessing and enhancing adaptive capacity*, another cross-cutting component, involves the integration of activities to better cope with climate change, including climatic variability, into national capacity strengthening efforts.

Handmer et al. (1999) argued that poorer economies find it difficult to adapt to climate change since they lack technical and financial abilities. This assertion corroborates Beg et al. (2002) who report that developing economies are more likely to face the most adverse effects of climate change and are less capable of coping with, or adapting to, such changes. The World Bank estimated that climate change adaptation measures within developing economies are likely to cost US$70-100 billion per year between 2010 and 2050 (World Bank, 2010). Table 1.2 presents the annual adaptation costs in developing economies. The estimates range from US$9 billion to $86 billion per year.

Table 1. 2: Annual Adaptation Costs in Developing Countries (US$ billion)

Institution	Assessment Year	Estimated Cost	Time Frame
UNDP	2007	$86	2015
UNFCCC	2007	$28-$67	2030
Oxfam	2007	$50	Present
World Bank	2006	$9-$41	Present

Sources: Human Development Report, UNDP (2007); Economic Aspects of Adaptation to Climate Change: Costs, Benefits, and Policy Instruments, OECD (2008)

Article 4.4 of the UNFCCC posits that developing economies should be supported by developed economies in meeting the costs of adapting to the adverse impacts of climate change. There has been a historical argument that funds for combatting climate change should be provided to developing economies in the form of Official Development Assistant (ODA). Nonetheless, developing countries have received less than 10 percent of the funds promised by developed countries to help them adapt to the impacts of climate change. The poorest economies have received the least, with Africa, the poorest continent, receiving less than 12 percent of all the climate change-related funds spent globally. Although developed countries have together pledged nearly US$18 billion in the last seven years, and despite many world leaders' assertions that the financing is essential, less than US$0.9 billion has been disbursed, and long delays are afflicting many funds. A summary of the UNFCCC climate-related funding mechanism is presented in Table 1.3.

Table 1. 3: Summary of the UNFCCC climate-related funding mechanisms

UNFCCC Climate-Related Funding Mechanism	Brief Description of Instruments Related to Adaptation	Convention Guidance for Integrating Adaptation into Climate Change and Sustainable Development
Global Environment Facility Trust Fund (GEFTF)	Contributions from donor states based on replenishment and established guidelines for project funding through Council meetings based on the general criteria of providing "global environmental benefits." The GEF TF includes the Strategic Priority for Adaptation (SPA), a Pilot program, operational in July 2004, of US $50 million.	Establishes pilot and demonstration projects to show how adaptation planning and assessment can be built into projects that provide real benefits and be integrated into national policy and sustainable development planning (UNFCCC Decision 6/CP.7).
Least Developed Countries Fund (LDCF)	Voluntary contributions by donor nations for adaptation in least developed countries. Policies and procedures established by the LDCF/Special Climate Change Fund (SCCF) Council. The LDCF supports a work program to assist least-developed countries with carrying out, inter alia, the preparation and implementation of National Adaptation Programs of Action.	Promotes the integration of adaptation measures in national development and poverty reduction strategies, plans or policies (UNFCCC Decision 3/CP.11).
Special Climate Change Fund (SCCF)	Two active financing windows to which nations voluntarily contribute, for the benefit of developing countries. The SCCF finances projects relating to four areas: (a) adaptation; (b) technology transfer and capacity building; (c) energy, transport, industry, agriculture, forestry, and waste management; and (d) economic diversification. Policies and procedures established by the LDCF/SCCF Council.	Contributes to the integration of climate change considerations into development activities (UNFCCC Decision 5/CP.9).
Adaptation Fund	Two percent of Clean Development Mechanism (CDM) projects funding supports the Adaptation Fund. Policies and procedures for funding established by the Adaptation Fund board.	

Source: Biagini et al. (2014).

In response to the climate change adaptation funding difficulties for developing countries, at the 7th CoP at Marrakech in 2001, three new funds known as the Marrakech Funds were established. These funds are the Least Developed Country Fund (LDCF), the Special Climate Change Fund (SCCF) and the GEF Adaptation Fund. The LDCF was established to support at least 49 LDCs to design National Adaptation Programmes of Action (NAPAs) over the period 2003–2005. The SCCF was established to aid a comprehensive mandate of adaptation covering areas such as technology transfer, transport, industry, natural resource and waste management, and health issues to help developing countries expand their economies. This fund became operational from 2005 with voluntary contributions. The LDCF and the SCCF are being operated by the GEF Adaptation Fund which was instituted under the Kyoto Protocol to support concrete adaptation projects and programmes. It was financed by a 2 percent levy on Clean Development Mechanism (CDM) projects. Since this financing was reliant upon the CDM generating funds, it seemed doubtful. Mitigation funding is still higher than the international funding for adaptation measures though it has increased in absolute terms recently (Schalatek et al., 2012). Table 1.4 outlines the UNFCCC Adaptation Funds in operation.

Table 1. 4: UNFCCC Adaptation Funds in Operation (US$ Million)

Fund	Total Pledged	Total Received	Project Approvals
LDCF	180	91.8	36.79
SCCF	90	59.9	67.6
GEF Trust Fund	50	50.0	50
Total	320	201.7	154.39

Source: GEF (2008)

Most of the climate change adaptation funding has gone towards impact assessments, capacity building, technical assistance for pilot demonstration activities, and knowledge transfer to facilitate the inclusion of adaptation concerns in development planning. "In November 2008, developed economies promised to contribute US$6.3 billion to the Climate Technology Fund and the Strategic Climate Fund also known as the Climate Investment Funds (CIF) of the World Bank, which also includes an adaptation component called the Pilot Program for Climate Resilience (PPCR). However, only US$240 million has been pledged specifically for the PPCR" (Lagos and Wirth, 2009, p. 26).

There are also dedicated bilateral sources of funding[3]. These are the:

International Climate Initiative (ICI) of Germany: In 2008, the German government auctioned 8.8 percent of its allowable emission permits to businesses, setting aside approximately 30 percent of the revenue to finance domestic and international climate action. Of the resulting €400 million per year, €120 million per year was earmarked for developing countries and countries in transition. Of this, half was intended for adaptation and biodiversity projects.

Adaptation to Climate Change Initiative of Australia: Australia invested AUD$150 million over three years, including AUD$35 million in 2008-2009, to meet high-priority climate adaptation needs in vulnerable countries. The program focused on the Pacific Island countries and East Timor but also included targeted assistance to other countries.

Cool Earth Partnership of Japan: Japan has pledged US$10 billion (JPY 1,250 billion) over five years to support developing countries that are already making efforts to reduce greenhouse gas emissions. Up to US$2 billion (about US$400 million per year) was provided for adaptation and improved access to clean energy in developing countries that are vulnerable to the adverse effects of climate change (e.g., African and Pacific Island countries).

Some developing economies that are already experiencing climate change impacts have begun to establish their own funds for adaptation. For example, Bangladesh has allocated US$40 million from its national budget to create a Trust Fund on Climate Change, inviting donors to contribute. The UK government has pledged US$132 million to this fund. The money will support measures such as protecting houses, schools and farms against flooding, and introducing new crop strains. A similar fund is under development in Bolivia. In addition, Sri Lanka has put in place an environmental levy that will, in part, fund climate change adaptation. Other proposals have been made for new sources of funding for climate change adaptation. Table 1.5 presents these proposed funding sources.

---

3 Facilitating an International Agreement on Climate Change: Adaptation to Climate Change www.globalclimateaction.org. June 2009.

Table 1. 5: Proposed Sources of New Funding for Climate Change Adaptation

Source of Funds	Estimated Revenue	Proposing Parties
Auctioning Permits and Extending Adaptation Levy to the International Emissions Trading Scheme (IET) of the Kyoto Protocol.	US$15 to $25 billion per year. A share of this revenue would be earmarked for adaptation.	Norway
International Air Passenger Adaptation Levy (IAPAL)	US$8 to $10 billion per year in the first five years and more in the longer term.	Maldives on behalf of LDCs.
International Maritime Emission Reduction Scheme (IMERS)	US$10 billion per year from 2013 at the price of US$15 per metric ton of carbon dioxide. A share of this revenue would be earmarked for adaptation.	Endorsed by Norway and some other developed countries; consistent with proposals by India, LDCs, Mexico and Nicaragua (on behalf of Guatemala, Dominican Republic, Honduras and Panama).
World Climate Change Fund (Green Fund)		
Additional 'Earmarked' Official development assistance (ODA)	US$185 billion per year. A share of this revenue would support adaptation.	China
Auction of Emissions Allowances	€50 billion by 2020 (US$80 billion). US$22 billion in 2010. A share of this revenue would support adaptation.	European Union
Global Carbon Adaptation Tax	US$48.5 billion per year in 2010, of which US$18.4 billion would be allocated to adaptation.	Switzerland
Global Climate Financing Mechanism (GCFM)		European Commission and World Bank
Burden Sharing Mechanism	US$39.6 million per year.	Tuvalu

Source: Global Leadership for Climate Action, June 2009
www.globalclimateaction.org

Few studies have investigated the costs of adaptation. The IPCC's Fourth Assessment Report (AR4) noted that there exist little literature on adaptation costs and benefits and what does exist is disjointed (IPCC, 2007). Table 1.6 presents some of the literature on climate change adaptation costs.

Table 1. 6: Global Studies of Climate Change Adaptation Costs

Study	Cost of Adaptation	Regional coverage	Time frame	Sectors	Comments
World Bank (2006)	$9 billion to $41 billion/year.	Developing countries.	Present	Unspecified	Based on OECD and World Bank (WB) analysis of flow exposed to climate change. Costs of reducing climate risks (climate proofing).
Stern Review (2006)	$4 billion to $37 billion/year.	Developing countries.	Present	Unspecified	Update with modification of the WB Study.
Oxfam (2007)	At least $50 billion/year.	Developing countries.	Present	Unspecified	WB study plus extrapolation from NAPA's and NGO estimates.
UNDP (2007)	$86 billion to $109 billion/year.	Developing countries.	2015	Unspecified	WB study, plus costing of targets for adapting poverty reduction programs and strengthening of disaster responses.
UNFCCC (2007)	$28 billion to $67 billion/year.	Developing countries.	2030	Agriculture, forestry, fisheries, water supply, health, coastal zones and infrastructure.	Analysis of specific adaptations.
UNFCCC (2007)	$44 billion to $166 billion/year.	Global	2030	Agriculture, forestry, fisheries, water supply, health, coastal zones and infrastructure.	

Source: Agrawala and Fankhauser. 2008. Presented at the OECD workshop on Economic Aspects of Adaptation.

Numerous policies have been implemented to support adaptation to climate change worldwide. Adaptation became a priority globally at the very start of international climate efforts. In the UNFCCC, all stakeholders generally committed to adopt national adaptation measures and to collaborate in preparing for the impacts of climate change. Countries reached an agreement to manage climate change as a global environmental issue through the UNFCCC on 9 May 1992. The

Convention entered into force on 21 March 1994 with the objective of stabilization of greenhouse gas concentrations in the atmosphere at a level that would prevent dangerous anthropogenic impacts on the global climate system. Table 1.7 outlines a selection of adaptation policies instituted in developing economies (such as Africa and Asia) across the globe.

Table 1. 7: Adaptation Initiatives in Selected Developing Countries

Region/Country	Climate Stress	Adaptation Practices
Africa		
Egypt	Sea-level rise	Adoption of National Climate Change Action Plan integrating climate change concerns into national policies; adoption of law 4/94 requiring Environmental Impact Assessments for project approval and regulating setback distances for coastal infrastructure; installation of hard structures in areas vulnerable to coastal erosion.
Sudan	Drought	Expanded use of traditional rainwater harvesting and water conservation techniques; planting of shelter belts and windbreaks to improve the resilience of rangelands; monitoring of the number of grazing animals and trees cut; establishing revolving credit funds.
Botswana	Drought	National government programs to re-create employment options after drought; capacity building of local authorities; assistance to small subsistence farmers to increase crop production.
Asia		
Bangladesh	Sea-level rise, saltwater intrusion	Consideration of climate change in the National Water Management Plan; building of flow regulators in coastal embankments; use of alternative crops and low-tech water filters.
Philippines	Drought, floods	Adjustment of silvicultural treatment schedules to suit climate variations; shift to drought-resistant crops; use of shallow tube wells; rotation method of irrigation during water shortages; construction of water impounding basins; construction of fire lines and controlled burning; adoption of soil and water conservation measures for upland farming.
Nepal	Glacial lake expansion	The Tsho Rolpa risk reduction project in Nepal is an example of adaptation measures being implemented to address the creeping threat of glacial lake outburst flooding because of rising temperatures, potentially causing serious damage to a dam and to livelihoods downstream.
Small Islands	Sea-level rise	Capacity building for shoreline defense system

(Somoa, Tonga, Cook Islands)		design; introduction of participatory risk assessment; provision of grants to strengthen coastal resilience and rehabilitation of infrastructures; construction of cyclone-resistant housing units; retrofitting of buildings to improved hazard standards; review of building codes; reforestation of mangroves.
	Drought, saltwater intrusion	Rainwater harvesting; leakage reduction; hydroponic farming; bank loans allowing for purchase of rainwater storage tanks.

Source: IPCC WGII Fourth Assessment Report – Final Draft for Government Review, Chapter 17: Assessment of Adaptation Practices, Options, Constraints and Capacity.

Nonetheless, the Convention was incompetent in dealing with climate change and countries adopted the Kyoto Protocol in December 1997. As a result, advanced economy countries or Annex 1 Parties were given lawfully obligatory emission targets to meet. The Kyoto Protocol came into being on November 18, 2004, after 55 countries party to the Convention approved it. These countries comprised many developed nations which have specific emission targets, encompassing 55 percent of their greenhouse gas emissions in 1990 (UN). Annex 1 Parties were given legally binding targets to reduce their greenhouse gas emissions to their 1990 levels by the year 2000. They were also asked to make available national communications, promote and facilitate the transfer of climate-friendly technologies, and financially support emerging economies through the GEF[4]. The Kyoto Protocol comprises the first international environmental agreement that developed market-based instruments to determine cost-efficient responses to the undoubted need for greenhouse gas abatement.

Annex 11 countries under the Kyoto Protocol were also instructed to reduce their greenhouse gas emissions, notably carbon dioxide (CO2) that emanates from fossil fuel combustion. According to the UNFCCC (1998), Annex 11 nations committed themselves to decreasing their greenhouse gas emissions by 5.2 percent on average lower than aggregate 1990 emission levels during the pledge period 2008-2012. Developed economies were given a leadership role on climate change by the UNFCCC and Kyoto Protocol. The principal limitation that bedeviled the

[4] ftp://ftp.zew.de/pub/zew-docs/dp/dp0361.pdf: The Kyoto Protocol: A Review and Perspectives

Kyoto Protocol was that it declined to include the United States. Nonetheless, the United States was a party to the UNFCCC[5].

The Paris Agreement is believed to be all-inclusive since all principal emitters of greenhouse gases, of which the United States is of no exception, are involved in the agreement. In terms of adaptation, the Paris Agreement is not too different from the UNFCCC and the Kyoto Protocol. It was also intended to help countries adapt to the unavoidable adverse impacts of climate change and develop techniques that can aid in increasing adaptation to climate change effects. The Paris Agreement was instituted successfully in December 2015. It was adopted as a decision of the Twenty-First Conference of the Parties and attached as an Annex. By 2020, the agreement is projected to take the place of the Kyoto Protocol. The agreement employed a bottom-up technique in terms of what countries can do on climate change. This is contrary to the top-down approach adopted by the Kyoto Protocol. The Paris Agreement is known to be the first worldwide consensus on climate change that covers policy obligations for all participating countries. Bodansky (2011) assert that the Paris Agreement adopts both top-down and bottom-up methods to international climate governance. The agreement is principally founded on voluntary commitments made by countries. Bodansky (2016) reports that the features of the agreement are very robust, due to its lawfully binding status, its worldwide coverage, and because it contains processes that take stock of progress on emissions reductions. Clemencon (2016) considers it to be better than no agreement at all.

The Paris Agreement also captures developed economies' climate change adaption actions. Nations are not only encouraged to collaborate on adaptation via measures such as exchanging information, institutional arrangements and scientific knowledge, there is also a recognition that adaptation needs to follow a country-driven approach, considering the

---

[5] **Annex 1 Parties:** include the industrialized countries that were members of the OECD in 1992, plus countries with economies in transition (the EIT Parties), including the Russian Federation, the Baltic States, and several Central and Eastern European States.

Annex 11 Parties: consist of OECD members of Annex 1, but not the EIT Parties. They are required to provide financial resources to enable developing countries undertake emissions reduction activities under the Convention and to help them to adapt to adverse effects of climate change. In addition, they have to "take all practical steps" to promote the development and transfer of environmentally friendly technologies to EIT Parties and developing countries. Funding provided by Annex 11 Parties is channeled mostly through the Convention's financial mechanism.

vulnerable communities within societies and their traditional knowledge and practices. For instance, Malaysia has been highlighted for its approach in handling climate change through its National Climate Change Policy, which indicates that both mitigation and adaptation must be given a high level of attention (UNFCCC, 2015). In fact, as Malaysia is experiencing the negative impacts of climate change, quite a number of adaptation measures have been undertaken. Flood mitigation programmes have been amongst the most important adaptation actions taken by Malaysia.

In Malaysia, the government has developed an adaptation plan known as Tenth Malaysia Plan (2011-2015). In this plan, there are 194 flood mitigation projects and 34 local-scale flood hazard maps developed to facilitate disaster prevention. It has been estimated that the cost of managing floods in Malaysia has decreased from "RM3 million per year during the Second Malaysia Plan (1971-1975) to RM1.3 million per year during the Tenth Malaysia Plan (2011-2015). Efforts to address floods through integrated flood management integrated flood forecasting and early warning systems are necessary (UNFCCC, 2015). Another adaptation measure that has received the Malaysian's Government attention is water resource management. While 97 percent of the water supply in Malaysia is derived from water in rivers and reservoirs, management of these resources to ensure adequate water supply continues to be given a high level of attention by the Government[6].

To reinforce adaptation attempts, the Paris Agreement spelt out that there should be a set of collective aims to reduce climate vulnerability and strengthen the climate resilience of ecosystems, economies and societies. The agreement commits all parties to prepare, implement and report on national adaptation efforts, with support provided to vulnerable countries with the least capacity to adapt. It also provides the opportunity, through regular stocktaking, to assess adaptation progress and priorities, and to exchange experiences and lessons learned.

> "The Paris Agreement is fair and just, comprehensive and balanced, highly ambitious, enduring and effective, and with legally binding force"[7].
>
> *China's Closing Statement at COP21, December 12, 2015*

---

[6] Essential Elements of the Paris Climate Agreement: December 2015 http://www.c2es.org/ international/2015-agreement

[7] The Paris Agreement on Climate Change: Behind Closed Doors: Radoslav S. Dimitrov (2016)

In continuance of the moves to climate change adaptation policy, the Cancun Adaptation Framework (UNFCCC, 2010) which sought to improve adaptation attempts through, inter alia, greater international cooperation and coherence, was instituted. This framework urges states to strengthen, consolidate and enhance the sharing of relevant information, knowledge, experiences and good practices, at the local, national, regional and international levels. To promote the implementation of enhanced action on adaptation in a coherent manner under the UNFCCC, the Adaptation Committee was set up and a three-year work plan was established in 2012 (UNFCCC, 2015). Table 1.8 presents some possible scenarios for an adaptation outcome in Cancun.

Table 1. 8: Possible Scenarios for an Adaptation Outcome in Cancun

Cancun adaptation scenario	**Elements contained and missing**
1. No agreement on adaptation.	Text not agreed, because of a) disagreement on substance, b) disagreement over trade-offs between controversial issues, c) non-adoption for tactical reasons; some support will flow bilaterally but with an unbalanced distribution of resources.
2. Agreement on a weak adaptationtext.	Only "low-hanging fruits" included (e.g., regional centers, adaptation committee, etc.), but more controversial issues (insurance mechanism, loss and damage, finance link) postponed.
3. Adaptation framework without strong financial mechanism.	Adaptation framework also contains more controversial approaches, such as insurance mechanism and loss and damage, but lack of reliable finance mechanism makes the agreement largely meaningless.
4. Adaptation framework with an overall strong financial mechanism.	A comprehensive, ambitious adaptation framework linked to a reliable and predictable longer-term finance mechanism.
5. A fully comprehensive post-2012 agreement.	In addition to scenario 4, an overall fair, ambitious and binding deal is adopted, delivering mitigation and the associated support for developing countries at a level which puts the world on a path below 2°C rise in temperature.

Source: Germanwatch & WWF International, Briefing Paper June 2010

In 2014, 46 new laws and policies on climate change adaptation were issued by 34 countries. Seventeen (17) of these countries were developing economies and 17 developed countries. Of the new laws, 21 were legislative (passed by parliaments) and 25 were executive (enacted by governments). The new climate change adaptation laws and policies are presented in Table 1.9. A framework law is defined as a law or regulation with equivalent status, which serves as a comprehensive, unifying basis for climate change policy, addressing multiple aspects and issues of climate change mitigation or adaptation (or both) in a holistic, overarching manner. Fankhauser et al. (2014) argued that not all climate laws are equal in importance or scope; however, framework laws and policies are very essential in propelling climate policy.

Table 1. 9: New Climate Change Adaptation Laws and Policies Passed in 2014

Country	Law/Policy	Legislative/ Executive
Algeria	Regulator Order on electricity generated from facilities using photovoltaics.	Executive
Angola	Presidential Decree 85/14 approving the Statute of the Ministry for Environment (MINAMB);	Executive
	Presidential Decree 17/14 approving the Modernization Programme of the National Institution of Meteorology and Geophysics (INAMET).	Executive
Australia	Carbon Farming Initiative Amendment Bill.	Legislative
Austria	Energy Efficiency Act.	Legislative
Bangladesh	Climate Fiscal Framework.	Legislative
Belgium	Special Act to reform the finance of the Communities and the Regions; the extension of the tax autonomy of the Regions; and the finance of new competencies.	Legislative
Bolivia	Law 602 on Risks Management.	Legislative
Bulgaria	Climate Change Mitigation Act.	Legislative
Chile	Law No. 20780 on environmental taxation (carbon tax);	Legislative
	National Climate Change Adaptation Plan.	Executive
China	Energy Development Strategy Action Plan (2014-2020);	Executive
	National Plan for Tackling Climate Change (2014-2020).	Executive
Colombia	Law 1715, regulating the integration and promotion of non-conventional renewable energy to the national energy system.	Legislative
Democratic Repu blic of Congo	Law 14/003 on the Protection of the Nature;	Legislative
	Law 14/011 on the Electricity Sector.	Legislative
Denmark	Climate Change Act 2014.	Legislative
EU	2030 Framework for Climate and Energy Policies;	Legislative
	European Energy Security Strategy;	Legislative

	Regulation 517/2014 on fluorinated greenhouse gases.	Legislative
France	Farming, forest and alimentation Framework Policy 2014-1170.	Legislative
Germany	Action Programme on Climate Protection 2020.	Executive
Greece	Ministerial Decision 21906 on the compensation to carbon-intensive industries (sectors and sub-sectors) exposed to carbon leakage caused by the indirect costs of the EU ETS.	Executive
Indonesia	National Medium Term Development Plan (2015-2019);	Executive
	New Geothermal Law 21/2014.	Legislative
Ireland	The National Energy Efficiency Action Plan; Offshore Renewable Energy Development Plan.	Executive
Italy	Law Decree 145/20131 that amended the incentive regime to electricity plants fuelled by renewable energy sources other than photovoltaic plants;	Legislative
	Law 116 on urgent provisions for the agricultural sector, environment, company development and, inter alia, reduction of energy bills.	Legislative
Mongolia	Green Development Policy.	Executive
Morocco	Dahir 1-14-09, which promulgates the Framework Law 99-12, concerning the national charter of the environment and sustainable development.	Executive
Pakistan	Pakistan 2025: One Nation, One Vision.	Executive
Peru	Directive on the Commercialization of Fees Generated by Ecosystem Conservation Projects in Nationally Protected Areas;	Executive
	Law 30215 on Mechanisms of Compensation for Services to Ecosystems.	Legislative
Philippines	Executive Order 174, institutionalizing the Greenhouse Gas Inventory Management and Reporting System.	Executive
Poland	Strategy for Energy Security and Environment.	Executive
Romania	Law on energy efficiency 121/2014;	Legislative
	Government Decision 1026/ 2014 on the reorganization of the National Commission on Climate Change.	Executive
Russia	Decree 321, approving the State Programme on Energy Efficiency and Energy Development.	Executive
Senegal	Decree 2014-880 on the powers of the Minister of Environment and Sustainable Development.	Executive
Slovakia	Resolution 148/2014 establishing the National Adaptation Strategy.	Executive
Spain	Royal Decree 413/2014, regulating the production of electricity generation from renewable energy, cogeneration and waste;	Executive
	Royal Decree 525/2014, on the subsidization of the Incentives Programme for Efficient Vehicles.	Executive
Tanzania	Agriculture Climate Resilience Plan.	Executive
Zimbabwe	National Climate Change Response Strategy.	Executive

Source: Germanwatch & WWF International, Briefing Paper June 2010.

References

Adger, W., Khan, S., and Brooks, N. (2003). Measuring and Enhancing Adaptive Capacity. The Energy and Resources Institute. New Delhi: TERI. In Dalei, N.N. (2016). *Do we need a Climate Change Adaptation Policy in Agriculture Sector in Indian State of Odisha*? Conference proceedings on Advances in the Field of Health, Safety, Fire, Environment, Allied Sciences and Engineering (HSFEA 2016) University of Petroleum and Energy Studies, Dehradun, INDIA.

Agrawala, S., and Fankhauser, S. Eds. 2008. *Economic Aspects of Adaptation to Climate Change: Costs, Benefits, and Policy Instruments.* Paris: OECD.

Beg, N., Morlot, J. C., Davidson, O., Okesse, Y.A., Tyani, L., Denton, F., Sokona, Y., Thomas, J.P., La Rovere, E.L., Parikh, J.K., Parikh, K., and Rahman, A.A. (2002) Linkages between climate change and sustainable development, *Climate Policy*, 2 (2-3),129–144.

Biagini, B., Bierbaum, R., Stults, M., Dobardzic, S., and McNeeley, S.M. (2014) A typology of adaptation actions: A global look at climate adaptation actions financed through the Global Environment Facility, *Global Environmental Change*, 25, 97–108.

Bodansky, D. (2011) A Tale of Two Architectures: The Once and Future U.N. Climate Change Regime. *Arizona State Law Journal* 43 (3), 697–712.

Bodansky, D. (2016) The Paris Climate Change Agreement: A New Hope? *American Journal of International Law*, 110(2), 288-319.

Carter, T.R., Parry, M.L., Harasawa, H., and Nishioka, S. (1994) Intergovernmental Panel on Climate Change Technical Guidelines for Assessing Climate Change Impacts and Adaptation. University College London and Center for Global Environmental Research, London.

Clemencon, R. (2016). Sustainable Development, Climate Politics and EU Leadership: Who else can lead the world? *European Journal of Sustainable Development*, 5(1), 125-144.

Cohen, S., Demeritt, D., Robinson, J., and Rothman, D. (1998) Climate Change and sustainable development: towards dialogue, *Global Environmental Change*, 8(4), 341–371.

Dimitrov, R.S. (2016). The Paris agreement on climate change: behind closed doors, *Global Environmental Politics* 16:3, doi:10.1162/GLEP_a_00361.

Downing, T.E., Ringius, L., Hulme, M., and Waughray, D. (1997) Adapting to climate change in Africa, *Mitigation and Adaptation Strategies for Global Change*, 2, 19-44.

Ebi, K., and Semenza, J. (2008) Community-based Adaptation to the Health Impacts of Climate Change, *Am J Prev Med*; 35(5), 501-507.

Fankhauser, S., Gennaioli, C., and Collins, M. (2014) *Domestic dynamics and international influence: What explains the passage of climate change legislation?* Working Paper. [pdf] Grantham Research Institute on Climate Change and the Environment and Centre for Climate Change Economics and Policy. [pdf] Available at: http://www.lse.ac.uk/GranthamInstitute/wp-content/uploads/2014/05/Wp156Domestic-dynamics-and-international-influence-what-explains-the-passage-of-climate-change-legislation.pdf [Accessed 7 May 2015]

Fussel, H.M., and Klein, R.J.T. (2002) Assessing the Vulnerability and Adaptation to Climate Change: An Evolution of Conceptual Thinking, Paper presented at the UNDP Expert Group Meeting on Integrating Disaster Reduction and Adaptation to Climate Change, Havana, Cuba, 17–19th June 2002.

GEF (Global Environment Facility). (2008) *GEF Funding for Adaptation to Climate Change.* PPt. for 13th Poverty and Environment Partnership Meeting, Manila, June 9-11.

Germanwatch, and WWF International, Briefing Paper June 2010.

Handmer, J., Dovers, S., and Downing T.E. (1999) Societal vulnerability to climate change and variability, *Mitigation and Adaptation Strategies for Global Change*, 4 (3, 4), 267–281.

Huq, S., Reid, H., Konate, M., Rahman, A., Sokona, Y., & Crick, F. (2004). Mainstreaming Adaptation to Climate Change in Least Developed Countries (LDCS). *Climate Policy*, 4(1), 25-43.

IPCC (2001) Climate Change 2001: Impacts, Adaptation and Vulnerability, McCarthy, J.J., Canziani, O.F., Leary, N.A., Dokken, D.J., and White, K.S., (eds.), Cambridge: Cambridge University Press.

IPCC (2007) *Mitigation of Climate Change*. Assessment Report 4, Working Group III, Inter- Governmental Panel on Climate Change.

Kates, R.W. (2000) Cautionary Tales: adaptation and the global poor, *Climatic Change*, 45 (1), 5–17.

Lagos, R., and Wirth, T. (2009). Facilitating an international agreement on climate change: Adaptation to climate change. Climate Change: Adaptation to Climate Change. Retrieved from www.globalclimateaction.org.

Lim, B., Spanger-Siefried, E., Burton, I., Malone, E., and Huq, S. (2004) Adaptation Policy Frameworks for Climate Change: Developing Strategies, Policies and Measures. United Nations Development U N Programme, Cambridge University Press.

OECD (2008) "Economic Aspects of Adaptation to Climate Change: Costs, Benefits, and Policy Instruments." Paris: OECD.

Oxfam (2007) Adapting to climate change. What is needed in poor countries and who should pay? Oxfam Briefing Paper 104, Oxfam, Oxford, UK.[Google Scholar]

Parry, J.E., Hammill, A., and Drexhage, J. (2005) Climate Change and Adaptation, IISD.

Pielke, R.A. (1998) Rethinking the role of adaptation in climate policy, *Global Environmental Change*, 8, 159–170.

Schalatek, L., Nakhooda, S., Caravani, A., and Barnard, S. (2012) Climate Finance Regional Briefing: Middle East and North Africa. Climate Finance Fundamentals No. 8. London/Washington, DC, Overseas Development Institute (ODI)/Heinrich Böll Stiftung North America.

Smit, B., and Pilifosova, O. (2001) Adaptation to climate change in the context of sustainable development and equity, in Climate Change 2001:

impacts, adaptation and vulnerability, Chapter 18, Cambridge: Cambridge University Press.

Smit, B., Burton, I., Klein, R.J.T., and Wandel, J. (2000) An anatomy of adaptation to climate change and variability. *Climate Change* 45, 223–251.

Smit, B., and Skinner, M.W. (2002) Adaptation options in agriculture to climate change: a typology. *Mitigation and Adaptation Strategies for Global Change* 7, 85–114.

Smithers, J., and Smit, B. (1997) Human adaptation to climatic variability and change, *Global Environmental Change*, 7(2), 129–146.

Stern, N. (2006) *Stern Review on the Economics of Climate Change.* London: HM Treasury Review, 30.

Tompkins, E.L., Adger, W.N., Boyd, E., Nicholson-Cole, S., Weatherhead, K., and Arnell, N. (2010) Observed adaptation to climate change: UK evidence of transition to a well-adapting society. *Global Environmental Change,* 20(4),627–635.

UKCIP (2005) Measuring Progress: Preparing for Climate Change through the UK Climate Impacts Programme. UK Climate Impacts Programme (UKCIP) (technical report).

UNDP (2007) Human Development Report 2007/2008, Fighting Climate Change: Human Solidarity in a Divided World. UNDP, New York.

UNDP (United Nations Development Programme) (2004) *Adaptation Policy Frameworks for Climate Change.* Cambridge, UK: Cambridge University Press.

UNDP/GEF (2003). *Capacity Development Indicators,* New York, US. UNDP/GEF Resource Kit (No.4), Page 4. http://www.undp.org/gef/undp-gef_monitoring_evaluation/sub_undpgef_monitoring_evaluation_documents/CapDevIndicator%20Resource%20Kit_ Nov03_Final.doc.

UNFCCC (2010) *Report of the Conference of the Parties on its sixteenth session, held in Cancun from 29 November to 10 December 2010.* Montreal, Canada, United Nations Framework Convention on Climate Change.

UNFCCC (United Nations Framework Convention on Climate Change) (1998) *Kyoto Protocol to the United Nations Framework Convention on climate change.* Available at http://unfccc.int/resource/docs/convkp/kpeng.pdf (Accessed: 10 October 2017).

UNFCCC (United Nations Framework Convention on Climate Change) (2015, December 12) Paris Agreement: FCCC/CP/2015/L.9/Rev.1.Retrieved from https://unfccc.int/documentation/documents/advanced_search/items/6911.php?priref=60000831 (Accessed: 14 September 2017).

UNFCCC (2007) Investment and financial flows to address climate change. Bonn, Germany: United Nations Framework Convention on Climate Change. Available at: http://unfccc.int/files/cooperation_and_support/financial_mechanism/application/pdf/ background_paper.pdf.

Wilbanks, T.J., and Kates, R.W. (1999) Global change in local places: how scale matters. *Climatic Change* 43, 601–628.

World Bank (2006) Managing climate risk: Integrating adaptation into World Bank Group Operations, Washington, DC: World Bank. [Google Scholar].

World Bank (2010) World Development Report: Climate Change and Development. World Bank, New York, NY.

CHAPTER 2

Assessing the vulnerability of smallholder women rice farmers to climate variability in the Northern Region of Ghana: The livelihood vulnerability index approach

Suhiyini Issah Alhassan, Yaw Bonsu Osei-Asare, John K.M. Kuwornu

Abstract

This study examined the vulnerability of smallholder women rice farmers to climate variability in the Savelugu/Nanton Municipality and West Mamprusi District in the Northern Region of Ghana using the Livelihood Vulnerability Index (LVI). Primary data was collected from two-hundred and sixty (260) randomly selected women rice farmers. Time series data on rainfall and temperature were sourced from the Ghana Meteorological Service. The results revealed that Savelugu/Nanton Municipality was more vulnerable to natural disasters and climate variability, socio-demographic profile, health and water, while West Mamprusi District was more vulnerable to livelihood strategies, social network and food. West Mamprusi District was also more vulnerable in terms of adaptive capacities, while Savelugu/Nanton Municipality was more vulnerable in terms of exposure and sensitivity. In all, a majority of the sampled women farmers (i.e., 58%) were highly vulnerable to climate variability, with farmers in both districts showing lower adaptive capacities relative to their exposure to climatic stresses. The implications of the study are presented.

Keywords: Climate variability; Ghana; rice farmer; vulnerability

2.1 Introduction

The adverse impacts of climate variability on agriculture, especially women farmers, is increasing globally and has in recent times attracted the attention of academics, policymakers and non-governmental organizations (NGOs). Climate variability is a major global issue given its impacts on the environment and agriculture, especially smallholder farmers who earn their livelihood through rain-fed agriculture (Morton, 2007; IPCC, 2014). The growing conditions of crops and livestock are affected by climatic conditions. Poverty and food insecurity in northern Ghana continue to be high as a result of climate change and variability (Amikuzuno and Donkoh, 2012).

Compared to male farmers, the agricultural activities of female farmers are generally more constrained due to a lack of resources. This difference in resource endowments adversely affects efficient agricultural investments by female farmers, thereby affecting their resilience to climate variability (FAO, 2011). In the Northern Region of Ghana, just as in many other parts of Ghana, relative to male-headed households, female-headed households have less tenure security and access to land including other supporting production resources, making them more vulnerable (Blackden and Wodon, 2006; Doss and Morris, 2000; Koru and Holden, 2008). Research has revealed that only one out of every forty-five female farmers owns land for rice cultivation in the Northern Region (Asare, 2000). This, coupled with the fact that women in the region are marginalized in certain aspects of vital decision-making including responses to climate variability, makes women more vulnerable and susceptible to the impacts of climate variability (Ministry of Environment, Science, Technology and Innovation (MESTI), 2013). Yet, females constitute about 50 percent of the region's population (Ghana Statistical Service (GSS), 2012).

According to Asante and Amuakwa-Mensah (2014), rice and root crop production are likely to reduce due to climate variability. The concern about the potential effects of climate variability includes its impact on women and children, who are especially vulnerable due to their high poverty levels and household responsibility to search for water, food and fuel for cooking (MESTI, 2013). Supporting female farmers is therefore vital in ensuring gender equity.

The objective of this study was to assess the vulnerability of smallholder women rice farmers to climate variability. Unlike previous studies (e.g., Etwire et al., 2013; Van Der Geest, 2004) which assessed the vulnerability levels of all smallholder farmers, this study focused on assessing the vulnerability of only women rice farmers who are the most susceptible to climate variability. This study also employs more vulnerability indicators than the earlier studies. The current study tested for significant statistical differences in the vulnerability levels of farmers from two districts, which has been neglected in the literature. Thus, the findings of this study bring to light the significant contributors to vulnerability and the degree of female farmers' vulnerability to climate variability, while also providing female-focused recommendations for reducing their vulnerability levels. These recommendations can assist in redirecting the focus of both on-going and future intervention projects on climate variability and agriculture towards achieving Ghana's National Climate Change Policy objective on gender by adhering to the guiding principle of improving gender equity and sensitivity.

2.2 Materials and Methods

2.2.1 Conceptualizing Vulnerability to Climate Variability

The term 'vulnerability' has no generally accepted definition due to its wide usage in different disciplines. The conventional meaning of the term is the capacity to be distressed (Turner et al., 2003). That is, the extent of injury likely to be caused to a system as a result of its exposure to a hazard (Turner et al., 2003). Scientifically, vulnerability was originally used in geography and natural hazards fields of research. Vulnerability refers to the ability to forestall, manage with, resist and recuperate from the effect of natural disasters (Madhuri et al., 2014). Also, vulnerability has been defined to mean the likelihood of farming households and communities suffering detriment including a reduced capacity to respond to stresses ensuing from the effects of climate-related events (Juana et al., 2013).

In the views of Cutter et al. (2008) and Nelson et al. (2009), vulnerability could also be the predisposition of any group of people, location, or system to disorders determined by exposure and sensitivity to distress, including their adaptive capacity. The Intergovernmental Panel on Climate Change (IPCC) defined vulnerability as a system's susceptibility or incapacity to cope with the negative effects of climate change, climate variability and other extremes.

Common among these views on vulnerability is that the term concerns itself with the susceptibility of a system to a hazard resulting from exposure to the threats or drivers of climate variability, but not the estimation of harm itself. Vulnerability, as employed in this study, refers to the degree to which smallholder women rice farmers are susceptible to, or unable to adapt to, the negative impacts of climatic stresses (i.e., mainly floods, bushfires and droughts). Thus, vulnerability is composed of the type, level and frequency of climate variation to which a smallholder woman rice farmer is exposed or sensitive to, and her adaptive capacity to these climate extremes. This conceptualization of vulnerability is consistent with that of Adger (1999), Brooks (2003), Fussel (2006), IPCC (2007, 2014), Kelly and Adger (2000) and Turner et al. (2003). According to the IPCC (2007) and Fussel and Klein (2006), exposure is the extent to which a system is unshielded from major climatic impacts; sensitivity is the extent to which a system is affected by climate-related stimuli either positively or negatively, covertly or overtly; while adaptive capacity is a system's ability to adjust itself to withstand the likely harsh effect of climate variability by employing prevailing opportunities.

2.2.2 Approaches to Measuring Vulnerability

Deressa et al. (2009) distinguished between two main approaches to measuring vulnerability to climate variability; namely, econometric and indicator techniques. The econometric approach employs regression analysis methods, but this approach is inhibited by the limitation of testing several econometric assumptions on hypothesis, confidence intervals and standard errors including attributing causation without stringent assumptions. The indicator approach involves choosing indicators (i.e., components) that a researcher considers to largely account for vulnerability (Deressa et al., 2009). The major limitation of the indicator approach is the biasness or subjectivity on the part of the researcher in choosing the various indicators to be included in computing the index. A myriad of indices under the indicator approach has been developed by various authors in assessing vulnerability to climate variability. Deressa et al. (2009) developed the Vulnerability as Expected Poverty (VEP) index to measure farmers' vulnerability to drought, floods and other climatic extremes. The VEP index estimates the probability of a household's consumption falling below a minimum level due to the occurrence of a climatic shock. The main limitation of this index is that it measures future poverty (i.e., vulnerability) as a result of climatic extremes rather than current vulnerability. Another framework is the Umbrella Model developed by the Livelihood and Food Security Trust (2010) to measure households' vulnerability to shocks and hazards. The Umbrella Model draws on Moser's Assets Vulnerability Framework (Moser, 1998) and thus uses ten assets-based factors to measure and plot households' vulnerability. This model has been criticized for including too many people and hence, lacks specificity. Gbetibouo et al. (2010) also propounded a vulnerability index to measure the vulnerability of nine (9)

South African Provinces. A Principal Component Analysis (PCA) was used to weigh indicators of vulnerability. The greater likelihood of assigning paradoxical weight to indicators due to poor data structure has been the limitation to the use of this vulnerability index. Another useful tool for assessing spatio-temporal scale differences in vulnerability is the Climate Vulnerability Index (CVI), developed by Pandey and Jha (2012). This index is suitable only for mountainous regions, and data sensitivity due to the weighing of different sub-components is a major setback in applying the CVI.

The most widely used index in assessing farmers' vulnerability to climate variability is the Livelihood Vulnerability Index (LVI) (Hahn et al., 2009). This study employed the LVI as an indicator approach to measuring the vulnerability of women rice farmers to climate variability. The computation of the LVI is relatively simple compared to other indices since it relies more on primary data from farmers, aside from data on precipitation and temperature (Hahn et al., 2009). According to Hahn et al. (2009), the LVI not only determines the tendency of household vulnerability to droughts, bushfires and floods, it also considers existing vulnerability which is important for present planning.

Hahn et al., (2009) employed the LVI to assess vulnerability levels of households in Moma and Mabote Districts in Mozambique to climate change and variability. The results showed that Mabote was more vulnerable than Moma in terms of socio-demographic profile, social networks, natural disasters and climate variability. When all sub-components were aggregated, Mabote appeared more vulnerable (0.326) than Moma (0.316). Shah et al., (2013) also employed the LVI to compare the impact of climate change on agricultural and natural resource-dependent communities in Trinidad and Tobago. Findings from this research revealed that Nariva community was more vulnerable in terms of natural disaster and climate variability, socio-demographics, health and water indicators, while Caroni community was more vulnerable in terms of the food security indicator. Etwire et al., (2013) also applied the LVI to assess the vulnerability of farming households to climate change and variability in northern Ghana. The empirical results showed that the Northern Region was most exposed to climatic stresses, the Upper West Region was the most sensitive to climate change, while the Upper East Region had the least adaptive capacity to climatic stresses. Using the overall LVI_{IPCC}, the Northern Region (0.004) was most vulnerable to climate change and variability, whilst the Upper East (-0.007) was more vulnerable than the Upper West (-0.015) Region[1]. Simane et al., (2016) also used the LVI_{IPCC} to analyze the vulnerability to climate change of mixed crop farming households in communities on the Choke Mountains along the Blue Nile Highlands of Ethiopia. The results revealed that households in high elevation sloping and low steep agro-ecological areas had lower adaptive capacities, rendering them more vulnerable than households in midland agro-ecological areas. All of these studies have failed to test for significant differences in the vulnerability levels of the study population, a research gap which the current study has addressed.

2.2.3 Study Area

The study was conducted in the Savelugu/Nanton Municipality and West Mamprusi District in the Northern Region of Ghana. This region is semi-arid and within the Guinea Savannah agro-ecological zone with its vegetation comprising mainly of grassland areas, interspersed with Guinea and Sudan savannah woodland characterised by such drought-resistant trees as the baobab, shea nut, acacia, dawadawa, mango and neem. The average minimum and maximum temperatures of the region are 14^0C at night and 40^0C during the day (GSS, 2014). The region experiences two seasons: the dry season (i.e., November to April) and the wet season (i.e., May to October), with an average annual rainfall of 750mm – 1050mm (GSS, 2014).

The region occupies a land area of 70, 384 km^2 with 50.4 percent of its population (i.e., 2, 479, 461) being female (GSS, 2012). West Mamprusi District is the third largest district in the Northern Region with a total population of 168,011, of which 50.06% are female, whilst the Savelugu/Nanton Municipality is the fourth largest district in the region with a population of 139,285, of which 51.51% are female. The agricultural sector employs 73.11% of the region's economically active population, with 43.06% of these being female. The main crops cultivated in the region include maize, rice, yam, cassava, soy beans, millet and sorghum (GSS, 2012). Figure 2.1 shows a map of the study area indicating the locations of the study communities.

---

[1] LVI_{IPCC} is the Livelihood Vulnerability Index computed based on the IPCC framework.

Figure 2. 1: Map of Savelugu/Nanton Municipality and West Mamprusi District showing the study communities

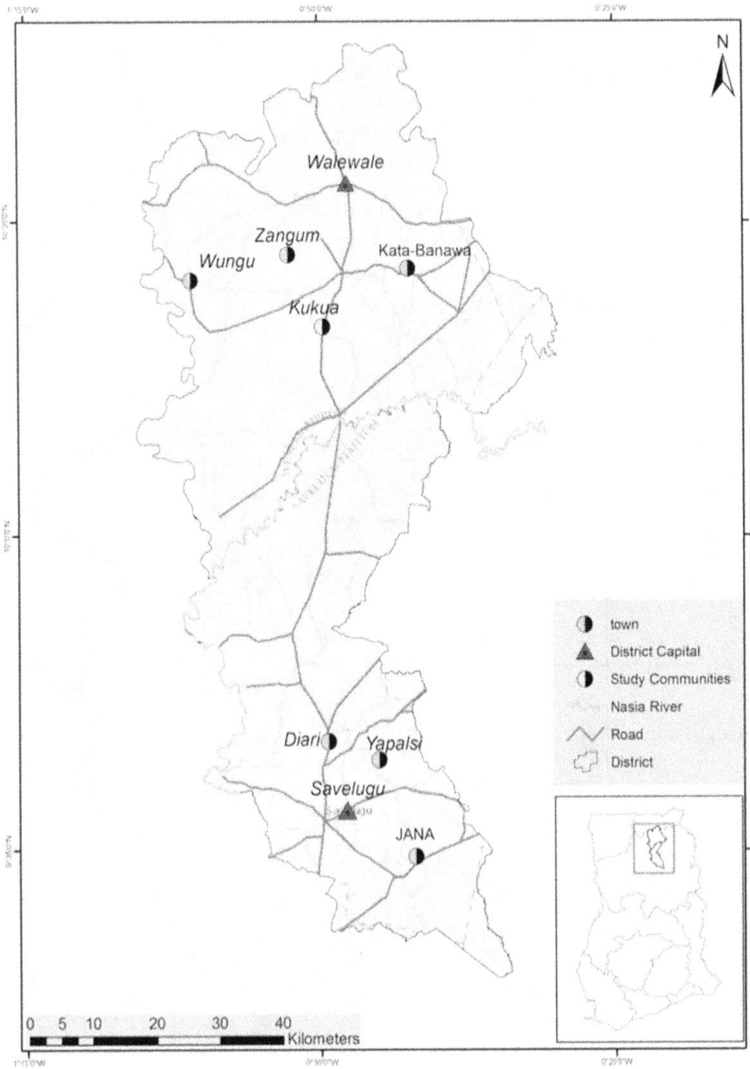

2.2.4 Sources of Data and Sampling Procedures

A multi-stage sampling technique was used. The Northern Region was purposively selected for this study because it is in a Guinea Savanna ecological zone located in a semi-arid climatic region (GSS, 2012) and is the nation's highest producer of rice (Ministry of Food and Agriculture

(MoFA), 2015). It is also Ghana's vulnerable region to climate variability (Etwire et al., 2013). According to the 2015 report of the Statistics, Research and Information Directorate of the Ministry of Food and Agriculture, Ghana (SRID of MoFA), West Mamprusi District and Savelugu/Nanton Municipality were purposively selected among other districts because they are the top two rice producing districts in the Northern Region and also have the most erratic rainfall patterns, according to rainfall data from the Ghana Meteorological Service.

Based on population and land size, four communities in West Mamprusi District and three communities in the Savelugu/Nanton Municipality were randomly selected. Simple random sampling was used to proportionally select a total of two hundred and sixty (260) smallholder women rice farmers from all communities. Details of the sample distribution are presented in Table 2.1.

Table 2. 1: Sample Size Distribution by Communities and Districts

District	Community	Estimated number of women rice farmers	No. of Respondents Sampled	
			Community	District
Savelugu/Nanton	Diare	108	38	110
	Jana	103	36	
	Yapalsi	103	36	
West Mamprusi	Zangum	106	37	150
	Kata-Banawa	108	38	
	Kukua	100	35	
	Wungu	115	40	
Total Sample Size		743		**260**

Primary data was collected from the 260 smallholder women rice farmers using a questionnaire. Data was collected in February 2016. Secondary data on temperature and rainfall in each district were collected from weather stations of the Ghana Meteorological Service.

2.2.5 Measuring Vulnerability to Climate Variability

The LVI is a commonly used framework in the assessments of vulnerability to climate variability. The framework makes it possible to analyze both the essential components that constitute livelihood and the contextual factors influencing these factors. The LVI assumes equal weights for all major components and sub-components (Sullivan et al., 2002, cited in Hahn et al., 2009).

The nature and magnitude of smallholder women rice farmers' vulnerability to climate-related events (i.e., bushfires, floods and drought) were examined at two levels: the community and district levels. In pursuance of this, two sets of analysis have been undertaken for each of the communities and the two districts under study, namely (1) estimation of the LVI based on a balanced weighted average, and (2) estimation of the LVI_{IPCC} based on the Intergovernmental Panel on Climate Change (IPCC) vulnerability framework.

2.2.5.1 Estimating LVI

Hahn et al., (2009) used seven major indicators to estimate the LVI. These include socio-demographic profile, livelihood strategies, social networks, health, access to food, access to water, and natural disasters and climate variability. Each major indicator consists of several sub-components known as indicators. Each of these indicators is measured on a different scale. It is, therefore, necessary to standardize each indicator as an index using the UNDP (2007) Life Expectancy index, given by equation (2.1):

$$Index_{sc/di} = \frac{S_{c/d} - S_{min}}{Smin_{max}} \tag{2.1}$$

Where $S_{c/d}$ is the observed sub-component indicator for community *c* or district *d*, $S_{min}$and $S_{max}$are the minimum and maximum values for each sub-component, respectively, determined using data from both districts.

The sub-component indicators are now averaged using equation (2.2) to obtain the index of each major component:

$$M_{c/di} = \frac{\sum_{i=1}^{n} Index_{sc/di}}{n} \tag{2.2}$$

Where $M_{c/di}$ is one of the seven major components [Socio-Demographic Profile (SDP), Livelihood Strategies (LS), Social Network (SN), Health (H), Food (F), Water (W), or Natural Disaster and Climate Variability (NDCV)] for community *c* or district *d*; $index_{sc/di}$ represents the sub-components, indexed by *i*, that make up each major component, and *n* is the number of sub-components in each major component. Once values for each of the seven major components for each community and district are calculated, they are also averaged using equation (2.3) to obtain the community level and district-level LVIs:

$$LVI_{c/d} = \frac{\sum_{i=1}^{7} w_{Mi} M_{c/di}}{\sum_{i=1}^{7} w_{Mi}} \quad (2.3)$$

Explicitly, equation (2.3) can be rewritten as equation 2.4, below:

$$LVI_{c/d} =$$

$$(wSDP_{c/d}SDP_{c/d} + wLS_{c/d}LS_{c/d} + wH_{c/d}H_{c/d}$$

$$+wSN_{c/d}SN_{c/d} + wF_{c/d}F_{c/d} + wW_{c/d}W_{c/d} + wNDCV_{c/d}NDCV_{c/d})$$

$$/(SDP_{c/d} + LS_{c/d} + H_{c/d} + SN_{c/d} + F_{c/d} + W_{c/d} + NDCV_{c/d}) \quad (2.4)$$

Where wM_i, the weights of each major component, is a function of the number of sub-components that make up each major component and are included to ensure that all sub-components contribute equally to the overall LVI.

2.2.5.2 Estimating LVI$_{IPCC}$

Considering the IPCC definition of vulnerability, Hahn et al. (2009) then used equations (2.1) – (2.3) and computed a new variable, $LVI_{IPCC_{c/d}}$. The $LVI_{IPCC_{c/d}}$ differs from the LVI when the major components are combined. Instead of merging the major components into the LVI using equation (2.3), the major components are first combined into three categories namely exposure, adaptation capacity and sensitivity using equation (2.5):

$$CF_{c/d} = \frac{\sum_{i=1}^{n} w_{Mi} M_{c/di}}{\sum_{i=1}^{n} w_{Mi}} \quad (2.5)$$

Where $CF_{c/d}$, is an IPCC-defined contributing factor (i.e. exposure, sensitivity or adaptive capacity) for community *c* (in the case of communities) or district *d* (in the case of districts), $M_{c/di}$ are the major components for the community or district indexed by *i*, wM_i is the weight of each major component, and *n* is the number of major components in each contributing factor. Once exposure, adaptive capacity and sensitivity are estimated, the three contributing factors are combined using equation (2.6):

$$LVI_{IPCC_{c/d}} = (E_{c/d} - A_{c/d})S_{c/d} \tag{2.6}$$

Where $LVI_{IPCC_{c/d}}$ is the LVI for community c or district d, expressed using the IPCC vulnerability framework, $E_{c\,/\,d}$ denotes the calculated exposure score for community c or district d (equivalent to the natural disaster and climate variability major component), $A_{c\,/\,d}$ denotes the calculated adaptation capacity score for community c or district d (weighted average of socio-demographic, livelihood strategies, and social networks major components), and $S_{c\,/\,d}$ denotes the calculated sensitivity score for community c or district d (weighted average of the health, food and water major components). The $LVI_{IPCC_{c/d}}$ is scaled between -1 (least vulnerable) to 1 (most vulnerable).

The major limitation of the LVI approach to measuring vulnerability is the equal weights of the sub-indicators assumption. Practically, the sub-indicators do not contribute equally to vulnerability and future studies should consider weighting the sub-indicators for more realistic outcomes. Table 2.2 presents the major components and sub-components used to measure LVI in previous studies.

Table 2. 2: Major Components and Sub-components for Estimating the Livelihood Vulnerability Index (LVI)

Major Component	Sub-component	Explanation	Measurement	Source of Adoption
Socio-demographic profile	Dependency ratio.	Ratio of the population under 15 and over 65 years of age to the population between 15 and 64 years of age.	Ratio	Domestic Household Survey (DHS) (2006). Measure DHS: Model Questionnaire with Commentary.
	Percentage of farmers from female-headed households.	Percentage of farmers from households where the primary adult is female. If a male head is away from the home >6 months per year, the female is deemed as the household head.	Percent	DHS (2006)
	Average age of female farmers.	Average of ages of all female farmers.	Years	Etwire et al. (2013)
	Percentage of farmers whose households are without latrines.	Percentage of farmers who live in households without latrine for defecation.	Percent	Developed by author for purposes of this research.
	Percentage of farmers who have not attended school.	Percentage of farmers who report to have attended 0 years of school.	Percent	DHS (2006)
	Percentage of farmer who live in households with orphans.	Percentage of farmers who have at least 1 orphan living in their home. Orphans are children <18 years old who have lost one or both parents.	Percent	DHS (2006)
	Average number of persons per room.	Average number of persons who sleep in a room within a household with the farmer.	Heads	Developed by author for purposes of this research.

Livelihood Strategies	Percentage of farmers who worked in a different community.	Percentage of farmers who report to have worked outside the community at least once as their primary work activity in the past 12 months.	Percent	World Bank (1997). Household Questionnaire: Survey of Living Conditions, Uttar Pradesh and Bihar.
	Percentage of farmers depending solely on agriculture as their source of income.	Percentage of farmers who report only agriculture as a source of income.	Percent	World Bank (1997)
	Percentage of farmers who do not own their farmlands.	Percentage of farmers who report not owning the farmland on which they cultivate.	Percent	Developed by author for the purposes of this research.
	Average Agricultural Livelihood Diversification Index.	The inverse of the number of agricultural livelihood activities +1, as reported by a household.	1/(# of agricultural livelihoods +1).	DHS (2006)
Health	Average time to health facility (minutes).	Average time it takes a farmer to get to the nearest health facility.	Minutes	World Bank (1997)
	Percentage of farmers with a family member with chronic illness.	Percentage of farmers who report at least 1 family member with a chronic illness. Chronic illness was defined subjectively by respondents.	Percent	DHS (2006)
	Percentage of farmers who had to miss work or school in the last 2 weeks due to illness.	Percentage of farmers who report to have missed school or work due to illness in the last 2 weeks.	Percent	World Health Organization/Roll Back Malaria (2003).
	Average Malaria Exposure*Prevention Index (range: 0–12).	Months reported exposure to malaria*Owning at least one bednet indicator (have bednet = 0.5, no bednet = 1).	Month*Bednet indicators.	Malaria: WHO/RBM (2003). Bednets: DHS (2006).

Social Networks	Average Receive: Give ratio.	Ratio of the number of types of help received by a farmer in the past 12 months + 1 to the number of types of help given by a farmer to someone else in the past 12 months + 1.	Ratio	DHS (2006)
	Average Borrow: Lend Money (range: 0.5–2).	Ratio of farmer's borrowing money in the past 12 months+1 to farmer lending money in the past 12 months +1.	Ratio	World Bank (1997)
	Percentage of farmers who have not gone to their local government for assistance in the past 12 months.	Percentage of farmers who reported that they have not asked their local government for any assistance in the past 12 months.	Percent	WHO/RBM (2003)
	Percentage of farmers who belong to a farmer-based organization (FBO).	Percentage of farmers who are members of a social group such as FBO.	Percent	Developed by author for the purpose of this study.
Food	Percentage of farmers depending on own farm for food.	Percentage of farmers who get their food primarily from their personal farms.	Percent	Hahn et al. (2009).
	Average number of months farmer struggle to find food (range: 0–12).	Average number of months farmers struggle to obtain food.	Number of months.	World Bank (1997)
	Average Crop Diversity Index (range: >0–1).	The inverse of the number of crops grown by a farmer +1.	1/# crops	World Bank (1997)
	Percentage of farmers who do not save crops.	Percentage of farmers who do not save crops from their harvest.	Percent	Hahn et al. (2009)

	Percentage of farmers that do not save seeds.	Percentage of farmers who do not save seeds from their harvest.	Percent	Hahn et al. (2009)
Water	Percentage of farmers reporting water conflicts.	Percentage of farmers who report having heard about conflicts over water in their community.	Percent	Hahn et al. (2009)
	Percentage of farmers that utilize a natural water source.	Percentage of farmers who report a creek, river, rain, lake, pool, or hole as their primary water source.	Percent	DHS (2006)
	Average time to water source (minutes).	Average time it takes a farmer to travel to their primary water source.	Minutes	DHS (2006)
	Percentage of farmers who do not have a consistent water supply.	Percentage of farmers who report that water is not available at their primary water source every day.	Percent	World Bank (1997)
	Inverse of the average number of litres of water stored per farmer (range:> 0–1).	The inverse of the average number of litres of water stored by each farmer + 1.	1/litres	Hahn et al. (2009).
Natural Disasters and Climate Variability	Average number of bushfires, flood and drought events between 2001 and 2015.	Average number of bushfires, floods and droughts reported by households between 2001 and 2015.	Count	City of Williamsburg Virginia's Colonial Capital Emergency Management (2004). Household Natural Hazards Preparedness Questionnaire.
	Percentage of farmers who lost crops as a result of bushfires, floods or drought between 2001 and 2015.	Percentage of farmers who reported losing part of their produce due to bushfires, floods or droughts.	Percent	Developed by author for the purposes of this research.

	Percentage of farmers with an injury or death as a result of the most severe natural disaster between 2001 and 2015.	Percentage of farmers who reported either an injury to or death of one of their family members as a result of the most severe bushfire, flood or drought between 2001 and 2015.	Percent	Hahn et al. (2009)
	Mean standard deviation of the daily average maximum temperature by month.	Standard deviation of the average daily maximum temperature by month between 1987 and 2015.	Celsius	Instituto Nacional de Estatistica (2007)
	Mean standard deviation of the daily average minimum temperature by month.	Standard deviation of the average daily minimum temperature by month between 1987 and 2015.	Celsius	Instituto Nacional de Estatistica (2007)
	Mean standard deviation of the average precipitation by month.	Standard deviation of the average monthly precipitation between 1987 and 2015.	Millimeters	Instituto Nacional de Estatistica (2007)

2.2.5.3 Testing for Difference in Means

An independent two samples students' *t*-test (2-tailed) was employed to test for the significant difference in the means of the LVI's major components, the overall LVIs, the IPCC vulnerability contributory factors, and the overall LVI_{IPCC} computed for the two districts. According to Ruxton (2006), the independent two sample students' *t*-test is most applicable when the two samples are randomly selected (between groups), have an interval or ratio scale dependent variable, and the sample size is large (N≥30).

Specification of Hypothesis

The null (H_o) and alternative (H_a) hypotheses are stated as:

H_o: There is no significant difference between the LVI means of Savelugu/Nanton

Municipality (μ_1) and West Mamprusi District.

H_a: There is a significant difference between the LVI means of Savelugu/Nanton
Municipality (μ_1) and West Mamprusi District.

2.3 Results and Discussion

Though the ideal value of the LVI ranges between 0 (i.e., least vulnerable) and 1 (i.e., most vulnerable), the computed indices for the major components in this study ranged from 0.28 (i.e., least vulnerable) to 0.82 (i.e., most vulnerable). The computed vulnerability indices for the major components and sub-components are presented in Tables 2.3 and 2.4, respectively. The two sample *t*-test results revealed that there are significant differences in the levels of vulnerability of women rice farmers in Savelugu/Nanton Municipality and West Mamprusi District to climate variability (Table 2.5).

Table 2. 3: Computed Major Component Indices and LVIs for Study Communities and Districts

Major Component	Index (M_i)								
	Community (M_{ci})							Districts (M_{di})	
	Diare	Jana	Yapalsi	Zangum	Kata-Banawa	Kukua	Wungu	Savelugu/ Nanton	West Mamprusi
Socio-Demographic Profile	0.43	0.41	0.50	0.48	0.47	0.43	0.47	0.43	0.45
Livelihood Strategies	0.59	0.48	0.59	0.57	0.59	0.51	0.62	0.56	0.57
Social Network	0.49	0.50	0.55	0.82	0.65	0.55	0.54	0.51	0.65
Health	0.43	0.44	0.53	0.38	0.30	0.28	0.36	0.47	0.32
Food	0.50	0.62	0.61	0.66	0.57	0.59	0.62	0.57	0.61
Water	0.64	0.60	0.50	0.60	0.37	0.35	0.60	0.58	0.43
Natural Disaster and Climate Variability	0.61	0.54	0.59	0.52	0.52	0.49	0.56	0.57	0.52
LVI	0.53	0.51	0.55	0.58	0.49	0.46	0.54	0.53	0.51

Source: Author's computation from field survey, 2016.

The first major component is socio-demographic profile comprising of seven sub-components. West Mamprusi District had a higher percentage of farmers from households with orphans (35.3%) than Savelugu/Nanton Municipality (27.3%). Savelugu/Nanton Municipality had a higher dependency ratio (0.82) than West Mamprusi District (0.79). Jana in the Savelugu/Nanton District had the highest dependency ratio (0.92) among the study communities. The national dependency ratio is 0.67. A higher dependency ratio implies many economically inactive persons are dependent on the working class. On average, each economically active person in Ghana supports an additional person (GSS, 2008). The results revealed that West Mamprusi District had more farmers from female-headed households (34.2%) than Savelugu/Nanton District (21.8%). A female is deemed as the household head if she is the head of the household or if the substantive male household head is away for more than six months, irrespective of the reason. About 47 percent of sampled farmers are from households headed by females. Female-headed households are usually marginalized in major decisions in the community including responses to climate variability. The average age of female farmers in the region was 50.8 years, almost the same as the 50.5 years for rural savannah areas (GSS, 2012). The percentage of farmers from households without pit latrines are higher in both Savelugu/Nanton District (70%) and West Mamprusi District (54 %) than the national figure of 19.3 percent (GSS, 2012). Yapalsi had the highest number of households (83.3%) without pit latrines. The study revealed that West Mamprusi District had a higher percentage of farmers who have never attended school (79.3%) than Savelugu/Nanton Municipality (77.3%), with Yapalsi having the highest percentage. Overall, about 78.3 percent of sampled farmers reported to have never attended school. This is higher than the 67.4 percent of 15 years and older persons in the Northern Region reported to have never attended school (GSS, 2012). Education (either formal or informal) assists farmers to read and easily comprehend issues affecting them, including to quickly adapt to variability in climatic conditions to reduce their vulnerability. The findings revealed that West Mamprusi District (2.5) had a higher average number of persons per room than Savelugu/Nanton Municipality (1.9). Among the study communities in the Northern Region, Jana (2.85) had the highest average number of persons living in a room. Higher average numbers of persons living in a room is a sign of inadequate rooms, overcrowding and insufficient privacy for farmers, and is a catalyst for the spread of communicable diseases in times of outbreak. The overall socio-demographic profile of the major component index indicates that whereas West Mamprusi District (0.45) is more vulnerable than Savelugu/Nanton Municipality (0.43), Yapalsi (0.50) and Jana (0.41), both in the Savelugu/Nanton Municipality, are the most and least vulnerable communities, respectively (Table 2.4).

The second major component of the LVI is livelihood strategies and it consists of four sub-components. West Mamprusi District recorded a higher percentage (47%) of farmers working outside their communities as a primary source of income within the past twelve months, with Wungu being the community with the highest percentage among the studied communities. About 42.8 percent of sampled farmers in the pooled data had worked outside their communities within the period under study. Declining soil fertility and erratic rainfall patterns were reported to be responsible for the migration of youth to Southern Ghana in search of non-existent jobs. This out-migration of the active labour force tends to make farmers more vulnerable as they may learn certain negative attitudes and contract deadly diseases from the cities. The findings revealed that about 66.8 percent of sampled farmers depend solely on agriculture as their source of income, with farmers in Savelugu/Nanton Municipality (71%) being the most dependent on agriculture and least diversified (0.34), making it the most susceptible to climatic stresses. Between the two districts, Savelugu/Nanton had the higher percentage of farmers not owning their farmland (60%). Lands are vested in the hands of chiefs and 'Tindanas' in the Northern Region, making it difficult for non-royal members to own farmland. The situation for female farmers is worse, to the extent that they cannot rotate farmlands during infertile periods to increase crop yields, making them more vulnerable. Aggregating the four sub-components of the livelihood strategies major component makes West Mamprusi District (0.57) more vulnerable than the Savelugu/Nanton Municipality (0.56) (Table 2.4).

Social network is the third major component of the LVI and consists of four sub-components. Farmers in the West Mamprusi District received more assistance (8%) than they gave (5%), compared to farmers in the Savelugu/Nanton District. Zangum is the community with the highest percentage of farmers who received more assistance than they gave within the past twelve months. The assistance is usually in the form of money, care during sickness including providing medicine, sales of crops, and remittances (Table 2.4).

Table 2. 4: Computed Sub-component Indices for the Study Communities and Districts

Major Component	Sub-Component of the Livelihood Vulnerability Index	Index									
		Community (S_{ci})							District(S_{di})		
		Diare	Jana	Yapalsi	Zangum	Kata-Banawa	Kukua	Wungu	Savelugu/ Nanton		West Mamprusi
Socio-Demographic Profile	Dependency ratio.	0.12	0.13	0.10	0.12	0.11	0.10		0.11	0.12	0.11
	Percentage of farmers from female-headed households.	0.21	0.33	0.11	0.35	0.24	0.31		0.45	0.22	0.34
	Average age of female farmers.	0.68	0.56	0.70	0.80	0.89	0.61		0.66	0.65	0.64
	Percentage of farmers from households without latrines.	0.74	0.53	0.83	0.76	0.61	0.46		0.63	0.70	0.61
	Percentage of farmers who have never attended school.	0.71	0.69	0.92	0.84	0.74	0.83		0.78	0.77	0.79
	Percentage of farmers with orphans in their households.	0.34	0.19	0.58	0.19	0.37	0.43		0.43	0.27	0.35
	Average Number of persons per room.	0.23	0.42	0.27	0.33	0.32	0.26		0.28	0.28	0.29
Livelihood Strategies	Percentage of farmers who have worked in a different community.	0.53	0.17	0.44	0.43	0.37	0.51		0.58	0.38	0.47
	Percentage of farmers dependent solely on agriculture as a source of income.	0.76	0.67	0.69	0.70	0.71	0.43		0.65	0.71	0.63
	Percentage of farmers who do not own their farmland.	0.55	0.58	0.69	0.57	0.68	0.40		0.68	0.62	0.59
	Average Agricultural Livelihood Diversification Index.	0.51	0.52	0.55	0.56	0.58	0.69		0.56	0.53	0.60

Social Network	Average Receive: Give ratio.	0.49	0.50	0.55	0.82	0.65	0.55		0.54	0.51	0.65
	Average Borrow: Lend Money (range: 0.5–2).	0.45	0.35	0.47	0.50	0.48	0.45		0.55	0.42	0.50
	Percentage of farmers who have not gone to their local government for assistance in the past 12 months.	0.92	1.00	1.00	0.97	1.00	1.00		1.00	0.97	0.99
	Percentage of farmers who belong to an FBO.	0.32	0.31	0.25	0.13	0.48	0.49		0.25	0.29	0.31
Health	Average time to health facility (minutes).	0.40	0.58	0.80	0.40	0.30	0.14		0.38	0.59	0.31
	Percentage of farmers with a chronic illness.	0.45	0.17	0.33	0.41	0.29	0.49		0.53	0.32	0.39
	Percentage of farmers who had to miss work or school in the last 12 weeks due to illness.	0.58	0.69	0.58	0.51	0.34	0.29		0.28	0.62	0.35
	Average Malaria Exposure* Prevention Index (0–12).	0.31	0.31	0.41	0.19	0.26	0.22		0.26	0.34	0.23
Food	Percentage of farmers dependent on personal farm for food.	0.42	0.58	0.64	0.73	0.50	0.83	0.75	0.55		0.70
	Average number of months farmers struggle to find food (range: 0–12).	0.22	0.27	0.12	0.40	0.39	0.25	0.46	0.20		0.38
	Average Crop Diversity Index (range: 0–1).	0.42	0.70	0.61	0.37	0.32	0.29	0.42	0.57		0.35
	Percentage of farmers who do not save crops.	0.53	0.67	0.69	0.92	0.82	0.74	0.65	0.63		0.78
	Percentage of farmers who do not save seeds.	0.89	0.89	0.97	0.86	0.84	0.83	0.80	0.92		0.83
Water	Percentage of farmers reporting water conflicts.	0.97	0.67	0.44	0.78	0.29	0.49	0.80	0.70		0.59
	Percentage of farmers who	1.00	0.92	0.81	1.00	0.95	0.77	0.50	0.91		0.91

	utilize a natural water source.									
	Average time to water source (minutes).	0.37	0.46	0.38	0.54	0.16	0.14	0.83	0.43	0.20
	Percentage of farmers who do not have a consistent water supply.	0.87	0.92	0.83	0.68	0.42	0.31	0.83	0.87	0.44
	Inverse of the average number of litres of water stored per farmer (range: >0–1).	0.01	0.01	0.01	0.02	0.02	0.02	0.02	0.01	0.02
Natural Disaster and Climate variability	Average number of flood, bushfires and drought events in the past 15 years.	0.93	0.59	0.93	0.69	0.83	0.62	0.93	0.82	0.77
	Percentage of farmers who lost crops as a result of the floods, drought and bushfires.	0.76	0.36	0.69	0.81	0.42	0.60	0.65	0.61	0.62
	Percentage of farmers who did not receive a warning about the pending natural disasters.	0.61	0.39	0.50	0.49	0.42	0.34	0.43	0.41	0.42
	Percentage of farmers with an injury or death as a result of the most severe natural disaster in the past 6 years.	0.87	0.89	0.81	0.81	0.74	0.83	0.90	0.85	0.82
	Mean standard deviation of the daily average maximum temperature by month.	0.71	0.71	0.71	0.68	0.68	0.68	0.68	0.71	0.68
	Mean standard deviation of the daily average minimum temperature by month.	0.39	0.39	0.39	0.29	0.29	0.29	0.29	0.39	0.29
	Mean standard deviation of average precipitation by month.	0.79	0.79	0.79	0.68	0.68	0.68	0.68	0.79	0.68

Table 2. 5: Two Sample t-test Results on Differences in Means of Major LVI Components between Savelugu/Nanton and West Mamprusi

District	Obs.	Mean	Std Error	Std Dev	t-test	P - value	d.f.
Socio-demographic profile							
Savelugu/Nanton	110	0.448	0.004	0.040	-4.4723	0.000	258
West Mamprusi	150	0.465	0.002	0.021			
Pooled data	260	0.457	0.002	0.314			
Livelihood strategies							
Savelugu/Nanton	110	0.556	0.005	0.051	-2.6596	0.008	258
West Mamprusi	150	0.571	0.003	0.039			
Pooled data	260	0.565	0.002	0.045			
Social network							
Savelugu/Nanton	110	0.512	0.002	0.025	-11.533	0.000	258
West Mamprusi	150	0.639	0.009	0.114			
Pooled data	260	0.585	0.007	0.108			
Health							
Savelugu/Nanton	110	0.466	0.004	0.045	25.981	0.000	258
West Mamprusi	150	0.330	0.003	0.039			
Pooled data	260	0.388	0.005	0.079			
Water							
Savelugu/Nanton	110	0.579	0.006	0.062	7.696	0.000	258
West Mamprusi	150	0.482	0.010	0.122			
Pooled data	260	0.523	0.007	0.112			
Food							
Savelugu/Nanton	110	0.573	0.005	0.056	-6.405	0.000	258
West Mamprusi	150	0.608	0.003	0.031			
Pooled data	260	0.593	0.003	0.046			
Natural disaster & climate variability							
Savelugu/Nanton	110	0.580	0.003	0.032	15.954	0.000	258
West Mamprusi	150	0.524	0.002	0.023			
Pooled data	260	0.547	0.002	0.039			

Source: Authors' Computation from field survey, 2016.

Farmers who received assistance are better able to adapt to climatic stresses, thereby making them less vulnerable. West Mamprusi District had a higher percentage of farmers borrowing (26%) more than they lent (2%) within the last twelve months, with a borrow-lend ratio of 1.25. In comparison, 29 percent of farmers in the Savelugu/Nanton Municipality borrowed, with only 20 percent of farmers lending to other people, giving a borrow-lend ratio of 1.14. Overall, 27 percent of farmers in the Northern Region borrowed from friends and other relatives, while 9.6 percent lent money to friends and other relatives within the last twelve months. Farmers with a high borrow-lend ratio are more vulnerable. About 98.5 percent of sampled farmers in the Northern Region reported not going to their District assembly for assistance for the past twelve months. Farmers rather prefer soliciting assistance from friends and relatives than going to their local government. The study also revealed that about 70.9 percent of respondents in Savelugu/Nanton Municipality belonged to social groups (i.e., FBO), relative to 66.7 percent of respondents in West Mamprusi District. Memership of FBO is a major source of farming information to farmers. In terms of overall vulnerability to social network, West Mamprusi District was more vulnerable (0.65) than the Savelugu/Nanton Municipality (0.51) (Table 2.4).

The fourth major component is health, consisting of four sub-components. Farmers in Savelugu/Nanton Municipality on average spend more time to reach healthcare facilities than those in West Mamprusi District. Inaccessibility of health services reduces smallholder women rice farmers' health status, rendering them more vulnerable. Over 36 percent of sampled farmers in the Northern Region had a chronic disease, with West Mamprusi District having a higher percentage (39.3%) than Savelugu/Nanton Municipality (31.8%). A relatively larger percentage of farmers in the Savelugu/Nanton Municipality (61.8%) reported missing work or school in the past six months due to illness, compared to farmers in the West Mamprusi District (35.3%). The region in general recorded 46.5 percent of sampled farmers missing work or school as a result of sickness. Sickness increases the vulnerability level of farmers. This is, however, lesser than the 64.2 percent of people in rural savannah that halted their daily activities due to illness or injury two weeks prior to the fifth round of the Ghana living standards survey (GSS, 2008). Farmers in Savelugu/Nanton Municipality reported being more vulnerable to malaria (0.34) than those in West Mamprusi District (0.23). Among the study communities, Yapalsi is the most vulnerable to malaria. Averaging the four sub-components of the health major components finds Savelugu/Nanton

District (0.47) is more vulnerable than West Mamprusi District (0.32) (Table 2.4).

The fifth food major component is food. West Mamprusi District had a higher percentage of sampled farmers depending solely on personal farms (70%) than farmers in Savelugu/Nanton Municipality (54.6%). Farmers who depend solely on their personal farm as a source of food are more vulnerable to climatic stresses than those who have other sources of food. The average number of months farmers struggle to find food was higher in West Mamprusi District (2.27), making smallholder women rice farmers in the district more vulnerable to climate-related events than farmers in Savelugu/Nanton Municipality (1.22). These months are mostly between planting (i.e., June) and harvest (i.e., October). Farmers in West Mamprusi District were found to grow more diversified crops (2.5) than farmers in Savelugu/Nanton Municipality (1.9). Farmers with less crop diversification are more vulnerable to climatic stressors. About 62.7 percent and 91 percent of farmers in Savelugu/Nanton Municipality did not save crops and seeds, respectively, compared to 78 percent and 83.3 percent of farmers in West Mamprusi District. Farmers who save crops and seeds get higher income from the sales of their farm produce, are more food secure and do not struggle for seeds in subsequent farming seasons than farmers who do not. West Mamprusi District, with an LVI of 0.61, appeared to be more vulnerable in terms of food than Savelugu/Nanton Municipality, with an LVI of 0.57 (Table 2.4).

Water is the sixth major component of the LVI, with five sub-components. Savelugu/Nanton Municipality had a higher percentage of farmers reporting water conflict (70%) than those in West Mamprusi District (59.3%), indicating their higher level of vulnerability. Conflict is a disintegrative social phenomenon and increases vulnerability. Almost 91 percent of sampled farmers in both districts utilise water from natural sources such as rain, natural dams, streams, lakes and rivers. This is greater than the percentage of households in the rural savannah who source water from natural sources. The utilisation of natural water sources has a greater likelihood of increasing farmers' vulnerability to waterborne diseases such as bilharzias, and to water shortage due to the unreliable nature of natural water sources resulting from erratic rainfalls. The majority of sampled farmers in Savelugu/Nanton Municipality (87%) reported not having a consistent water supply compared to their counterparts in West Mamprusi District (44%). The situation is worst in the dry season with the drying up of the few natural wells, lakes and streams which serve as the main source of water for farmers in the

Municipality, thereby making their lives more precarious. A minority of farmers in both districts store water, with Savelugu/Nanton Municipality being the least (0.01). The overall index of the water major component indicates that Savelugu/Nanton Municipality (0.58) is more vulnerable than West Mamprusi District (0.43) (Table 2.4).

The last but not the least major component is natural disasters and climate variability, which also consists of seven sub-components, as presented Table 2.4. Savelugu/Nanton District is more vulnerable to floods, droughts and bushfires (0.82) than West Mamprusi District (0.77). Almost half of the respondents in both Savelugu/Nanton Municipality (40.9%) and West Mamprusi District (42.0%) did not receive warnings of impending natural disasters, thereby resulting in most farmers (60.9% and 62.0%, respectively) losing crops as a result of floods, droughts and bushfires. Savelugu/Nanton Municipality had a higher percentage of farmers (85.5%) with natural disaster-related injuries than West Mamprusi District (82.0%). Though both districts are in a semi-arid climatic region, an analysis of the meteorological data revealed that Savelugu/Nanton received more precipitation over the past twenty-nine years than West Mamprusi District. However, there is very little difference between the two districts in terms of average maximum and minimum temperatures over the same period. Thus, when all seven sub-components of the natural disaster and climate variability are aggregated, Savelugu/Nanton Municipality (0.57) was more vulnerable than West Mamprusi district (0.52) (Table 2.4).

The vulnerability levels of the communities in the two districts are as follows: Diare was the most vulnerable community in terms of water and natural disasters and climate variability; Yapalsi was the most vulnerable community to health and socio-demographic profile; Wungu was most vulnerable in terms of livelihood strategies; Kata-Banawa was most vulnerable to social network; while Zangum is the most vulnerable community to food (Figure 2.2). For the districts, Savelugu/Nanton Municipality is more vulnerable in terms of health, water, and natural disasters and climate variability, while West Mamprusi District is more vulnerable in terms of socio-demographic profile, livelihood strategies, social network and food (Figure 2.3). Averaging all seven major components, farmers in Savelugu/Nanton Municipality were more vulnerable (0.53) than those in West Mamprusi District (0.51).

Figure 2. 2: Vulnerability spider diagram of study communities

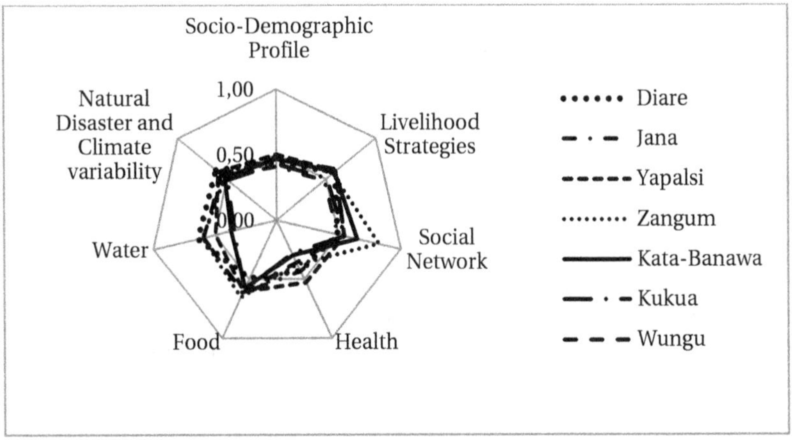

Source: Authors' Computation from field survey, 2016.

Figure 2. 3: Vulnerability Spider Diagram of Study Districts in Northern Region

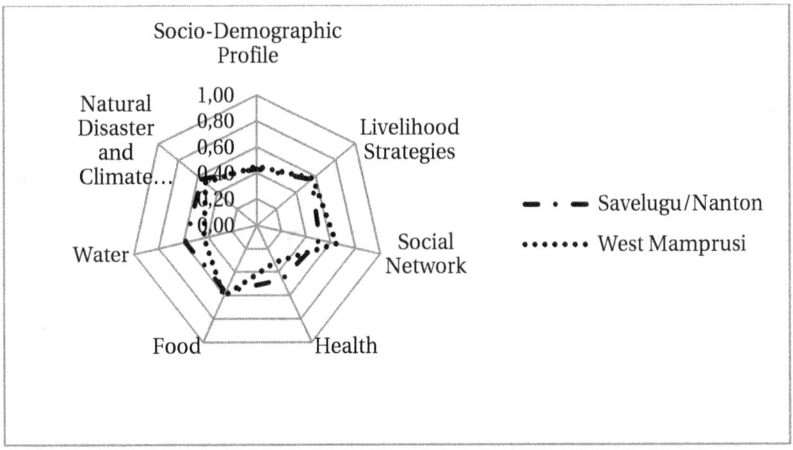

Source: Authors' Computation from field survey, 2016.

The IPCC definition of vulnerability takes into consideration adaptive capacity, sensitivity and exposure of a system to climatic stress. The LVI_{IPCC} was computed by further grouping the seven major components into these three main contributory factors, namely adaptive capacity, exposure and sensitivity.

The computed contributory indices revealed that Savelugu/Nanton Municipality is more sensitive (0.55) and exposed (0.32) to climate variability, while West Mamprusi District is more vulnerable in terms of farmers' adaptive capacity (0.53). The LVI_{IPCC} for Savelugu/Nanton District and West Mamprusi District are -0.09 and -0.11, respectively. This suggests that the adaptive capacities of farmers in both districts fall short of their exposures to climatic stresses (Table 2.6).

Farmers' adaptive capacities in both districts fall short of the extent to which they are exposed to climate variability and are thus more sensitive to climatic stresses. Using the LVI_{IPCC}, farmers in Savelugu/Nanton Municipality are more vulnerable to climate-related events such as floods, drought and bushfires than those in West Mamprusi District. For the communities, women rice farmers in Zangum have the least adaptive capacity, while farmers in Diare are the most exposed to climatic stressors. Jana and Zamgum are the most sensitive communities to climate variability (Table 2.6).

A comparison of the IPCC vulnerability indices was conducted and the results of the independent two sample *t*-test showed that there are no significant differences in vulnerability of farmers in the two districts in terms of the IPCC vulnerability contributory factors ($p<0.01$) (Table 2.7).

Table 2. 6: Computed Vulnerability Contributory Factors and LVI_{IPCC} Indices

Contributory Factor	Index ($CF_{c/di}$)								
	Community (CF_{ci})							Districts(CF_{di})	
	Diare	Jana	Yapalsi	Zangum	Kata-Banawa	Kukua	Wungu	Savelugu/ Nanton	West Mamprusi
Adaptive capacity	0.49	0.45	0.54	0.58	0.54	0.48	0.53	0.48	0.53
Sensitivity	0.53	0.56	0.55	0.56	0.42	0.41	0.54	0.55	0.46
Exposure	0.61	0.33	0.32	0.31	0.27	0.26	0.28	0.32	0.30
LVI_{IPCC}	0.06	-0.07	-0.12	-0.15	-0.11	-0.09	-0.14	-0.09	-0.11

Source: Authors' computation from field survey, 2016.

Table 2. 7: Two Sample t-test Results of Differences in Means of the IPCC Vulnerability Contributory Factors for Savelugu/Nanton and West Mamprusi

District	Obs.	Mean	Std Err	Std Dev	t-test	P - value	d.f.
Adaptive capacity							
Savelugu/Nanton	110	0.492	0.003	0.037	-8.759	0.000	258
West Mamprusi	150	0.532	0.002	0.036			
Pooled data	260	0.516	0.003	0.041			
Sensitivity							
Savelugu/Nanton	110	0.545	0.001	0.012	9.823	0.000	258
West Mamprusi	150	0.481	0.005	0.004			
Pooled data	260	0.509	0.004	0.058			
Exposure							
Savelugu/Nanton	110	0.531	0.002	0.017	3.0636	0.000	258
West Mamprusi	150	0.517	0.004	0.044			
Pooled data	260	0.523	0.002	0.358			

2.4 Conclusions

From the findings of the study, the following conclusions can be drawn. Overall, most farmers (58%) in the Northern Region of Ghana are highly vulnerable to climate variability. Farmers in Savelugu/Nanton Municipality are significantly more vulnerable to climate variability than those in West Mamprusi District. Farmers in Savelugu/Nanton are more vulnerable in terms of health, water, and natural disasters and climate variability than farmers in West Mamprusi District who are more vulnerable in terms of socio-demographic profile, livelihood strategies, social network and food. Farmers in Savelugu/Nanton Municipality are more exposed and sensitive to climatic stresses than farmers in West Mamprusi District who have the least adaptive capacity. The adaptive capacities of smallholder women rice farmers in both districts fall short of the extent to which they are exposed to natural disasters and climate variability. These women farmers are thus highly sensitive to climatic stresses.

Based on the findings, the following informed policy recommendations are made to help reduce the vulnerability of women farmers in the Northern Region: The Ministry of Health and other health-based NGOs should build Community-based Health Planning and Services (CHPS) compounds for communities in the Savelugu/Nanton Municipality to enhance access to healthcare and reduce the time spent in reaching healthcare centers. The Ministry of Water Resources and Ministry of Local Government through the Municipal assembly including NGOs involved in water and sanitation management should provide communities in the Savelugu/Nanton with household latrines and portable water sources such as boreholes and mechanized wells. This will reduce the time women spend in search of water, the prevalence of waterborne diseases contracted using contaminated water, and open defecation with its associated health hazards. West Mamprusi District should be given priority in terms of government and NGO's social interventions in the form of developmental and income generation projects. Such projects should involve the community senior high schools and the Youth Enterprise Support program to increase livelihood diversification and reduce out-migration of the economically active population in the district.

Acknowledgements

This study was partly funded by the A. G. Leventis Foundation Scholarship Scheme, College of Basic and Applied Science, University of Ghana, Legon.

References

Adger, W. N. (1999) Social Vulnerability to Climate Change and Extremes in Coastal Vietnam. *World Development,* 27(2), 249-269.

Amikuzuno, J., and Donkoh, S. A. (2012). Climate variability and yield of major staple food crops in Northern Ghana. *African Crop Science Journal,* 20(2), 349-360.

Asante, F. A., and Amuakwa-Mensah, F. (2014) Climate change and variability in Ghana: Stocktaking. *Climate , 2015(3),* 78-99.

Asare, A. (2000) *Operational guidelines on community forest (management) committees.* ITTO/FSD Collaborative Off-Reserve Forest Management Project.

Blackden, M., and Wodon, Q. (Eds.). (2006) Gender, Time use, and Poverty in Sub-Saharan Africa. *World Bank Working Paper No. 73.*

Brooks, N. (2003) Vulnerability, risk and adaptation: A conceptual framework. *Tyndall Center for Climate Change research, Working paper 38.*

City of Williamsburg Virginia's Colonial Capital (2004). Williamsburg Emergency Management. Virginia, USA.

Cutter, S. L., Barnes, L., Berry, M., Burton, C., Evans, E., Tate, E., et al. (2008) A place-based model for understanding community resilience to natural disasters. *Global Climate Change, 18,* 598-606.

Deressa, T. T., Hassan, R.M., Ringler, C. Alemu, T. and Yesuf, M. (2009) Determinants of farmers' choice of adaptation methods to climate change in the Nile Basin of Ethiopia. *Global Environmental Change, 19*(2), 248-255.

DHS (Demographic Health Survey) (2006) Measure DHS: model questionnaire with commentary. Basic Documentation, Number 2.

Doss, C. R., and Morris, M. L. (2000) How does gender affect the adoption of agricultural innovations? The case of improved maize technology in Ghana. *Agricultural Economics, 25,* 27-39.

Etwire, P. M., Al-Hassan, R. M., Kuwornu, J. K. M., and Osei-Owusu, Y. (2013) Application of the Livelihood Vulnerability Index in Assessing Vulnerability to Climate Change and Variability In Northern Ghana. *Journal of Environment and Earth Science, 3(2),* 157-170.

Food and Agricultural organisation, FAO. (2011) *FAO-Adapt Framework Programme on climate change Adaptation.* Rome, Italy.

Fussel, H. M., and Klein, T. R. (2006) Climate change vulnerability assessment: an evolution of conceptual thinking. *Climate Change, 75,* 301-329.

Fussel, H.M. (2006) Vulnerability: A generally applicable conceptual framework for climate change research. *Global Environmental Change, 17,* 155-167.

Gbetibouo, G. A., Ringler, C., and Hassan, R. (2010) Vulnerability of the South African farming sector to climate change and variability: an indicator approach. *Nat. Resource Forum, 34,* 175-87.

Ghana Statistical Service, GSS (2008) *Ghana Living Standards Survey.* Report of Fifth Round (GLSS 5).

Ghana Statistical Service GSS. (2012) *2010 Population and Housing Census.* Analytical Report, Northern Region.

Ghana Statistical Service GSS (2014) *Ghana Living Standards Survey Round 6 (GLSS 6).* Poverty Profile in Ghana (2005-2013), Accra, Ghana.

Hahn, M. B., Riederer, A. M., and Foster, S. O. (2009) The Livelihood Vulnerability Index: A pracgmatic approach to assessing risks from climate variability and change - A case study in Mozambique. *Global Climate Change, 88,* 74-88.

Instituto Nacional de Estatistica, Mocambique (NationalInstituteof Statistics, Mozambique) (2007) Base de dados – Clima (Database – Climate). http://196.22.54.6/ pxweb2007/Database/INE/01/13/13.asp (accessed 23 December 2007).

IPCC (2007) *Climate Change 2007: Sythesis Report.Contribution of Working Groups I, II and III to the Fourth Assessment Report of the Intergovernmental Panel on Climate Change.* In: IPCC, 104.

IPCC (2014) *Climate Change: Impact, Adaptation and Vulnerability. Contributions of Working Groups I, II and III to the Fourth Assessment Report.* Cambridge University Press, Cambridge, UK.

Juana, J. S., Kahaka, Z., and Okurut, F. N. (2013) Farmers' perception and adaptation to climate change in Sub-Sahara Africa: A synthesis of empirical studies and implications for public policy in African Agriculture. *Journal of Agricultural Science, 5(4)*, 121 - 135.

Kelly, P. M. and Adger, W. N. (2000) Theory and Practice in Assessing Vulnerability to Climate Change and Facilitating Adaptation. *Climatic Change, 47(4)*, 325-352.

Koru, B., and Holden, S. (2008) Differences in maize productivity between male- and female-headed households in Uganda. *Ethiopian Development research Institute.*

Madhuri, S. R., Divase, G. T., Devkar, T. B., and Tekade, A. R. (2014) Solubility and Bioavailability Enhancement of Poorly Aqueous Soluble Atorvastatin: In Vitro, and In Vivo Studies. *2014*. Hindawi Publishing Corporation.

Ministry of Environment, Science, Technology and Innovation, MESTI (2013) *Ghana National Climate Change Policy*. Accra, Ghana.

Ministry of Food and Agriculture, MoFA. (2015) Statistics, Research and Information Directorate, Accra, Ghana.

Morton, J. F. (2007) The impact of climate change on smallholder and subsistence agriculture. *Proceedings of the National Academy of Sciences, 104(50)*, 19680-19685.

Moser, C. O. (1998). The asset vulnerability framework: reassessing urban poverty reduction strategies. *World development*, 26(1), 1-19.

Nelson, R., Kokic, P., Crimp, S., Martin, P., Meinke, H., Howden, S. M., et al. (2009) The vulnerability of Australian rural communities to climate variability and change: Part II - Integrating impacts with adaptive capacity. *Environmental Science and Policy, 13(2010)*, 18-27.

Pandey, R. and Jha, S. (2012) Climate vulnerability index - measure of climate change vulnerability to communities: a case study of rural lower Hima-laya, India. Mitig. Adapt. Strategies. *Global Change*, 487-506.

Ruxton, G. D. (2006) *Unequal Varian test*. Oxford, UK: Oxford University Press.

Shah, U. K., Dulal, H. B., Johnson, C., and Baptiste, A. (2013) Understanding livelihood vulnerability to climate change: Applying the livelihood vulnerability index in Trinidad and Tobago*Regional Geoforum*, 47*(2013)*, 125-137.

Simane, B., Zaitchik, B. F., and Foltz, J. D. (2016) Agroecosystem specific climate vulnerability analysis:application of livelihood index to a tropical highland region. *Mitig Adapt strateg Glob Change, 21*, 39-65.

Sullivan, C., Meigh, J.R., and Fediw, T.S. (2002) Derivation and testing of the water poverty index phase 1, Final Report. Department for International Development, UK, 2002.

Turner, B. L., Kasperson, R. E., Matson, P. A., McCarthy, J. J., Corell, R. W., Christensen, L., et al. (2003) A framework for vulnerability analysis in sustainability science. 100(14), 8074-8079.

UNDP (2007) Human Development Report 2007/2008, Fighting Climate Change: Human Solidarity in a Divided World. UNDP, New York.

Van Der Geest K. (2004) "*We're managing!" Climate Change and Livelihood Vulnerability in Northwest Ghana,* African Studies Centre, Leiden, 288pp.

WHO/Roll Back Malaria, (2003) Economic impact of malaria: household survey.

World Bank (1997) Survey of living conditions: Uttar Pradesh and Bihar. Household Questionnaire, December 1997–March 1998.

WHO/RBM (2003) Position of WHO's Roll Back Malaria Department on Malaria Treatment Policy. November 2003. Geneva (http://www.who.int/malaria/treatment.html; accessed 28.07.2005).

CHAPTER 3

Food crop vulnerability to climate variability and change at the household level in Bui Division, Northwest Cameroon

Suiven John Paul Tume[1], Mbu Dora Nyuykighan[2], Moye Eric Kongnso[3], Bankui Andrew Dzeaye[4], Mairong Frederick Nsaikii[4], Njodzeka Gilbert Njodzeka[1]

[1]Green Care Association, Shisong, P.O. Box 141 Kumbo, Bui Division, Northwest Cameroon. [2]Department of English & Literary Studies, Federal University of Lokoja, Nigeria.

[3]Department of Geography, University of Dschang, Cameroon.

[4]Faculty of Agronomy & Agricultural Sciences, University of Dschang, Cameroon.

Abstract

Climate change and climate variability are major challenges to rural livelihoods and national economies in Sub-Saharan Africa, most of which are sustained by smallholder agriculture. These environmental challenges are confronting food crop production systems. There is a general consensus that the climate is changing and impacting both negatively and positively on different livelihood systems. Most often, the negative impacts tend to overshadow the positive outcomes. Individuals, households and communities, therefore, need to adapt to these environmental changes. Rainfall has been fluctuating, with a tendency towards a decrease at all the stations where data was collected. This is revealed by an average Rainfall Seasonality Index (SI) of 0.89 (i.e., markedly seasonal with a longer dry season). To ascertain the vulnerability of food crop production systems to climate variability and change, a crosssection of 200 households were surveyed in Bui Division of Northwest Cameroon. Household views revealed that there have been considerable drops in the main food crops (i.e., maize, beans and solanum potato) on which livelihoods depend. At the household and village levels, adaptation strategies include soil and water conservation, micro-irrigation, multiple farms, inter/relay cropping, mixed farming, organic manure usage and drought-resistant crops. These adaptations are constrained by limited know-how, gender imbalances in land management and other resources, cultural issues, and institutional failures amongst others. The low adaptive capacity of most rural households is due to extreme poverty and frequent irregular onset of the first rains. The full implications of climate change for food crop production are, however, currently not clearly understood by most rural people. Many still associate crop failures and food security threats to superstition. At the more local level, the poor often cannot adopt diversification as an adaptive strategy and often have very limited options available.

Keywords: Adaptation; Farming Households; Food Production Systems; Food Security; Limitations

3.1. Introduction

Climate change vulnerability is the degree to which a system is susceptible to the adverse effects of climate change, including climate variability and extremes (IPCC, 2007; Thorlakson, 2011). Africa is one of the most, if not the most, vulnerable continents to the effects of climate change (Patt and Winkler, 2007). This has to do not only with climatic and physical features, but also with social and political conditions. Besides a community's exposure to climatic disruptions, vulnerability is also linked to sensitivity of adaptive capacities (OECD, 2009). Climate change interacts with non-climate drivers and stressors to intensify vulnerability of agricultural systems (Niang et al., 2014). Increasing temperatures and changes in precipitation are very likely to reduce cereal crop productivity. This is a threat to food security. New evidence is emerging that high-value perennial crops could also be adversely affected by temperature rises (Smit and Pilifosova, 2001; Kolawole et al., 2014). Pest, weed and disease pressure on crops and livestock is expected to increase as a result of climate change combined with other factors. Moreover, new challenges to food security are emerging as a result of strong urbanization trends in Africa and increasingly globalized food chains, which require a better understanding of the multi-stressor context of food and livelihood security in both urban and rural areas in Africa (Niang et al., 2014). Urgent technical, financial and institutional support is needed to improve food security and make rural farmers' livelihoods resilient to climate change (Smit and Pilifosova, 2001; OECD, 2009; Harvey et al., 2014; Kolawole et al., 2014).

Increasing temperatures and declining rainfall patterns, as well as increasing frequency of extreme climate events (i.e. droughts and floods), are the expected future weather patterns within the tropics (Smit and Pilifosova, 2001; Kolawole et al., 2014, 2016; Luxon and Pius, 2014). Africa is projected to be the region most vulnerable to the impacts of climate variability and change especially in the agriculture and water sectors (Niang et al., 2014). This has dire consequences as many African people depend on agriculture for their livelihoods. The impacts of climate change and variability are felt locally at household and community levels, and regional and national levels (Kolawole et al., 2016). In most cases, adaptation occurs at a local level in ways that are usually unnoticed and unassisted by national governments (IPCC, 2014). Down-scaling the vulnerability to climate change and variability to the household and community levels is proactive to designing indigenous adaptation strategies (IPCC, 2014; Kolawole et al., 2016).

Adaptation reduces the impacts of climate stresses on human and natural systems (Abeysinghe and Prolois, 2016; Sharmais et al., 2016). It consists of a multitude of behavioural, structural and technological adjustments. Adaptation measures can vary in timing (i.e. anticipatory vs reactive; ex-ante vs ex-post), scope (i.e. short-term vs long-term; localised vs regional), purposefulness (i.e. autonomous vs planned; passive vs active) and adapting agent (i.e. private vs public; societies vs natural systems) (Smit & Pilifosova, 2001; Töpfer and Hunter, 2002; UNFCCC, 2006, 2011; IPCC, 2007, 2014). Examples of adaptation measures include changing crop varieties and altering farming practices, developing drought-resistant crops, diversifying livelihoods and land-use planning (OECD, 2009; Rhodes et al., 2014; IIED, 2015 a, b).

Climate variability and change is a major challenge to rural livelihoods and national economies in Sub-Saharan Africa. This is because most livelihoods and economies in this context are sustained by smallholder agriculture. Climate change (long-term) and variability (inter-annual, intra-annual) are among the major environmental challenges confronting food crop production. There is general consensus that the climate is changing and impacting both negatively and positively on different agricultural and livelihood systems. Most often, the negative impacts overshadow the positive outcomes. Individuals and communities, therefore, need to adapt to these environmental changes. Different farming landscapes and intricacies associated with people's livelihoods varying over time and space suggest a need for localized innovation to enhance and sustain productivity. Low adaptive capacity of most Indigenous communities is due in large part to their extreme poverty. Frequent irregular onset of rains further compound the issues, especially for agriculture which is heavily dependent on rainfall, as well as a range of macro- and micro-structural problems for rural areas. The full implications of climate change for development are, however, currently not clearly understood. At the local level, the poor often cannot adopt diversification as an adaptive strategy and often have limited diversification options available to them. Micro-financing, other social safety nets and social welfare grants, as a means of enhancing adaptation to current and future shocks and stresses, may be successful in overcoming such constraints if supported by local institutions on a sustainable basis. Local communities in Bui Division of Northwest Cameroon will, therefore, need to focus on increasing their adaptive capacity to climate variability and climate change over the long-term. Other solutions that could be considered include mainstreaming adaptation into national development processes. In this case, there are several opportunities to link disaster risk reduction, poverty and development. Where communities live with various risks, coupling risk reduction and development activities can provide additional adaptation benefits.

3.2 Materials and Methods

3.2.1 Study area

Bui is one of the seven divisions that make up the Northwest Region of Cameroon (Figure 3.1). It is located about 80 km from Bamenda (regional headquarters) on the ring road from Mezam through Ngoketunjia to Donga-Mantung. Bui Division is situated between latitudes 6°-6°20″N and longitudes 10°30″-10°60″E (Tume, 2008). This gives a typical *'Aw'* climate type (as per Köppen's 1936 classification). Bui is an orographic plateau within the Cameroon Volcanic Line (CVL) with spectacular mountainous landscapes like Mount Oku (3,011masl) and the Mbaw–Tikar Plains (about 900m asl). The division has a total surface area of about 2,795 km^2 (Divisional Delegation of Lands, Kumbo, 2008), with six sub-divisions—Kumbo, Jakiri, Nkum, Nkor–Noni, Mbiame and Oku. Fieldwork evidence gathered from July through October 2016 reveals that the area can be divided into seven agricultural basins (Table 3.1).

Figure 3. 1: Location of Bui Division

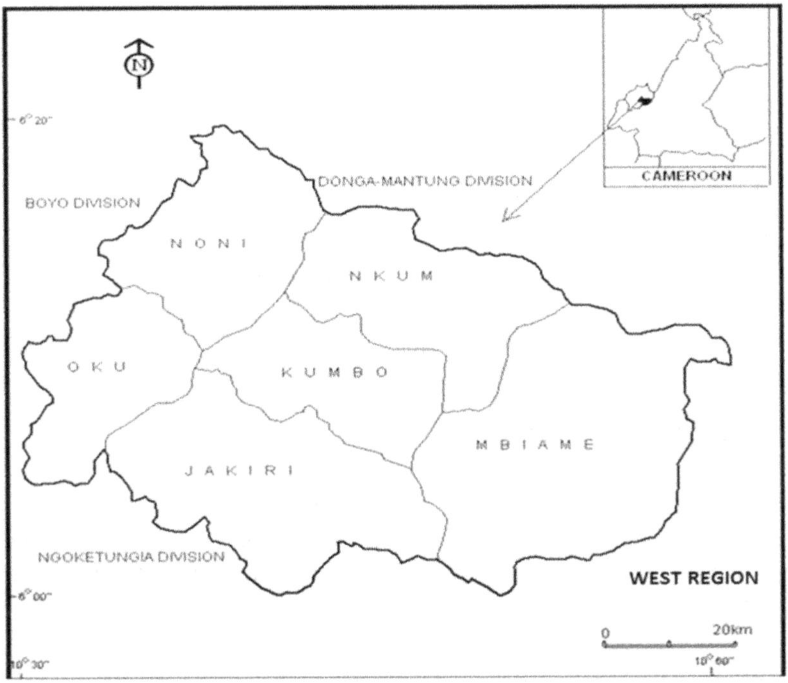

Source: Tume (2008).

Table 3. 1: Agricultural Basins of Bui Division

Agricultural Basin	Crops Cultivated
Wasi-Ber	Rice, maize, beans, cassava.
Mbaw Nso Plain	Rice, maize, soya beans, groundnuts, oil palm, cassava.
Vekovi-Simonkoh	Solanum potato, maize, beans, market gardening.
Mbam Valley (Nkum)	Market gardening, maize, beans, groundnuts, cassava, plantains.
Nkum Kov-Kumbo	Maize, solanum potato, beans.
Nkuv Kov Ndzeen-Gwarkang	Maize, cassava, beans, yams, cow pea, groundnuts.
Nkor-Lassin Plain	Groundnuts, maize, beans.

Source: Fieldwork (July-October 2016).

3.2.2 Data Collection

Field data collection was carried out using a questionnaire, a guided interview at the Divisional Delegation for Agriculture and Rural Development (Kumbo), and observations of physical manifestations of climate change in the food crop production sector. Two hundred (200) questionnaires were administered in the six sub-divisions that make up Bui Division. The questionnaire was structured into three main sections (i.e., agricultural vulnerability to climate change, climate change adaptation, limitations to adaptation). Responses from households were collated and analyzed in Microsoft Excel. These responses were presented in frequency tables and expressed as percentages. Another guided interview was administered at the Divisional Delegate of Agriculture and Rural Development (DDARD) to understand what this government service has been doing to assist rural farmers in adapting to climate change and their limitations in adaptation to climate change.

Rainfall data was collected from seven weather stations located across Bui Division, to clearly represent the spread over the agricultural basins. Three of the weather stations (i.e., Tatum, Jakiri and Shisong) have relatively longer periods of climatic data records. Rainfall anomalies were calculated for these stations to show trends. Decadal rainfall averages and percentage changes have also been calculated for each of the stations with longer periods of data collection.

The main indices used in this study to determine the impact of changing rainfall on food crop vulnerability are Mean Annual Rainfall, Seasonality Index, Standard Deviation and Coefficient of Variation. Rainfall Seasonality Index is a strong indicator of changing rainfall patterns in the tropics because it is the sum of the absolute deviations of mean monthly rainfalls from the overall monthly mean divided by the mean annual rainfall. This index is used to determine the rainfall regime. The rainfall regime is a good indicator of the current climatic conditions and it can support agricultural planning and the management of other related environmental issues.

Food crop output data was collected for Mbiame and Oku. To determine the impact of the changing climate and environmental conditions on crop outputs, anomalies were calculated and illustrated in trend graphs. These graphs represent the basic food crops produced and consumed across Bui Division at the household level.

3.2.3 Climate variability and change

The long-term mean monthly, seasonal or annual precipitation does not necessarily indicate the regularity with which given amounts of rainfall can be expected, especially in the low latitudes where rainfall is known to be highly variable from one year to another (Ayoade, 1998). Rainfall variability is an important consideration in the tropics, where rainfall tends to be more seasonally than annually variable. The less variable rainfall is, the more reliable it is, and the more variable rainfall is, the less reliable it is. This is because the index of variability is a measure of the degree of likelihood that the mean rainfall may be repeated each year, season or month, depending on the period under consideration. The distribution of rainfall over Bui Division does not only vary in space, but also with time (i.e., monthly, seasonal and annual variations). Rainfall data was collected from seven weather stations in Bui Division (i.e. Tatum, Shisong, Jakiri, Oku, Tobin, Kumbo and Takui). Mean annual rainfall (MAR), Seasonality Index (SI), Standard Deviation (SD) and Coefficient of Variation (CV) were calculated. The rainfall characteristics for Bui Division are summarized in Table 3.2.

Table 3. 2: Rainfall Characteristics for Bui Division

Station	MAR (mm)	SI	SD	CV (%)
Tatum	168.03	0.92	44.06	2.18
Shisong	173.06	0.91	2.51	1.49
Jakiri	174.77	0.91	1.13	0.65
Oku	179.46	0.92	96.38	28.84
Tobin	137.98	0.91	93.38	12.53
Kumbo	133.89	0.91	92.25	18.68
Takui	205.1	0.75	95.6	18.99

The MAR in Bui Division ranges from 133.89-205.1 mm. This is not a good indicator of change because it can be calculated from a small data set. However, rainfall standard deviations are 44.04mm (Tatum), 2.51mm (Shisong), 1.13mm (Jakiri), 96.38mm (Oku), 93.38mm (Tobin), 92.25mm (Kumbo) and 95.6mm (Takui). Despite the positive standard deviations, rainfall is increasingly unreliable in Bui Division and this is affecting the timing of sensitive phases of food crop production, from germination of crops to maturity.

The rainfall CV is also critical for water-dependent food crop production in Bui Division. Lower CV values reveal a higher reliability of rainfall, while high CV values indicate rainfall unreliability. Within the tropics, a CV threshold of 20% is critical for water-dependent activities. In Bui, the CV values were 2.18% (Tatum), 1.49% (Shisong), 0.65% (Jakiri), 28.84% (Oku), 12.53% (Tobin), 18.68% (Kumbo) and 18.99% (Takui). All but one of the stations still has reliable rainfall because the CV was less than 20%. Oku (28.84%) is a mountain community which has been greatly affected by unreliable rainfall over the past 20 years. This has altered the sub-division's agricultural calendar, thereby making food crop production more vulnerable to these changes.

Another indicator of changing climate in Bui Division is the SI. Walsh and Lawler (1981) and Olaniran (1983) stated that the seasonality of precipitation refers to the tendency for a location to receive more rainfall in certain months than in others. This makes use of the SI (Table 3.3).

Table 3. 3: Rainfall Seasonality Indices (SI) Regimes

Rainfall Regime	SI Class Limit
Very equable	≤0.19
Equable but with a definite wetter season	0.20–0.39
Rather seasonal with a short drier season	0.40–0.59
Seasonal	0.60–0.79
Markedly seasonal with long drier season	***0.80–0.99***
Most rain in three months or less	1.00–1.19
Extreme, almost all rain in one to two months	≥1.20

Source: Walsh and Lawler (1981).

The SI values in Bui are 0.92 (Tatum), 0.91 (Shisong), 0.91 (Jakiri), 0.92 (Oku), 0.91 (Tobin), 0.91 (Kumbo) and 0.75 (Takui). Six of the stations reveal that rainfall in Bui Division is markedly seasonal with a longer dry season (i.e., SI class limit 0.80-0.99). The only station with seasonal rainfall is Takui, with an SI of 0.75. The average SI for all of the stations is 0.89. Despite that the first rains are expected by mid-March in Bui Division, it has drastically altered over the past 20 years. Experience reveals that the first rains have been increasingly arriving in April and when they do come, they are not intensive enough to support food crops. Again, the inconsistent rainfall is marked by consistent droughts which extend until June. Tender food crops are often attacked by pests and diseases. This is a serious threat to food security in Bui Division. When heavy rains begin in July, they are marked by the other extreme of weather events (i.e., flooding), which destroys agricultural land and further complicates the threat to food security at the household level.

3.2.4 Rainfall variability over time

The rainfall CV only indicates the fluctuation of the mean and never gives a clear indication of the general trend whether rainfall is increasing or decreasing. Given the long rainfall data records for Tatum, Shisong and Jakiri, rainfall trends have been established to reveal the current prevailing conditions (Figures 3.2, 3.3 and 3.4).

Figure 3. 2: Slightly increasing rainfall trend, Tatum (1957-2015)

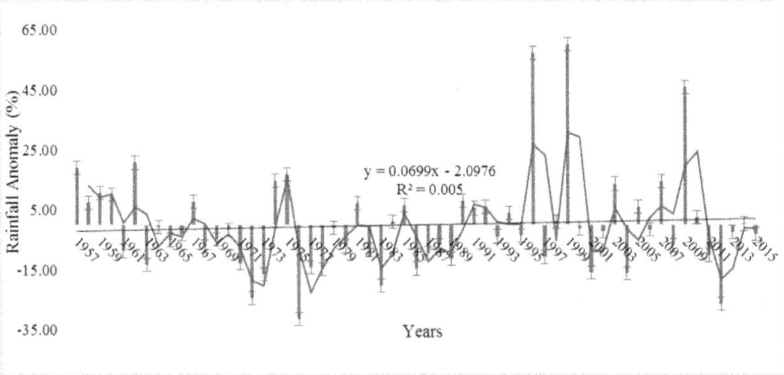

Figure 3. 3: Decreasing rainfall trend, Shisong (1975-2015)

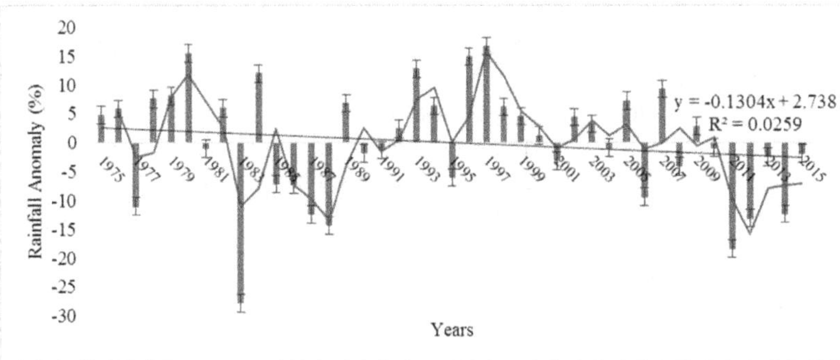

Figure 3. 4: Slightly increasing rainfall trend, Jakiri (1961-2006)

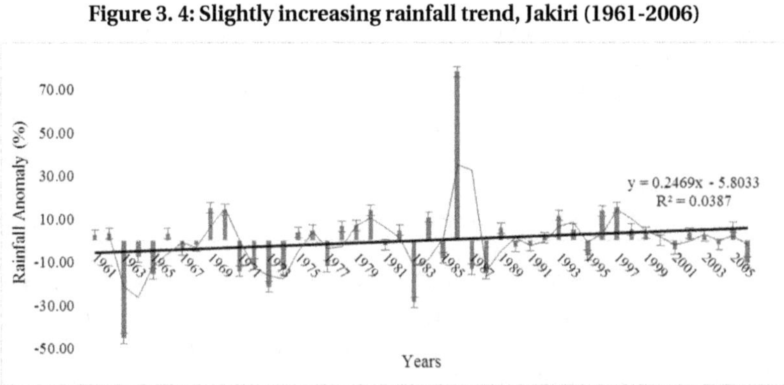

Figure 3.2 reveals an almost constant rainfall pattern for Tatum, while Figure 3.3 shows declining rainfall for Shisong and Figure 3.4 shows a slight increase in rainfall over Jakiri. Despite these disparities, the SI of over 0.9 for these stations reveals a prolonged dry season. This makes food crop production in these sub-divisions highly vulnerable to climate change.

The years 1957-1966 and 1967-1976 were marked by decreasing precipitation, while there was a slight increase in precipitation during the decades 1977-1986 and 1987-1996. Rainfall was again characterized by a downward trend during the periods 1997-2006 and 2007-2015. In Tatum, during the decades 1957-1966, 1967-1976, 1977-1996, 1997-2006, 2007-2015, rainfall variability was +3.61%, -6.03% (severe drought decade), -6.66% (severe drought decade), 4.41%, +3.18% and +1.48%, respectively. From 1957-1966, the rainfall percentage change was 3.63%. For the other decades, the percentage changes were as follows - 1967-1976 (-6%), 1977-1986 (-6.62%) 1987-1996 (4.43%), 1997-2006 (93.2%) and 2007-2015 (1.49%). This gives a cumulative rainfall reduction of 0.024% from 1957-2015. There have also been major changes to rainfall in Jakiri. From 1961-1970, the percentage change was 3.61%. For the other decades, the percentage changes were as follows -1971-1980 (3.61%), 1981-1990 (-0.98%), 1991-2000 (-4.69%) and 2001-2006 (0.78%). This gives a cumulative reduction of 2.02%. Decadal rainfall has also changed tremendously over time in Shisong. From 1975-1984, rainfall changed by +1.92%. For the other decades, the changes were as follows - 1985-1994 (-1.62%), 1995-2004 (4.07%) and 2005-2015 (-3.98%). This gives a rainfall reduction of 3.375% over the period 1975-2015.

3.2.5 Socio-economic characteristics of the respondents

Demographic characteristics of respondents revealed their susceptibility to declining food production. A total of 200 questionnaires were administered in the six sub-divisions. They were distributed as follows (Table 3.4): Kumbo (16.5%), Jakiri (17.5%), Nkum (18.5%), Oku (16.5%) and Nkor-Noni (14.5%). Among the respondents, 38.5% were male and 61.5% were female. Male-headed households were dominant (i.e., 85.5%), while 14.5% of households were headed by women. The respondents were mostly married (i.e., 69.5%), with 20.5% being single, 4.5% divorced and 5.5% widowed. Education attainment, which is a key factor in assessing household vulnerability to climate change, is as follows: no formal education (7.5%), primary education (50.5%), secondary school dropout (5.5%), General Certificate of Education, Ordinary Level (GCE O/L) (10.5%), General Certificate of Education, Advanced Level (GCE A/L) (4.5%), Teachers' Grade I (8.5%), ISTP (2.5%), HND/HPD (5.5%), BSc/BA (3.5%), MSc/MA (1.5%) and PhD (0%). A majority of the farmers had only primary education. Farmers with no formal education and those who dropped out of school were particularly reluctant to adopt innovations in their agricultural practices, thus, making them more susceptible to climate change. The outcome is declining food production, which is a major threat to food security.

Table 3. 4: Distribution of respondents

Bui Sub-Division	**Frequency**	**%**	
Kumbo	33	16.5	N: 200
Jakiri	35	17.5	Mean: 33.33
Nkum	37	18.5	
Mbiame	33	16.5	SD: 1.33
Oku	33	16.5	
Nkor-Noni	29	14.5	
Sex			
Male	77	38.5	
Female	123	61.5	
Household head			
Male	171	85.5	
Female	29	14.5	
Marital status			
Single	41	20.5	
Married	139	69.5	SD: 30.56
Divorced	9	4.5	
Widowed	11	5.5	
Education attainment			
No formal education	15	7.5	
Primary education	101	50.5	
Sec school dropout	11	5.5	
GCE O/L	21	10.5	
GCE A/L	9	4.5	SD: 14.07
Teachers' Grade I	17	8.5	
ISTP	5	2.5	
HND/HPD	11	5.5	
BSc/BA	7	3.5	
MSc/MA	3	1.5	
PhD	0	0	

Source: Fieldwork (July–October 2016).

3.3 Results and Discussion

3.3.1 Vulnerability of food crop production to climate variability and change

Basic foodstuffs that support people's daily lives have been declining in Bui Division. The division's food crop production system is rain-fed. Thus, when there is a drought, threats to food security and crop failure will be evident.

In Mbiame Sub-Division, the trend of food crop output is a dramatic reduction and the climate continues to be very uncertain. This has to do with the timing of the onset of the first rains and prevailing drought conditions throughout the growing season (Figure 3.5). Apart from maize, beans, and solanum potato, the rest of the crops have negative returns. These crops include soybeans, groundnuts, rice, cocoyam, plantains, onion and macabo.

Figure 3. 5: Decreasing food crop output trend, Mbiame Sub-Division (2006-2010)

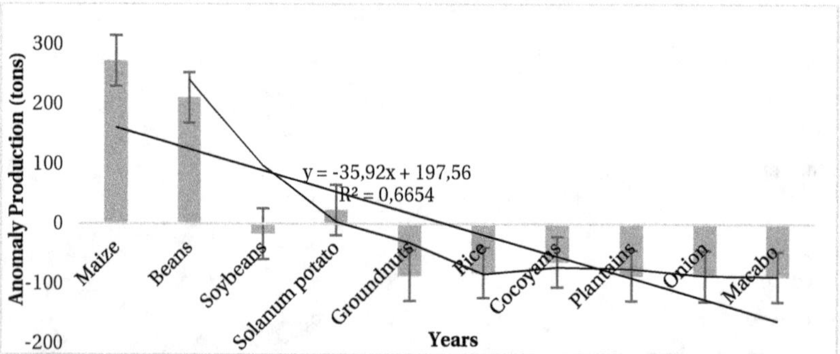

Data Source: Mbiame Council Development Plan (CDP), (2012)

In Oku Sub-Division, food production has been similarly responding to climate variability (in addition to other factors). Sources from *Mission de Developpement de la Nord-Oest* (Northwest Development Authority) (MIDENO) reveal a gradual decrease in outputs of maize and solanum potatoes from 1982-2005 (Figures 3.6 and 3.7). In contrast, the output of beans has been relatively constant over the same period (Figure 3.8).

Figure 3. 6: Relatively stable beans output, Oku

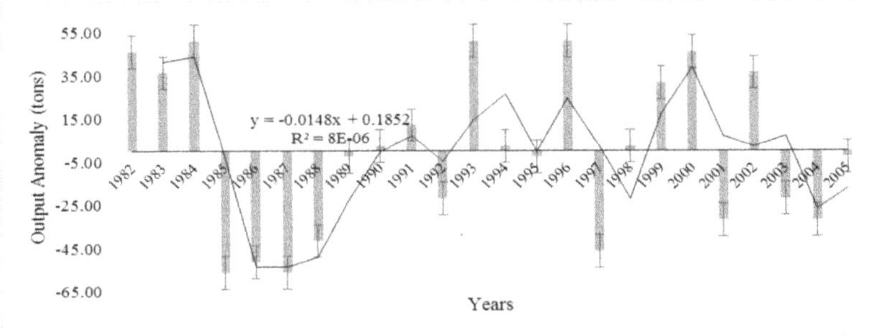

Data source: MIDENO

Figure 3. 7: Decreasing maize production trend, Oku

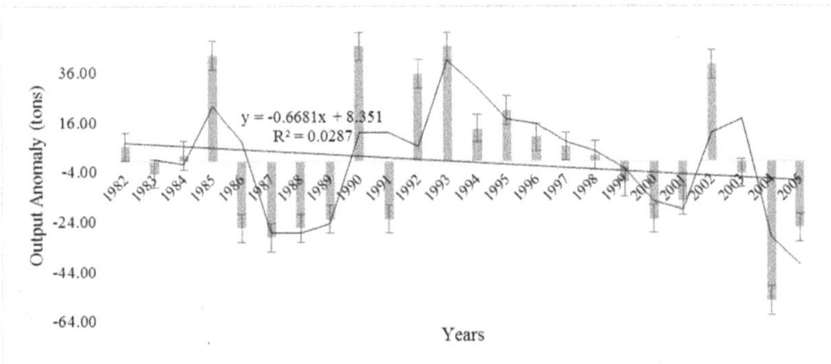

Data source: MIDENO

Figure 3. 8: Decreasing solanum potatoes production trend, Oku

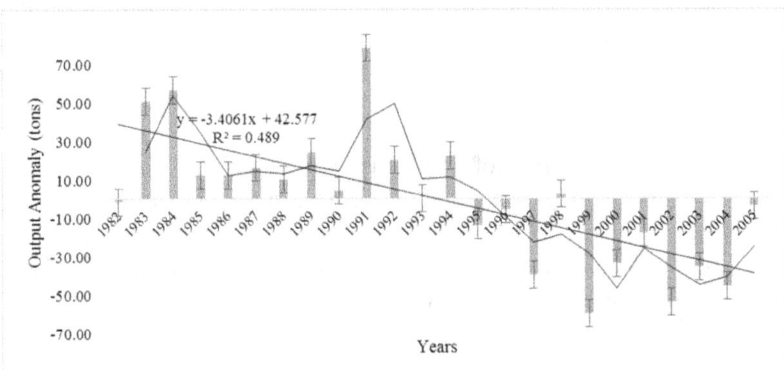

Data source: MIDENO

3.2.2 Discussion

The effects of climate change on crop and terrestrial food production are evident in several regions of the world. Negative impacts of climate trends have been more common than positive impacts (Porter et al., 2014). All aspects of food security are potentially affected by climate change, including food access and price stability (OECD, 2009; Patt and Winkler, 2007). In Bui, smallholder household food crop production is vulnerable to climate change as a result of the late onset of first rains to start the planting season, early first rains (i.e. in February), droughts, insect/pest attacks on crops, plant diseases, soil erosion, declining soil fertility, the lack of new farmland, a lack of fertilizers and inappropriate use of pesticides (Table 3.5).

Table 3. 5: Climate change vulnerability factors in crop production in Bui Division

Vulnerability factors	Frequency[1]	%
Late rains	160	80
Early rains	40	20
Droughts	200	100
Insect/pest attacks	200	100
Plant diseases	200	100
Soil erosion	133	66.5
Declining soil fertility	145	72.5
Lack of new farmland	166	83
Lack of fertilizers	153	76.5
Poor use of pesticides	167	83.5

Source: Fieldwork (July-October 2016).

The late and early onset of first rains is a serious threat to food crop production in this part of Cameroon because the agricultural system is rain-fed. Late rains are a testament to the fact that the dry season (drought season) is prolonged. On the other hand, early rains have proven to be very unreliable. This is because when the early rains come in February and farmers sow crops, the rains will then cease to fall. Such early rains are

[1] Frequency of the number of respondents that mentioned the factor.

often followed by several weeks of severe droughts, such that the tender crops wither and die. Hence, farmers are compelled to alter planting dates. This makes the agricultural sector increasingly vulnerable to uncertain rainfall patterns.

3.3.2.1 Adaptation of food crop production to climate variability and change

Rural communities have a long history of responding to climate variability and change, but with varying levels of success. These short-term coping strategies can form the basis for successful long-term adaptation strategies (OECD, 2009). However, care needs to be taken as some of these coping strategies could prove to be unsustainable over time as climate change progresses, leading to a greater risk of maladaptation (i.e., low adaptive capacity). In Bui, rural households adapt to climate change by using local seeds, pesticides, irrigation, altering planting dates, organic manures, mono-cropping, poly-cropping and incorporation of agroforestry (Table 3.6). The households have the perception that local seeds are more weatherproof than improved seeds. With the erratic onset of first rains, the farmers resort to altering planting dates. To enhance soil fertility, they have embarked on the use of organic manure and irrigation (during the dry season) for large-scale vegetable production (i.e., mono-cropping). Some farmers have also incorporated poly-cropping and agroforestry as an adaptation to climate change.

Table 3. 6: Food crop adaptation to climate change in Bui Division

Adaptation factors adopted	Frequency	%
Improved seeds	125	62.5
Local seeds	193	96.5
Use of pesticides	127	63.5
Irrigation	67	33.5
Altering planting dates	200	100
Organic manures	173	86.5
Mono-cropping	167	83.5
Poly-cropping	200	100
Agroforestry	163	81.5

Source: Fieldwork (July-October 2016).

Households are also diversifying their sources of livelihoods, rather than depending solely on agriculture. These supplementary activities include apiculture (i.e., bee farming), poultry farming, petty trading, livestock rearing and exploitation of raffia palm for wine and furniture (Table 3.7).

Table 3. 7: Supplementary livelihood activities in Bui Division

Supplementary activities	Frequency	%
Apiculture	31	15.5
Poultry farming	65	32.5
Petty trading	73	36.5
Livestock rearing	47	23.5
Exploitation of raffia palm	27	3.5

Source: Fieldwork (July-October 2016).

A considerable proportion of the population of Bui Division has embarked on small- to medium-scale livestock production (i.e., cattle). This has improved the protein intake of the people. However, this practice is not sustainable because the cattle numbers are already more than the carrying capacity of most rangelands. The outcome has been serious land degradation, competition between arable and grazing land, and frequent cultivator-grazier conflicts. This situation has accelerated environmental vulnerability to climate change.

3.3.2.2 Limitations of food crop production adaptation to climate variability and change

Limits to adaptation can emerge from the interactions between climate change, the biophysical environment and socio-economic constraints (IPCC, 2014). An adaptation limit occurs owing to the inability to avoid an intolerable risk to an actor's objectives and the sustainability of a natural system. Understanding of limits is informed by historical and recent experience where limits to adaptation have been observed, and by limits that are anticipated to arise as a consequence of future global change (Klein et al., 2014). The limiting factors to food crop production at the household level in Bui Division are presented in Table 3.8.

Table 3. 8: Limiting factors to climate change adaptation in the agricultural sector in Bui Division

Limiting factors	Frequency	%
Inadequate finances to buy farm inputs	173	86.5
Inadequate fertilizers	183	91.5
Inadequate pesticides	147	73.5
Inadequate farming land	166	83
Stringent land laws	177	88.5
Poor farm-to-market roads	154	77
Competing land use with graziers	57	28.5
Inadequate government assistance	187	93.5
Inadequate assistance from the council	153	76.5
Gender inequality in land matters	184	92
Poor land use planning	194	97
High labour burdens	198	99
Inadequate information & education	174	87
Farming on steep slopes	104	52
Farming in swampy land	68	34
Disease vectors (mosquitoes/malaria)	67	33.5
Infertile soils	145	72.5
Ineffectiveness of Agricultural Technicians & Extension field workers	168	84
Difficulties in understanding climate change & climate change adaptation	123	61.5
Inadequate farm tools	166	83

Source: Fieldwork (July-October 2016).

The problem of late rains and early rains has been lingering in Bui Division for the past 20 years. When the rains come early (i.e., usually in February or early March) and farmers sow seeds, thereafter the rainfall will not be consistent. Drought conditions will prevail, such that the tender crops wither and dry out. Farmers are compelled to wait for late rains to start planting (usually in April). Even when the rains begin in April, they support only the germination of crops. After germination, droughts will prevail until the end of May. It is during these phases of consistent droughts that pests like stem borers attack tender maize crops and blight destroys tender solanum potato and beans. As such, farmers adjust to these conditions by altering planting dates. This intensifies the vulnerability of subsistence households to climate change. As a result of these factors, there have been drops in the main staple food crops (i.e., maize, solanum potato and beans). This confirms the view that climate change is expected to lead to declining cereal production in developing countries, with the risk of decreasing yields particularly high in Africa (OECD, 2009; Patt and Winkler, 2007).

The DDARD provides improved seeds to local farmers to assist them to adjust their agricultural practices to prevailing climatic conditions. These improved seeds are weatherproof and serve as an option to adaptation to climate change. However, 63.5% of households surveyed are reluctant to change their practices and prefer using local seeds, which in most cases are not weatherproof. As such, adaptation becomes difficult. An indigenous strategy for building resilience in managing climate change is through planting seeds from recently harvested crops, especially beans and solanum potato. Rural farmers in Bui perceive that these recently harvested seeds are more drought and blight proof than old seeds.

In addition, the DDARD indicated that there are a number of support programs delivered by Non-Governmental Organisations (NGOs) that have been proactive in agricultural adaptation to climate change. The Programme for the Improvement of Competitiveness of Family Agro-pastoral Farms (ACEFA) provides counselling to producer groups and financing of projects for farm equipment and infrastructure. The ACEFA has constructed solanum potato and maize stores at Tadu, Takijah, Tatum, Nseh and Mbaw Nso, and has registered 361 farmer producers and 5 professional agro-pastoral groups in Bui. So far, only 59 of the 361 producers have submitted projects for financing worth $116,457. Another proactive program in food crop adaptation to climate change is the National Agricultural Extension and Research Programme (PNVRA). The PNVRA assists households in rural agricultural extension services. Also, the Upper Nun Valley Development Authority (UNVDA) promotes large-scale rice production using modern and indigenous methods. The UNVDA also provides training and inputs support to farmer groups, collection and processing of the paddy rice.

Other actors who have played prominent roles in climate change adaptation in the agricultural sector are the Green Care Association (GCA)-Shisong, Strategic Humanitarian Services (SHUMAS) and the Agricultural Training Programme (ATP) of the Catholic Diocese of Kumbo. The GCA is a local NGO that promotes climate-smart agriculture, apiculture, mushroom production, non-conventional livestock rearing (i.e., indoor animal rearing), watershed protection and analogue forestry. The SHUMAS promotes training on sustainable agriculture and constructs farm-to-market roads and bridges, while ATP offers training on climate-smart agriculture and organic manures. The response of households to these sustainable practices to supplement their food crop production is very slow.

The agricultural sector is further constrained by some cultural and traditional practices. These practices include weekly traditional holidays during which farming activities are suspended and when death occurs in the community. Each community in Bui Division has at least three days a week during which any farming activity is not carried out. This has gone a long way to limit efforts to adapt to climate change. These traditional holidays play a negative role during critical phases of food crop production like sowing of seeds, weeding, application of fertilizers and harvesting. When planting dates fall within those traditional holidays, it exposes the crops to attacks by wild rodents when planting takes place at a later date. Again, when a member of a community dies, farming activities are suspended till the mourning period is over. The duration of suspension varies. When the paramount traditional ruler (i.e., *Fon*) dies, farming activities are suspended for 8 weeks. If an adult dies, farming activities are suspended for 8 days and if a child dies, farming activities are suspended for 3 days. Other cultural practices that have frustrated adaptation efforts to climate change are the end of year festivities (i.e., cultural week) and periods when traditional authorities visit shrines to appease the *gods* of the land.

3.4 Conclusions

Since the 1980s, farmers have experienced a significant increase in variability to crop yields, prices and farm incomes. Over the same period, several crop pests and diseases have emerged and affected the output of beans, solanum potatoes and maize. Crop losses from extreme weather events such as droughts and heavy downpours have had substantial costs.

These negative trends have continued, exacerbated by warming temperatures and a more variable climate. The livelihoods of many farmers in Bui Division have been substantially altered because they are dependent on rain-fed agriculture. On the other hand, warming temperatures have encouraged the production of warm-season plants such as plantains and cassava around Bui. The impacts of these trends may lessen the comparative advantage that Bui Division now enjoys as one of the leading producers of solanum potatoes in the Northwest Region of Cameroon. The response of individual farmers to changes in the climate regime involves changes in the selection of crop species and in cultivation practices such as agro-forestry, cropping systems, irrigation and pest control. Besides crop and species diversification, farmers in this region are actively engaged in other activities such as bee-keeping, livestock rearing and forest exploitation. Adaptation in this area is still limited by poverty and lack of information. As such advances in climate forecasting may improve preparations and help prevent some of the projected losses. Ultimately, the ability of farmers to effectively adapt can decide the success or failure of individual farms, and by extension, can affect the entire Bui Division. Under progressively changing climatic conditions, adaptations would need to continuously evolve and may be increasingly difficult to plan.

References

Abeysinghe, A., and Prolois, C. (2016) Entry into force of the Paris Agreement: The legal process. Issue Paper, March 2016, International Institute for Environment and Development (IIED), London, 4-7.

Ayoade, J.O. (1998) Introduction to Climatology for the Tropics. Fourth edition, Spectrum Books Ltd., Ibadan.

Divisional Delegation of Lands, Kumbo (2008) Map of Kumbo. National Geographic

Institute, Yaounde, Cameroun.

Harvey, C.A., Rakotobe, Z.L., Rao, N.S., Dave, R., Razafimahatratra, H., Rabarijohn, R.H., Rajaofara, H, and Mackinnon, J. (2014) Extreme vulnerability of smallholder farmers to agricultural risks and climate change in Madagascar. Phil.Trans.R.Soc. B369, 20130089. (doi:10.1098/rstb.2013.0089)[Google Scholar].

Intergovernmental Panel on Climate Change (IPCC) (2007) Climate Change 2007: Impacts, Adaptation and Vulnerability. Contribution of Working Group II to the Fourth Assessment Report of the Intergovernmental Panel on Climate Change. Cambridge University Press, Cambridge UK, 976.

Intergovernmental Panel on Climate Change (IPCC) (2014) Climate Change 2014: Impacts, Adaptation and Vulnerability, Contribution of

Working Group II to the IPCC Fifth Assessment Report. Cambridge University Press, Cambridge, United Kingdom & New York, 1774.

International Institute for Environment and Development (IIED) (2015a) Vulnerable communities: getting their needs and knowledge into climate policy. IIED Briefing, November 2015, London, 1-4.

International Institute for Environment and Development (IIED) (2015b) Vulnerable communities: climate adaptation that works for the poor. IIED Briefing, November 2015, London, 1-4.

Klein, R.J.T., Midgley, G.F., Preston, B.L., Alam, M., Berkhout, F.G.H., Dow, K., and Shaw, M.R. (2014) Adaptation opportunities, constraints, and limits. In: Climate Change 2014: Impacts, Adaptation, and Vulnerability. Part A: Global and Sectorial Aspects. Contribution of Working Group II to the Fifth Assessment Report of the Intergovernmental Panel on Climate Change [Field, C.B., V.R. Barros, D.J. Dokken, K.J. Mach, M.D. Mastrandrea, T.E. Bilir, M. Chatterjee, K.L. Ebi, Y.O. Estrada, R.C. Genova, B. Girma, E.S. Kissel, A.N. Levy, S. MacCracken, P.R. Mastrandrea, & L.L. White (Eds.)]. Cambridge University Press, Cambridge, United Kingdom & New York, NY, USA, 899-943.

Kolawole, D.O., Wolski, P., Ngwenya, B., and Mmopelwa, G. (2014) Ethno-meteorology and Scientific Weather Forecasting: Small Farmers and Scientists' Perspectives on Climate Variability in the Okavango Delta, Botswana. *Climate Risk Management* (4–5), 43–58.

Kolawole, D.O., Motsholapheko, M.R., Ngwenya, B. N., and Ethakadu, O. (2016) Climate Variability and Rural Livelihoods: How Households Perceive and Adapt to Climatic Shocks in the Okavango Delta, Botswana. *American Meteorological Society* (8), 131-145.

Luxon, N., and Pius, C. (2014) Water and Agriculture Vulnerability to Climate Change and Adaptive Capacity in Southern Africa. International Water Management Institute, 1-8.

Mbiame Council, (2012). Council Development Plan. Mbiame, Cameroun.

Niang, I., Ruppel, O.C., Abdrabo, M.A., Essel, A., Lennard, C., Padgham, J., and Urquhart, P. (2014) Africa. In: Climate Change 2014: Impacts, Adaptation and Vulnerability. Part B: Regional Aspects. Contribution of Working Group II to the Fifth Assessment Report of the Intergovernmental Panel on Climate Change [Barros, V.R., C.B. Field, D.J. Dokken, M.D. Mastrandrea, K.J. Mach, T.E. Bilir, M. Chatterjee, K.L. Ebi, Y.O. Estrada, R.C. Genova, B. Girma, E.S. Kissel, A.N. Levy, S. MacCracken, P.R. Mastrandrea, & L.L. White (Eds.)]. Cambridge University Press, Cambridge, United Kingdom & New York, USA, 1199-1265.

Olaniran, O.J. (1983) The Monsoon factor and Seasonality of Rainfall Distribution in Nigeria. *Malaysian Journal of Tropical Geography* (7), 38-41.

Organization for Economic Co-operation and Development (OECD) (2009) Integrating Climate Change Adaptation into Development Co-operation, Policy Guidance. OECD, Paris, 193.

Patt, A.G., and Winkler, J. (2007) Applying Climate Information in Africa: An Assessment of Current Knowledge. University of Boston, 19.

Porter, J.R., Xie, L., Challinor, A.J., Cochrane, K., Howden, S.M., Iqbal, M.M., Lobell, D.B., and Travasso, M.I. (2014) Food Security and Food Production Systems. In: Climate Change 2014: Impacts, Adaptation, and Vulnerability. Part A: Global and Sectorial Aspects. Contribution of Working Group II to the Fifth Assessment Report of the Intergovernmental Panel on Climate Change [Field, C.B., V.R. Barros, D.J. Dokken, K.J. Mach, M.D. Mastrandrea, T.E. Bilir, M. Chatterjee, K.L. Ebi, Y.O. Estrada, R.C. Genova, B. Girma, E.S. Kissel, A.N. Levy, S. MacCracken, P.R. Mastrandrea & L.L. White (Eds.)]. Cambridge University Press, Cambridge, United Kingdom & New York, NY, 485-533.

Rhodes, E.R., Jalloh, A., and Diouf, A. (2014) Review of Research and Policies for Climate Change Adaptation in the Agriculture Sector in West Africa. Future Agricultures, Working Paper 090, 12-18.

Sharmais, A., Schwarteis, C., Mülleris, B., Abeysinghe, A., Barakat, S. (2016) Pocket Guide to the Paris Agreement. IIED Policy Paper G04042, European Capacity Building Initiative (ECBI), 44.

Smit, B., and Pilifosova, O. (2001) Adaptation to Climate Change in the Context of Sustainable Development and Equity. In Climate Change 2001: Impacts, Adaptation and Vulnerability, Cambridge: Cambridge University Press, 879-905.

Thorlakson, T. (2011) Reducing Subsistence Farmers' vulnerability to Climate Change. World Agroforestry Centre, Nairobi, Occasional Paper (16), 1-2.

Töpfer, K., and Hunter, J.W. (2002) Climate Change Information Kit. UNEP & UNFCCC.

Tume, S.J.P. (2008) Rainfall Variability and Implications on Water Resources Management: The Case of Bui Division, Northwest Province of Cameroon, MSc Thesis in Geography, University of Buea.

United Nations Framework Convention on Climate Change (UNFCCC) (2011) The Nairobi work program on impacts, vulnerability and adaptation to Climate Change; Climate Change and Freshwater Resources: a Synthesis of Adaptation actions undertaken by Nairobi work program partner organizations. UNFCCC Secretariat, Bonn, 10-11.

United Nations Framework Convention on Climate Change (UNFCCC) (2006) UNFCCC Handbook. UNFCCC Secretariat, Bonn.

Walsh, R.P.D., and Lawler, D.M. (1981) Rainfall Seasonality: Description, Spatial Patterns and Change through Time. Weather (36) July 1981, 201-208.

CHAPTER 4

Perceptions of climate change, traditional beliefs and determinants of farm households' adaptation decisions in Northern Togo

Ali Essossinam

Abstract

The increasing scale of potential climate change impacts on agricultural production must be addressed through mitigation and adaptation. Cross-sectional data from 704 farm households in Northern Togo are used to assess farmers' perceptions of climate change, their traditional beliefs about and barriers to adaptation, and the factors that affect their adaptation decisions. The results showed that climate change is well recognized, while a lack of information, capital and knowledge about adaptation were factors that have constrained farmers' adaptation to climate change. Cultural beliefs and modern approaches have been suggested as adaptation measures. Shifts in the rainy season, declines in livestock yields and food shortages have led farmers to pursue adaptations to climate change. Farm household socioeconomic characteristics such as the farm location, sex of household head, number of active household members, land devoted to cereal production, access to credit and climate information, government assistance, off-farm income and extension services were adaptation determinants. The interest rate applied to credit has a negative effect on farmers' decisions about adaptation to climate change. Adaptation policies should focus on improved extension services for better dissemination of knowledge about climate change adaptation strategies, accessibility of weather information, and possible synergies between traditional and modern beliefs about climate change adaptation.

Keywords: Adaptation Decisions; Climate Change; Northern Togo; Perceptions; Traditional Beliefs

4.1 Introduction

The effects of climate change on the agricultural sector in developing countries, characterized by weather-sensitive agricultural production systems and low adaptive capacities, is now recognized (World Bank, 2013; IPCC, 2014). The frequencies of floods and prolonged drought have also increased (IPCC, 2014). For example, McSweeney et al. (2010) have indicated that Togo has experienced sixty major floods between 1925 and 1992. The most damaging were those in 2007, 2008 and 2010. The African Development Bank has indicated that the flood events of 2008 destroyed 11,688 hectares of cultivated land and led to food shortages and price spikes throughout Togo (AfDB, 2011). The estimated agricultural loss from this event was approximately $22 million[1], with 83.8% of the country's food crops, 12.5% of its cotton crops and 3.7% of its cattle operations affected. The same source has indicated that the flood events in 2010 destroyed approximately 7,744 hectares of farms and 3,947 houses and also killed at least 21 people. According to the post-disaster Needs Assessment, the estimated losses resulting from the 2010 flood events were around $32 million.

McSweeney et al. (2010) have also documented temperature and precipitation changes in Togo. Temperature has increased on average about 1.1°C and at the rate of 0.24°C per decade, with a peak of 40°C now sometimes observed in Northern Togo. In terms of precipitation, the trend has been a decrease at the rate of 2.3mm per month or 2.4% per decade. In this context, climate change might constitute a threat to livelihoods, especially for rural populations that are characterized by a poor asset base and no use of irrigation practices. This is true for Togo where agriculture contributes about 42% of Gross Domestic Product (GDP) and employs 54.1% of the active population (FAO, 2015; AfDB et al., 2016). Food crops, particularly cereals, could be the most affected sub-sector, and reductions in these crops will increase food insecurity in Togo, given the high contribution of cereal crops to agricultural GDP (i.e., 68.5%).

There is a need for adaptation or mitigation strategies to cope with the induced effects of climate change. However, the effectiveness of current adaptation and climate risk management is questionable. Although some

---

[1] United States Dollars = 500 Franc CFA (FCFA)

farmers have realized the change in climatic conditions over the past decade, they have been unable to implement any adaptation measures to cope with these effects. Adapting to climate change by adopting new technologies or farming practices could reduce yield losses, decrease vulnerability and increase food security status (Di Falco et al., 2011). In the context of Africa and Asia, it has been documented that traditional knowledge, as well as modern conceptions of climate change mitigation, could play an important role (FAO, 2008; Berkes, 2009; Green and Raygorodetsky, 2010; Burkett, 2013; Ebge et al., 2014; Waapela, 2016). Also, the difference in farmers' socioeconomic and cultural endowments, as well as institutional factors, would influence farm households' decisions to adapt to new climate realities (Hassan and Nhemachena, 2008; Shongwe et al., 2014; Temesgen et al., 2014). Understanding farmers perceptions including their cultural conceptions of climate change, and their adaptation decisions could also be dependent on geographical location. It, therefore, becomes important to understand farmers' perceptions of adaptation to climate change and the barriers to their adaptation. It is also relevant to assess the factors influencing farm households' decision-making. Understanding these factors can help to reorient investment policies toward adoption of the best adaptation strategies to enhance agricultural productivity and reduce farm household vulnerability to food insecurity.

This paper aims to assess farm households' perceptions of climate change in Northern Togo. Specifically, it seeks to determine the barriers to farmers' adaptation, to shed light on their beliefs, and determine the factors that influence their decisions regarding adaptation to climate change.

4.2 Materials and Methods

4.2.1 Study area

The study was carried out in the three Northern Regions of Togo (i.e., Central, Kara and Savannah Regions). Based on the basic well-being indicators for Togo derived from the 2015 QUIBB survey, the populations in these regions are very poor (i.e. poverty index of 90%, 77.7% and 75% for Savannah, Central and Kara Regions, respectively) compared to the Southern Regions of the country (i.e. 69.4% and 56.2% in the Maritime and Plateaux Regions, respectively).

The climate in Southern Togo is characterized by two rainy seasons and two dry seasons, while the climate in the Northern Regions is characterized by one rainy season (from April to October) and one dry season (from November to March). The temperature sometimes reaches above 30°C in the Northern Regions and there has been an observed shift in the rainy season of up to 30 days beyond the norm (McSweeney et al., 2010). The Northern Regions of Togo are also characterized by three different agro-ecological zones: Northern Guinea Savannah, Southern Guinea Savannah and Derived Savannah (Figure 4.1).

Figure 4. 1: Study areas and different agro-ecological zones in Northern Togo

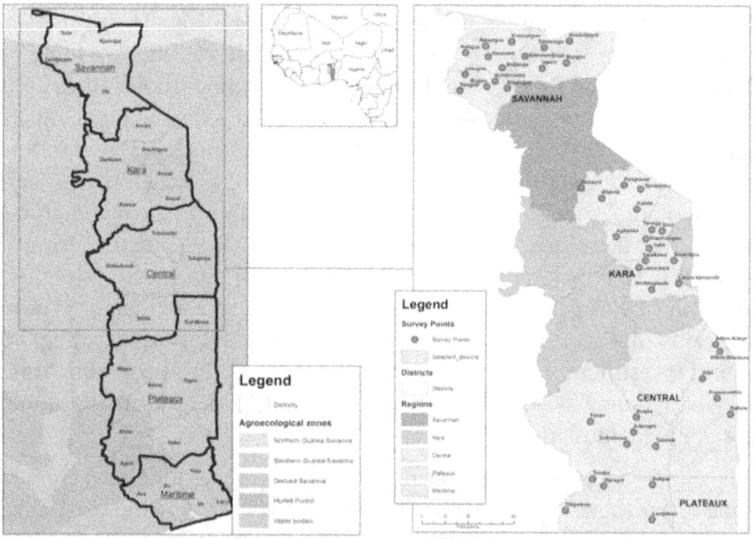

Source: Author, using WorldClim-Global Climate Data

These climatic and agro-ecological characteristics give the opportunity to produce diverse crops. Maize, sorghum and rice are the food crops that are mostly cultivated in these areas. Indeed, cereal production, especially maize, sorghum and rice production in the Northern Regions of Togo, represented on average 30%, 85% and 69% of national production, respectively, over the period 1972-2014[2]. Other major crops that are cultivated in Northern Togo are grain legumes (i.e., soya bean, beans and cowpea) and tuber crops (i.e., yam, cassava, sweet potatoes). Cotton and cashew are other cash crops produced in the area.

[2] Author, using data from the national statistics of crop production (the national agricultural database, information and documentation of Togo-DSID)

4.2.2 Sampling and data collection

Crosssectional data were used for this study. Data collection was undertaken in October, November and December 2015. A multi-staged sampling technique was used to collect data at the farm household level. The first stage was the selection of three regions among five administrative regions in Togo. The Central, Kara and Savanah Regions were purposely selected by considering the characteristics of the wet season and the poverty index in these regions. The second stage was the simple random selection of three districts in each region. The third stage was the selection of villages. Five villages were randomly selected in each district using a random number table. The lists of villages were available in each district head office. The last stage was the selection of the respondents. The respondents were selected randomly using a random number table. The survey was conducted at the household level.

The data were collected from 704 farm households using a structured questionnaire that was addressed to the head of the household. The structured questionnaire included open-ended questions and was initially pre-tested in Binah District, which is not of part of the selected districts considered in this study. The purpose of the testing was to check local farmers' understanding of the questionnaire, to improve its quality, simplicity and layout. The enumerators were periodically visited by the author, at least three times a day. The purpose of this was to check the consistency and accuracy of completed questionnaires and in the cases of inconsistency, the questionnaires were returned to the responding household for rechecking. The author also sought help from local authorities who facilitated the interaction between interviewers and respondents and assisted in identifying the locations of selected farm households.

The collected data covered farm household demographic and socioeconomic characteristics, farmers' perceptions of climate change and their observations of frequent climate change impacts over the past decade, and the farmers' access to weather information. Other sections of the survey covered the barriers to adaptation to climate change and asked the respondents to suggest strategies that could be implemented to reduce the impacts of climate change in their community. To determine the factors that influence farmers' climate change adaptation decisions, we asked farmers to indicate whether they had implemented any adaptation measures to respond to the effects of climate change and variability.

4.2.3 Qualitative and quantitative research approaches

Adaptation to climate change is a multi-scaled process (Bryant et al., 2000; Maddison, 2007; Deressa et al., 2009; Kubik and Maurel, 2016; Delaporte and Maurel, 2018). Various approaches have been applied in the literature when analyzing the determinants of farm households' adaptation to climate change by using mostly cross-sectional data. Qualitative approaches such as focus group discussions, consultation with Indigenous People and survey data using structured or semi-structured questionnaires can be used to analyze farmers' perceptions of climate change (Habiba et al., 2012). As for quantitative approaches, Logit and Probit models are widely used in decision or adoption studies for binary data, while a Tobit model is used when the dependent variable is continuous (Nichola, 1996; Pannell, 1999; Ransom et al., 2003; Fosu-Mensah et al., 2012; Tiwari et al., 2014; Abid et al., 2015; Ndamani and Watanabe, 2016). For instance, a Tobit model was used by Ransom et al. (2003) to determine factors that influence the use of improved maize varieties by farmers in Nepal. These authors found that the farmers' socioeconomic characteristics were determinants of adoption of the use of improved varieties, which was less than 60% in all village development committees.

Ndamani and Watanabe (2016) have used the weighted average index and logistic regression to analyze the determinants of farmers' adaptation to climate change in Ghana. The results from the study have indicated that dry spells and drought are major extreme natural events that have occurred more over the past 10 years than flood events. The socioeconomic characteristics of the respondents and institutional factors have been shown to be determinants of farmers' adaptations to climate change in the empirical model. The same approach was used by Tiwari et al. (2014) and Fosu-Mensah et al. (2012) to analyze the determinants of climate change adaptation in Nepal and Ghana, respectively, and the results from these studies were similar to those of Ndamani and Watanabe (2016). Assuming that different climatic settings could lead farmers to behave differently, Delaporte and Maurel (2018) have used the IV Probit approach to analyze climate change adaptation in Bangladesh and show that a 1% climate-induced reduction in farm-level income led farmers to pursue adaptation options. Maddison (2007) has used the Probit model to analyze the perceptions of and adaptations to climate change in eleven countries in Africa. But researchers must be careful when dealing with adoption studies. For instance, Nichola (1996) criticizes the use of Tobit and Probit models in adoption studies in the case where the agricultural technologies are scarce and suggests a double-hurdle approach that could take into account the scarcity of agricultural technologies. In this study, qualitative methods and a Probit model were used to analyze farmers' perceptions of climate change and the determinants of their adaptation decisions.

4.2.4 Method of analysis

A binary Probit model was used to assess the factors influencing farmers' climate change adaptation decisions. The decision of a farm household to adapt to climate change is assumed to be dependent on the expected utility from adaptation. This is called a latent variable (Gould et al., 1989; Bryan et al., 2009; Wooldridge, 2014). If Y^* denotes the latent variable of the adaptation decision, then its relationship with the potential explanatory variable could be noted as:

$$Y_i^* = X^T\beta + \mu_i \tag{4.1}$$

Where X is a vector of explanatory variables, β is a column vector of predictor coefficients and μ_i is an error term that is assumed to be normally distributed. The parameters β are estimated using maximum likelihood. The adaptation decision is noted in equation (4.2):

$$Y = \begin{cases} 1 & \text{if} \quad Y^* > 0 \\ 0 & \text{Otherwise} \end{cases} \tag{4.2}$$

Thus, the model takes the form:

$$P_r(Y = 1 \mid X) = \Phi(X^T\beta) \tag{4.3}$$

Φ stands for the cumulative normal distribution function. Equation 4.3 can be rewritten as:

$$\begin{cases} \Pr(Y_i = 1) = 1 - \Phi(-X_i^T\beta) \\ \Pr(Y_i = 0) = \qquad \Phi(-X_i^T\beta) \\ \text{where } \Phi(X) = \int_{-\infty}^{X} \frac{1}{\sqrt{2\pi}} \exp\left(-\frac{t^2}{2}\right) dt \end{cases} \tag{4.4}$$

The maximum likelihood estimation is given as:

$$\text{Log } L(\beta) = \sum_{i=1}^{n} (y_i \text{ Log } \Phi\left(X_i^T\beta\right) + (1 - y_i) \text{ Log } (1 - \Phi\left(X_i^T\beta\right)) \tag{4.5}$$

The marginal effects that show the real effects in terms of quantification of the explanatory variable on the adaptation decision can be derived using equation 4.5.

4.3. Results and Discussion

4.3.1 Descriptive statistics

On average, 36.5 percent of the surveyed farm households are from the Central Region, while 26.70 and 36.8 percent are from the Kara and Savannah Regions, respectively (Table 4.1).

Table 4. 1: The socioeconomic characteristics of the surveyed farm households in Northern Togo

Variables	Mean	Std. Dev	Min	Max
Age (Number of years)	43.81	11.77	20	90
Household size (Number of household members)	7.5	3.745	1	27
Active household members	4.48	2.75	1	22
Farming Experience (Number of years)	21.10	12.58	1	75
Cultivated land (ha)	2.75	1.45	0.25	10
Land for maize (ha)	1.05	0.670	0	5
Land for sorghum (ha)	0.65	0.609	0	5
Land for rice (ha)	0.17	0.290	0	2
Land for grain legume (ha)	0.63	0.689	0	4.75
Land for tuber crop (ha)	0.08	0.245	0	3
Land for cotton (ha)	0.21	0.525	0	5
Labour (Man-days)	131.70	82.85	6	722
Minimum consumption need ($)	1,027.78	618.5	120	5,534
Off-farm income ($)	163.11	278.79	0	2,500
Distance to the nearest input market (Km)	9.28	10.78	1	200
Interest rate (%)	2.759	6.0	0	50
		Proportion (%)		**Std. Err.**
• **Central Region**		36.50		0.018
• **Kara Region**		26.70		0.016
• **Savannah Region**		36.78		0.018
• **Male**		90.05		0.011
• **Female**		9.94		0.011
Education				
• **No formal education (%)**		15.48		0.014
• **Primary (%)**		46.87		0.019
• **Secondary I (%)**		33.24		0.017
• **Secondary II (%)**		3.55		0.006
• **University (%)**		0.85		0.003
Access to Finance (1=Yes, 0=No)[3]		36.3		0.018
Access to Extension services (1=Yes, 0=No)		31.5		0.017
Government assistance (1=Yes; 0=No)		32.6		0.017
Access to climate information (1=Yes, 0=No)		25.99		0.016
Access to Electricity (1=Yes; 0=No)		26.5		0.016
Adapt to climate change (Yes=1, No=0)		83.5		0.013

[3] For variables; access to finance, extension services, climate information, electricity, government assistance, and adapt to climate change, the proportions represent the percentage of those who answered in the affirmative. Thus, the percentage who answered 'yes'.

The results show that the surveyed population in the study areas is relatively young. The age distribution of the respondents ranges between 20 and 90 years, with a mean of 44 years (Table 1). The household size is relatively large (i.e., on average 8 people per farm household) and the study areas are dominated by small-scale farmers, operating on very limited farmland. The cultivated land in the study areas ranged from 0.25 to 10 hectares, with an average of 2.75 hectares.

Table 4.1 indicates that the average land devoted to the cultivation of maize, sorghum and rice in the study area was relatively small (i.e. 1.05, 0.65 and 0.17 hectares, respectively). The farm households' minimum consumption needs ranged from $120.00 and $5,334.00. The average income generated from off-farm activity is about $163.11, with the maximum reaching $2,500.00. Market accessibility is very important in the value chain of the products derived from farm households' production, especially in the distribution process. In the surveyed areas, the nearest input market, where farmers often buy their farming inputs, seems to be fair. For instance, on average, a farmer has to travel about 9.29 kilometers to reach the nearest input market but sometimes this distance is about 200 kilometers. Apart from the distance potentially leading to the loss of farming man-days, the quality of transportation infrastructure is also questionable. The road infrastructure in rural areas is often poor in developing countries like Togo, while the impact of increased market accessibility on rural development is indisputable (Terluin, 2003; Mu and Van de Walle, 2011).

Education is another important aspect of rural development. On average, about 15.48 percent of farmers did not have any formal education, while 46.87 percent had attended at least a primary school, and 33.22 and 3.55 percent had attended secondary I and secondary II, respectively. Only 0.85 percent of the sample had attended university. Education promotion in the study area is needed since it can help farmers in the adoption of best agricultural practices and thereby increase their farming productivity.

Access to finance is a common problem in the agriculture sector in developing countries (Rojas-Suarez, 2010). For instance, only 36.3 percent of respondents had access to financial products, such as credit. This low level of access to financial products could probably be due to the risks involved in agricultural activities. The expansion of extension services and access to and the use of climate information could help farmers cope with agricultural risks and increase their farming productivity. Unfortunately, the farm households' access to extension services and climate information is still low in the surveyed regions. Only 31.5 and 26 percent of the respondents have access to extension services and climate information, respectively (Table 4.1). However, a major proportion of farm households in the surveyed areas have implemented at least one adaptation measure in response to climate change.

4.3.2 Farm households' perceptions of climate change in Northern Togo

Climate change seriously matters in Northern Togo (Table 4.2). During the field survey, farmers were asked whether they have noticed any change in rainfall or temperature patterns, or whether they have observed any shift in the rainy season during the past decade. The results indicate that almost the majority of farmers in the sample had noticed a change in precipitation and temperature patterns as well as a shift in the rainy season over the past decade. Specifically, 99.15 percent of respondents had noticed a decrease in rainfall, while 99.57 percent had experienced an increase in temperature. Also, 88.49 percent of respondents had experienced that not only are the rainy seasons arriving later, but they are also finishing earlier than usual. In that situation, a farmer who does not have access to drought-tolerant seeds could see a reduction in their farming revenue due to decreased crop yields. Early interventions in the form of motivating the adoption of new technologies to adapt to the new climate realities should, therefore, attract policymakers' attention.

Table 4. 2: Farm households' perceptions of climate change

Farm households' perceptions of climate change	Yes	No	Percentage of respondents
Notification of change in rainfall pattern	704	0	100
• **Increase in rainfall**	698	06	99.15
• **Decrease in rainfall**	05	699	0.71
Notification of change in temperature pattern	704	0	100
• **Increase in temperature**	701	03	99.57
• **Decrease in temperature**	02	702	0.28
Shift in the rainy season	703	01	99.86
• **Rainy season comes later than usual**	694	10	98.88
• **Rainy season comes sooner than usual**	09	695	1.28
• **Rainy season finished earlier than usual**	623	81	88.49
• **Rainy season finished later than usual**	155	549	22.02

Drought	682	22	96.85
Floods	296	408	42.05
Off-seasonal rainfall	280	424	39.77
Too much rain	115	589	16.34
Too little rain	589	115	83.66
High temperature	693	11	98.44
High wind	514	190	73.01
Number of respondents			704

Other climate change indicators that have threatened agriculture in the Northern Region of Togo are frequent dry spells or droughts, flood events and high winds. From the sample, 96.85 percent of respondents had experienced a prolonged drought or dry spells, and this is observed almost every year. The flood events seem to be less frequent, but their effects still disastrous. One farmer in Natchabonga Village in the Kpendjal District (Savannah Region) testified as such: "*The rain is infrequent but when it comes, it is too much with violent winds destroying our crops. It often happens between crop flowering and harvesting periods, resulting in considerable harvest loss. It often removes our roofs and destroyed our houses. It happens almost each year. We cannot do anything about it. We cannot stop these violent winds. But from now, we only try to plant trees around our field and our houses. Meanwhile, this strategy is still costly and requires a long time to mature*".

Indeed, 73.01 percent of respondents have noticed the high winds during the past decade and these events now often occur up to three times per year.

4.3.3 Observed climate change impacts in Northern Togo

Table 4.3 indicates the climate change impacts that have been observed over the past decade in Northern Togo. The results show that climate change has threatened crop production in the region. On average, 96.59 percent of respondents claimed that climate change has reduced their crop yields, while 90.48 percent had noticed a decline in livestock yields. In recent years, climate change has led to the shortage of fodder, resulting in the death of livestock. Water scarcity has also increased, along with a decrease in water quality. This has increased food insecurity and households' vulnerability. It also constitutes a danger for food crop production, for both animal and human health.

Table 4. 3: Observed climate change impacts in Northern Togo

Observed climate change impacts	Yes	No	Percentage of respondents who answered 'yes'
Decline in crop yields	680	24	96.59
Increased in crop yields	24	680	3.41
Decline in livestock yield	637	67	90.48
Increased in livestock production	67	637	9.52
Death of livestock due to shortage of fodder and water	586	118	83.24
Food shortage	548	156	77.84
Increased weed and pest pressures	548	156	77.84
Communicable diseases	473	231	67.19
Decrease in water quantity and quality	598	106	84.94
High risk of crop damage due to prolonged dry spells or drought	656	48	93.18
Number of respondents			704

The evidence is that 93.18 percent of respondents have claimed that crop production is now high risk due to dry-spells that often last one month or more. An example is the dry spell in the 2014/2015 crop year. After planting, it took two months before the rains returned. Also, 77.84 percent of respondents had seen an increase in weed and pest pressures, while 66.7 percent confirmed an increase in communicable diseases. Policy interventions to deal with these effects of climate change could help to reduce food insecurity and household vulnerability.

4.3.4 Barriers to climate change adaptation in Northern Togo

Adaptation to climate change remains a challenge for farm households in Northern Togo. The results show that 83.52 percent of farm households have decided to adapt to the new climate realities by adopting one or more technologies or changing their farming strategy (Table 4.4). Many reasons could explain the farmers' choices or decisions. For example, 84.48 percent of the farmers who have not taken any adaptation measure lack information on available technology or new farming practices, while 79.31 percent do not have any knowledge about climate adaptation. Access to extension services could better inform these farmers about the available technologies and how they can be used to reduce yield losses.

Table 4. 4: Barriers to climate change adaptation in Northern Togo

Barriers to climate change adaptation	Yes	No	Percentage of respondents who answered 'yes'
Lack of information	98	18	84.48
Lack of capital	89	27	76.72
Lack of knowledge	92	24	79.31
Shortage of farming land	51	65	43.97
Not observing any climate problem	0	116	0
Number of observations			116

Training of extension service providers in climate-smart agriculture could be very helpful. Another barrier to climate change adaptation is the accessibility of financial products. The accessibility of financial products, especially credit, remains the main constraint in agriculture. This may be due to the risky nature of agriculture. Sometimes, the fear of going into debt leads farmers to not apply for formal financial credit. Even in the informal credit supply chain (i.e., credit supply through a wealthier farmer or relative), most of the farmers do not ask for it, often simply to preserve their dignity in their social environment. Only 43.97 percent of non-adapted farmers claimed they had a shortage of farming land.

4.3.5 Farm households' beliefs about and adaptation to climate change in the Northern Regions of Togo

During the field-work, farmers were asked to elucidate what they think should be done to reduce climate change impacts in their areas, given their experiences with changing weather patterns. The responses can be summarized into the following 21 categories: tree plantations, agro-forestry systems, not cutting trees, reducing charcoal production, praying to God or making sacrifices to "living gods" in farmers' communities, expansion of extension services, irrigation practices, change of crop varieties, adopting new technologies and farming practices, involvement in trade, use of fertilizers, practicing mixed crop/livestock systems, going back to traditional agriculture, less use of pesticides/herbicides, bio food production, food aids, fighting against bush fire, electricity for better information, waste and perishable plastics management, other (including intensive agriculture, changing of human behaviors, not using chemical fertilizers and providing agricultural insurance), and those who thought that nothing could be done about climate change (Figure 4.2).

Figure 4. 2: Farmers' suggestions to reduce climate change impacts in Northern Regions of Togo

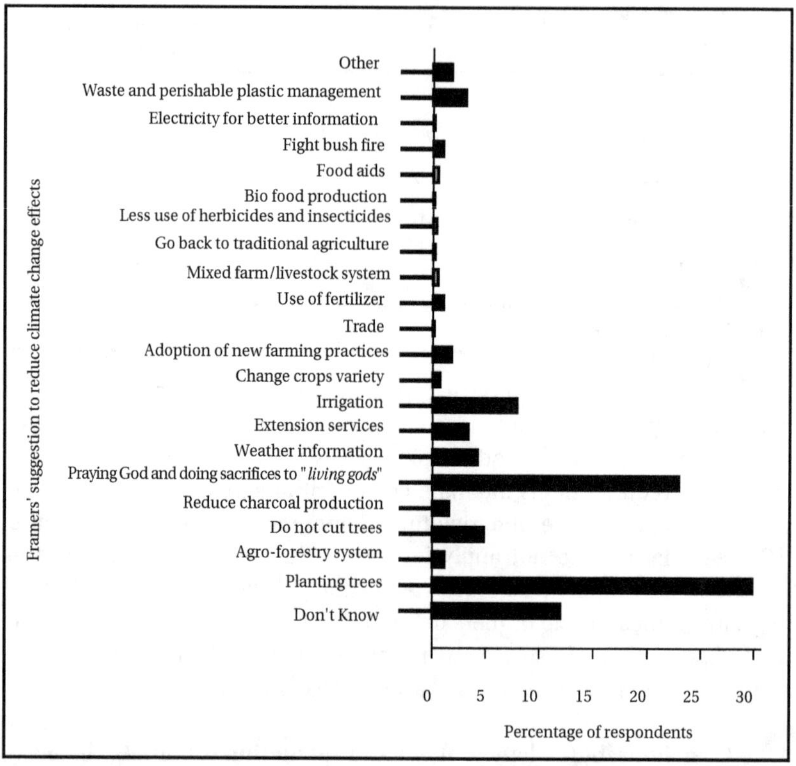

The results indicate that 30 percent of respondents have suggested tree plantations as the best way to reduce climate change impacts on farming activities, while 23 percent strictly believe that praying to god or by having ceremonies to "living gods" in their communities could help. Also, 8 percent of respondents suggested building dams and barrages for irrigation purposes, 4.26 percent suggested dissemination of weather information through an increase in weather stations and weather forecasts, 3.4 percent expect to enhance their capacity to adapt through extension services and farmer-to-farmer learning approaches, and 3 percent suggested the management of waste and perishable plastic for a better environment. Again, 6 percent of respondents suggested a need to stop destroying trees and reduce charcoal production, and only 12 percent of respondents said they did not know what to suggest as an alternative to reduce the effects of climate change. For these respondents, there was a belief that nothing could be done about climate change (Figure 4.2).

Extension services are an essential need in the study areas to let more farmers understand that it is possible to do something to reduce the effects of climate change on crop production. Table 4.5 presents the farmers' suggestions for what could be undertaken to reduce climate change effects. The results show disparities in suggestions between the study regions. The results show that most of the farmers who think that nothing can be done to reduce the impacts of climate change are mostly located in the Central and Savannah Regions (i.e., 19.46% and 10.04%, respectively).

The proportions of farmers who suggested tree plantations as an adaptation strategy are similar between regions. For example, 38.83 percent of the surveyed household heads in the Kara Region think that the warming effects of climate change can be reduced by planting trees. This proportion was 28.02 and 25.48 percent for the Central and Savannah Regions, respectively. The net difference equates to supernatural beliefs. On average, 42.55 and 25.48 percent farm households in the Kara and Savannah Regions, respectively, believe that praying to the gods and frequent social ceremonies for "living gods" of their communities could solve their climate change problems. For them, people have neglected their community ceremonies and the scarcity of rain, high temperatures, dry spells and high winds are seen as a form of punishment by the "gods".

Table 4. 5: Farmers' suggestions to reduce the effects of climate change in the Northern Regions of Togo by Regions[4]

Suggested strategies	Percentage of respondents		
	Central Region	Kara Region	Savannah Region
Planting trees	28.02	38.83	25.48
Agro-forestry systems	3.50	0	0
Do not cut trees	8.56	4.26	1.54
Reduce charcoal production	4.28	0	0
Praying to God and making sacrifices to "living gods"	6.23	42.55	25.48
Weather information	4.28	0.53	6.95
Extensions services	3.11	1.06	5.41

[4] Figure 4.2 presents the views of all the sampled respondents in Northern Togo, whereas Table 4.5 presents the views of the sampled respondents in the three specific regions in Northern Togo.

Irrigation practices	1.95	0.53	19.31
Changing crop varieties	1.17	0	0.77
Adoption of new farming practices	2.33	1.60	1.54
Trade	0	0	0.39
Use of chemical fertilizer	1.17	1.60	0.77
Mixed farm/livestock systems	1.17	0	0.30
Go back to traditional agriculture	0.78	0	0
Less use of herbicides/insecticides	1.17	0	0
Bio food production	0.39	0	0
Food aids	1.56	0	0
Fight bush fire	1.17	2.13	0
Electricity for better information	0.39	0	0
Waste and perishable plastic management	5.06	1.60	1.93
Other	4.28	0.53	0
Don't Know	19.46	4.79	10.04
Total Number of respondents	**257**	**188**	**259**

According to the respondents, praying or making sacrifices to the "gods" as well as planting trees are sure ways to mitigate climate change in the Kara and Savannah Regions. As for the Central Region, planting trees or reforestation, and better management of waste and perishable plastics are the best hopes for responding to climate change. A possible close collaboration between Indigenous People and scientists could lead to a better understanding of how to cope with climate change by sharing traditional and modern knowledge about weather information (Berkes, 2009; Williams and Hardison, 2013; Waapela, 2016).

4.3.6 Determinants of farm households' climate change adaptation decisions

Results of the Probit regression model in Table 4.6 show the factors that have influenced the farmers' climate change adaptation decisions in Northern Togo.

Table 4. 6: Probit regression results of the factors influencing the farmers' climate change adaptation decisions in Northern Togo

Variables	Probit		Marginal effects	
	Coef.	Std. Error	Coef.	Std. Error
Savannah	0.410**	0.206	0.070**	0.032
Kara	0.751***	0.194	0.111***	0.024
Sex (Male=1, Female=0)	0.571***	0.194	0.135**	0.056
Age (Number of years)	0.002	0.008	0.001	0.002
Education (1=Formal, 0=Informal)	0.121	0.187	0.023	0.037
Farming Experience (Number of years)	0.004	0.009	0.001	0.002
Family size (Household members)	-0.098***	0.026	-0.018***	0.004
Active household members	0.092**	0.038	0.017**	0.007
Log labour (man-days)	0.127	0.132	0.023	0.024
Log consumption expenditure (FCFA)	0.105	0.134	0.019	0.024
Log distance to the market (km)	-0.107	0.071	-0.019	0.013
Land under maize, sorghum and rice cultivation (hectares)	0.183**	0.081	0.033**	0.014
Land under grain legumes, tuber and cotton cultivation (hectares)	0.104	0.087	0.019	0.016
Dry spell (1=Yes, 0=No)	0.161	0.345	0.032	0.075
Flood (1=Yes, 0=No)	0.142	0.151	0.025	0.026
Decrease in water quantity and quality (1=Yes, 0=No)	-0.200	0.202	-0.033	0.030
Decline in livestock yield (1=Yes, 0=No)	0.613***	0.191	0.148***	0.056
Food shortage (1=Yes, 0=No)	0.342*	0.180	0.070*	0.040
Raining season gone earlier than usual (1=Yes, 0=No)	0.386*	0.194	0.084*	0.048
Off-farm income (FCFA)	1.17×10^{-6}*	6.52×10^{-7}	2.14×10^{-7}*	0.000
Access to finance (1=Yes, 0=No)	0.307*	0.174	0.053*	0.029
Interest rate (percentage)	-0.024*	0.012	-0.004*	0.002
Access to extension services (1=Yes, 0=No)	0.6472**	0.284	0.103**	0.038
Government assistance (1=Yes, 0=No)	0.701**	0.274	0.111***	0.037
Access to climate information (1=Yes, 0=No)	0.570***	0.207	0.123**	0.052
Access to Electricity (1=Yes, 0=No)	0.090	0.201	0.016	0.034
Intercept	-4.062**	1.717		

Log pseudolikelihood=-250.42776
Number of observation= 704
Wald chi2(26) = 107.46***
Pseudo R2 = 0.2051

*** p<0.01, ** p<0.05; * p<0.10

The farm households' location, its socioeconomic characteristics such as the gender of the household head, family size, land devoted to the main crops (i.e., maize, sorghum and rice), and the revenue generated from off-farm activities, are determinants of the climate change adaptation decisions (Table 4.6). The induced effects of climate change, notably food shortages, the shift in rainy seasons over the years, the decline in livestock yields as well as institutional factors, such as access to finance, interest rates applied to credit, access to and use of climate information in farm household decision-making, extension services and the assistance of government were also determinants of climate change adaptation decisions (Table 4.6).

4.3.7 Discussion

Adaptation to climate change is dependent on regional endowments in the Northern Region of Togo. The study results indicate that the coefficients of regional dummy variables are positive and significant, at least at the 5% level (Table 4.6). The results showed that the farm households that are located in the Kara and Savannah Regions are more likely to adapt to climate change than farmers in the Central Region. Indeed, the Central Region is in transition to a Southern Guinean climate and the type of tropical savannah that covers the Kara and Savannah Regions. The Kara and Savannah Regions are the driest areas; hence farmers in these regions are pursuing more adaptation strategies than those in the Central Region.

The coefficient of the sex of the household head is positive and significant at the 1% level, implying that men are more likely to adapt than women. This may be because women operate on only a single piece of land and may, therefore, prefer to undertake other activities such as small business trading. Expanding cultivated cereal land, especially for maize, sorghum and rice, will increase the probability of farm households to adapt to climate change. Certainly, maize, sorghum and rice are more sensitive crops to climatic conditions than grain legumes (i.e., soya beans and beans), tuber crops and cotton in the study area. Climate change-induced effects such as declines in livestock yields, food shortages and shifts in the rainy season have all led farmers pursue adaptation strategies.

The effects of climate change have induced farmers to undertake off-farm activities. The results have indicated that increasing earnings from off-farm activities could increase farm households' probability of pursuing climate change adaptation strategies. Increased opportunities for off-farm employment in the Northern Regions of Togo could, therefore, reinforce climate change adaptation by farmers. Institutional factors such as access to credit, the interest rate policy, government assistance, extension services and accessing climate information have played an important role in farm households' adaptation decisions in Northern Togo. Increasing farm households' access to financial products such as credit will increase the probability of adaptation to climate change. Also, high-interest rates will decrease the probability of the farm households' adaptation decision. A financial policy that grants a special interest rate for farming activities with flexibility in repayments could help farmers in their climate-coping strategies. For example, the Togo government has launched a financial program that seeks to support farming or other types of activities by giving the applicants a small amount of credit (i.e., 60 United States Dollars on average). At the end of the year, the applicant will pay this amount plus interest at the rate of 5 percent. There is no official documentation requirement. The applicants are only asked to be organized by groups. In cases of non-payment by one member of the group, the rest are required to pay for the one member's non-payment. The 5 percent interest rate may be reasonable for a farmer with a low asset base, compared to the rates required by banks and micro-finance providers. Indeed, sometimes the banks or micro-finance providers supply credit with interest rates of more than 18 percent, and this becomes a great challenge for farmers during repayments. The results have indicated that those farmers with access to low-interest credit products are more likely to adapt to climate change. Increasing the number of beneficiaries as well as the amount of credit could motivate more farmers to pursue adaptation strategies.

Access to extension services also remains an important criterion in the household's adaptation decision-making. The extension services coefficient is positive and significant at the 5% level, meaning that increasing farmers' access to extension services could shed light on their perceptions and increase their knowledge about climate conditions and how to better cope with climate change. Also, access to climate information has increased the probability of farmers pursuing climate change adaptation strategies. Expanding the number of weather stations and increasing both meteorological and extension services could help farmers to integrate meteorological information into their farming decision-making and reduce the pronounced effects of climate change on crop production in Northern Togo.

4.4 Conclusions

This study used cross-sectional survey data from 704 farm households in Northern Togo to examine farmers' perceptions of and barriers to climate change adaptation and to analyze the factors that have influenced their climate change adaptation decisions. The results indicate that farmers have perceived an increase in temperatures, while precipitation is perceived to have decreased. Also, shifts in the rainy season have been observed, with frequent dry spells and high winds. These effects have led to decreased crop and livestock yields and increased livestock deaths and incidences of communicable diseases. Changes in the climate have also resulted in a decrease in the quantity and quality of drinking water, and a food shortage in the study area.

The lack of information about adaptation strategies and the lack of capital and knowledge are major barriers to climate change adaptation in the study area. To reduce climate change impacts in the Northern Regions of Togo, based on their experiences in the field, farmers have suggested various measures such as tree plantations, building dams and barrages for irrigation, better waste and perishable plastics management for a cleaner environment, adoption of new technologies (i.e. drought-tolerant crops, use of chemical fertilizer, adoption of new farming practices), reductions in tree cutting and charcoal production, agricultural insurance as well as praying to God or making regular sacrifices to the "living gods" of their communities. This latter suggestion implies a need for more research into the cultural beliefs of climate change.

Results of the Probit model have shown that the adaptation decision was dependent on the geographical location of the surveyed farm households. The farm households' climate change adaptation decisions were influenced by the sex of the household head, the family size and the number of active household members. Also, the results showed that the more farmers grow maize, sorghum and rice, the more they are likely to adapt to climate change. Yet, the decrease in livestock yields, food shortages and shifts in the rainy season in the area due climate change have pushed farmers to pursue adaptation strategies. Access to financial products such as credit with low-interest rates will increase the chances of adapting to climate change. Also, access to extension service could increase farmers' awareness of climate change, while increasing the availability of weather information could help farmers in their decision-making.

Government policies should target disaster risks preparedness and management, to help increase livestock production, decrease the incidence and spread of communicable diseases, and ensure food security in the study area. There is also an urgent need for water management policy in Northern Togo, to increase access for food production and household use. Government assistance through providing specialized financial products to farmers, and the review of financial policies vis-à-vis farming activities and especially the interest rate policy, is needed to enhance investments in farming activities and increase food production as both are already threatened by climate change. The Togo government must also look at the development of weather services so that farmers can access accurate weather information to support their decision-making. The possible synergy between traditional and modern conceptions of climate change adaptation and mitigation could help to make better use of meteorological information. Encouraging farmers' education through increased access to extension services or adult education could help farmers in the adoption of new technologies and farming practices to deal with the effects of climate change.

Acknowledgements

The authors acknowledge the financial support of the Alliance for Green Revolution in Africa (AGRA), International Development Research Centre (IDRC), the West and Central Africa Council for Agricultural Research and Development (CORAF/WECARD) and the International Institute of Tropical Agriculture (IITA) through the grant for thesis writing. However, any opinions, findings, and conclusions or recommendations expressed in this material are those of the authors and do not necessarily reflect the views of the AGRA, IDRC, the CORAF/WECARD and the IITA. The authors would like to thank Prof Al-Hassan Wayo SEINI from University of Ghana, Mr. Bamba Zoumana and the capacity building office of IITA for their support and comments.

References

Abid, M., Scheffran, J., Schneider, U., and Ashfaq, M., (2015) Farmers' perception of and adaptation strategies to climate change and their determinants: the case of Punjab province, Pakistan. *Earth System Dynamics*. 6, 225-243.

AfDB, OECD, and UNDP (2016) African Economic Outlook (2016) Cities and Structural transformation. 15th Edition, African Economic Outlook. DOI: http://dx.doi.org/10.1787/aeo-2016-en

AfDB (2011) Togo country strategy paper, 2011-2015. African Development Bank, p.36.

Berkes, F. (2009) Indigenous ways of knowing and the study of environmental change. *Journal of Royal Society of New Zeland*. 39(4), 151-156.

Bryan, E., Deressa, T., Gbetibouo, G., and Ringler, C. (2009) Adaptation to climate change in Ethiopia and South Africa. Options and Constraints. *Environ. Sci. Policy*, 12(4), 413-426

Bryant, C., Smit, B., Brklacich, M., Johnston, T., Smithers, J., Chiotti, Q., and Singh, B. (2000) Adaptation in Canadian agriculture to climatic variability and change. *Climatic Change*, 45, 181-201.

Burkett, M. (2013) *Indigenous environmental knowledge and climate change adaptation.* In abate RS, Kront EA (eds) Climate change and indigenous peoples: the search for legal remedies. Edward Elgar, Cheltenham, pp.96-118.

Delaporte, I., and Maurel, M. (2018) Adaptation to climate change in Bangladesh. *Climate Policy,* 18 (1), 49-62.

Deressa, T., Hassan, M., Ringler, C., Alemu, T., and Yesuf, M. (2009) Determinants of farmers' choice of adaptation methods to climate change in the Bassin of Ethiopia. *Global Environmental Change,* 19, 248-255.

Ebge, C., Yaro, M., Okon, A., and Bisong, F. (2014) Rural Peoples' Perception to Climate variability/Change in Cross River State-Nigeria. *Journal of Sustainable Development,* 7(2), 25-36.

Di Falco, S., Veronesi, M., and Yesuf, M. (2011). Does adaptation to climate change provide food security? A micro-perspective from Ethiopia. *American Journal of Agricultural Economics,* 93(3), 829-846.

FAO (2015) *FAO Statistical Pokectbook: World food and Agriculture.* Food and Agriculture Organization of the United Nations, Rome. ISBN 978-92-5-108802-9; pp.236

FAO (2008) *Climate change and food security: A framework Document.* Food and Agricultural Organization of United Nation, Rome, Italy. pp.107

Fosu-Mensah, B., Vlek, P., and MacCarthy, D. (2012) Farmers' perception and adaptation to climate change: A case study of Sekyedumase district in Ghana. *Environment, Development and Sustainability,* 14(4), 495-505.

Gould, B., Saupe, W., and Klemme, R., (1989) Conservation Tillage: The role of farm and operator characteristics and the perception of soil erosion. *Land Economics,* 65(2), 167-182.

Green, D., and Raygorodetsky, G. (2010) Indigenous knowledge of changing a climate. *Climatic Change,* 100, 239-242.

Habiba, U., Shaw R., and Takeuchi Y. (2012) Farmers' perception and adaptation practices to cope with drought: Perspectives from Northwestern Bangladesh. *International Journal of Disaster Risk Reduction,* 1, 72-84.

Hassan R., and Nhemachena, C., (2008) Determinants of African farmers' strategies for adapting to climate change Multinomial Choice analysis. *African Journal of Agricultural and Resource Economics,* 2(1), 83-104.

IPCC (2014) Climate change impacts, adaptations and vulnerability, contribution of working group 2 to the fifth assessment report of the Intergovernmental Panel on Climate Change. Cambridge University Press, Cambridge

Kubik, Z., and Maurel, M. (2016) Climate variability and migration: Evidence from Tanzania. *Journal of Development Studies,* 52(5), 665-680.

Maddison, D. (2007). The perception of and adaptation to climate change in Africa. World Bank, Policy Research Working Paper No.4308, Washington D.C.

Maurel, M., and Tuccio, M. (2016) Climate instability, Urbanization and international migration. *Journal of Development Studies,* 52(5), 735-752.

McSweeney, C., New, M., Lizcano, G., and Lu, X. (2010) The UNDP Climate Change Country Profiles Improving the Accessibility of observed and projected Climate Information for Studies of Climate Change in Developing Countries. *Bulletin of the American Meteorological Society.* 91, 157-166.

Mu, R., and van de Walle, D. (2011) Rural roads and local market development in Vietnam. *The Journal of Development studies,* 47(5), 709-734.

Ndamani, F., and Watanabe, T. (2016) Determinants of farmers' adaptation to climate change: A micro level analysis in Ghana. *Scientia Agricola,* 73(3), 201-208.

Nichola, T., (1996). The decision to adopt and the intensity of adopting technology: A double hurdle model application in the adoption of a sorghum hybrid. *Journal of Studies in Economics and Econometrics,* 20, 49-59.

Pannell, D. (1999) Economic, Extension and the adoption of land conservation innovation in agriculture. *International Journal of Social Economics,* 26, 999-1012.

Ransom, K., Paudyal, K., and Adhikari, K. (2003) Adoption of improve maize varieties in the hills of Nepal. *Agricultural Economics,* 1(1), 299-305.

Rojas-Suarez, L. (2010) Access to financial services in emerging powers: facts, obstacles and policy implications. Global Development Background Paper. OEDC Global Development Centre, Washington, D.C.

Shongwe, P., Masuku, B., and Manyatsi, A. (2014) Factors influencing the choice of climate change adaptation strategies by households: A case of Mpolonjeni area deprogram in Swaziland. *Journal of Agricultural Studies,* 2(3), 86-98.

Temesgen, D., Yehualashet, H., and Rajan, D. (2014) Climate change adaptation of smallholder farmers in Southern Ethiopia. *Journal of Agricultural Extension and Rural Development,* 6(11), 354-36.

Terluin, I. (2003) Differences in economic development in rural regions of advanced countries: an overview and critical analysis of theories. *Journal of rural studies,*19(3), 327-344.

Tiwari, K., Rayamajhi, S., Pokharel, R., and Balla, M. (2014) Determinants of climate change adaptation in rural farming in Nepal Himalaya. *International Journal of Multidisciplinary and Current Research,* 2, 234-240.

Waapela, N. (2016). African Myth on Climate Change and Environmental Degradation and Challenges of Development in Africa. *Global Journal of Human-Social Science,* 16(1), 7-16.

Williams, T., and Hardison, P. (2013) Culture, law, risk and governance: context of traditional knowledge in climate change adaptation. *Climatic Change,* 120, 531-544.

World Bank. (2013) *Turn down the heat: Climate Extremes, Regional impacts, and the case for Resilience.* A report for the World Bank by Potsdam Institute for Climate Impact Research and Climate Analytics. Washington, D.C, World Bank, 20433.

Wooldridge, J. (2014) Introduction to Econometrics. Europe, Middle East and Africa. Edition, Cengage Learning.

CHAPTER 5

Social capital and climate change adaptation among smallholder farmers in the Central Region of Ghana

Sampson Osei and Abdulrazak Karriem

Abstract

The impact of climate change on food crop production is likely to worsen rural household food security and livelihoods in Ghana and many other parts of the developing world. Adaptation and mitigation are important avenues through which smallholder farmers can reduce the impacts of climate change. This study analysed the role of social capital on climate change adaptation mechanisms among smallholder farmers in the Central Region of Ghana. We utilized a mixed methodology that drew on a multi-stage sampling technique to administer questionnaires to 225 respondents from six communities in three different agro-ecological districts in the region. Using k-means cluster analysis, four individual social capital variables were constructed from name and position generator instruments. The results showed that smallholder farmers in the study area have established relationships with people of different socio-economic status in the domain of climate change. The strength of these networks, however, differs. The study also revealed that smallholder farmers have adopted 'indigenous' and 'introduced' adaptation strategies. Smallholder farmers' establishment of networks with relevant occupational prestige aids in the flow of information and behavioural change to adapt to the changing climate. The study recommends, among others, that the transfer of climate change adaptation techniques to smallholder farmers should be considered by increasing contact with a wide variety of local actors who can provide information and resources for agricultural production.

Keywords: Cluster analysis; Ghana; Smallholder agriculture; Social capital

5.1 Introduction

Climate change is arguably the most environmentally persistent threat to global stability in the world today. Projections during and beyond the 21st century indicate that terrestrial and freshwater species of plants and animals will face an increased risk of extinction (i.e., 20-30 percent) if the average global temperature exceeds 1.5 – 2.5°C (IPCC, 2007). In Ghana, as in most developing countries, the impact is likely to affect many sectors of the economy. In the agricultural sector, the most obvious direct impact will be on food crop production (i.e., crop failure, poor and unpredictable yields), which is likely to worsen rural household food security and livelihoods (FAO, 2009). Agriculture plays a significant role in the Ghanaian economy: it comprises 22 percent of the nation's GDP, contributes 20 percent of export earnings, and employs 42 percent of the workforce aged 15 years and above (ISSER, 2014). Smallholder farmers who cultivate 1-2 ha plots of land produce 80 percent of the country's agricultural outputs (MoFA, 2013). The smallholder agriculture sector is thus crucial to the livelihoods of rural households in Ghana. Climate change will not only lead to a decline in production but is also likely to lead to an increase in poverty among farming households. The adverse impacts of climate change, however, will fall most heavily on poor smallholder farmers who rely heavily on rainfall for farming activities.

Farmers, particularly smallholder farmers, are especially vulnerable to shocks such as seasonal variation in rainfall and increases in temperature (IPCC, 2007). To ensure that production of staple food staple foods and the overall food security of Ghana is not undermined, there is the need to develop innovative technologies to help mitigate the adverse effects of climate change in agriculture. State interventions to enhance adaptation capacity in the various sectors of the economy will require, among other means, diffusing information on the available adaptation mechanisms (e.g., improved drought-resistant seed varieties and breeds) in Ghana. The process of adaptation, therefore, involves the interdependence of agents (e.g., extension officers and farmers) through their relationships with each other. This includes the institutions in which the agents reside and the resource base on which farmers depend (Adger, 2003). The resources embedded in such relationships have been termed social capital.

Social capital – understood broadly as the features of social structure that facilitate action – has informed many studies, including those addressing farmer-related problems such as climate change (Bourdieu, 1986; Adger, 2001; Ngigi et al., 2012). The major constraints facing smallholder farmer adaptation to climate change have been poor extension services, and a lack of information on adaptation strategies and weather forecasts (Ozor et al., 2010). Adger (2001) and Deressa et al. (2009) have shown that social capital influences access to agricultural information, loan lending and capacity building to adapt to climate change. The adoption of technology by farmers, according to Casey and Lynne (1999), is more effective if the spread of knowledge is shared through social interactions. According to Fafchamps (2005), the interactions among individual farmers help in the exchange of ideas and information which enhances their capacity to respond to climate-related impacts and risks. This aspect of interaction is widely discussed in the literature under network social capital, which is often termed individual social capital (Lin, 2001).

Adaptation and mitigation are important avenues through which smallholder farmers can reduce the impacts of climate change. Numerous studies have investigated the determinants of climate change adaptation strategies among farmers (Mandleni and Anim, 2011; Tazeze et al., 2012). However, these studies have focused extensively on perceptions of climate change, incentives and the ability to adapt, and other environmental factors. The role of social capital as a determinant of household decision-making to adapt to climate change has not been comprehensively investigated. This study assesses the role of individual social capital on climate change adaptation strategies among smallholder farmers in the Central Region of Ghana. In doing so, the study seeks to answer the following research questions: What types of social relations exist among smallholder farmers that facilitate climate change adaptation, and what are the strengths of these relations? What are the various adaptation mechanisms used by smallholder farmers to mitigate the effects of climate change? and, Does the individual's social relations (i.e., social capital) influence climate change adaptation? The study employed both quantitative (i.e., household survey questionnaires) and qualitative (i.e., focus group discussions and key informant interviews) techniques to answer these research questions.

The remainder of the chapter is organised as follows: In section 2, we briefly present the background of the study with a specific focus on how social capital has been used to promote climate change adaptation strategies among smallholder farmers. Then, in section 3, we present the theoretical and conceptual framework underpinning the research project. The research methodology is presented in section 4. In the fifth section, we analyse and discuss the findings of the research project. Finally, in section 6, we conclude the chapter by summarising the main findings and providing recommendations.

5.2 Materials and Methods

5.2.1 Literature review

Research shows that individuals and communities through social relationships, networking and 'social learning' contribute greatly to environmental management and building adaptive capacity (Adger, 2003). Despite the debates around the definition of social capital (see Coleman, 1988; Sobel, 2002; Newton, 2001; Putnam, 1995), there are several elements that have been found to relate to adaptive capacity. These include social trust, reciprocity and interpersonal relationships (Pelling and High, 2005).

These divergent approaches to social capital have informed several empirical studies on climate change adaptation. Using the Mekong River Delta in Vietnam as a case study area, Nam (2011) assessed the role of social capital on adaptation to climate change. Nam constructed four social capital indicators: formal institutional index, informal institutional index, trust index and a cooperativeness index. Using a multinomial probit model, the study revealed that the formal institutional index positively influences the adaptation of different crop varieties and positively influences changing planting dates.

In Kenya, Ngigi et al. (2012) investigated the role of social capital on climate change adaptation. These authors argued that local groups and organizations play a significant role in economic, social and political outcomes. Drawing on the new institutional economics, they focused on the group dimension of social capital and constructed one index using principal components analysis. The study found that social capital positively influences the adoption of new agricultural technologies.

In the American state of Georgia, household survey data collected by National Agricultural Statistics was used to study the effect of social capital on the form of community involvement in the adoption of sustainable agricultural practices. Munasib and Jordan (2011) measured social capital in the form of associational membership. The authors used the number of associations a farmer was a member of to construct a social capital index. The study revealed that social capital not only has a positive causative effect on the adoption of pest control practices, grazing practices and soil management practices but also on the extent to which farmers adopt these practices.

In Kirundo Province of Northern Burundi, Minani et al. (2007) investigated how social capital improves family agriculture and resilience to climate change. Social capital indicators were constructed using local institutional networks. Thus, three groups of farmers were constructed. The first group of farmers comprised of local associations with experience in agriculture and livestock rearing and who received training from formal institutions. The second group of farmers comprised of different organisations, but without experience and support. The last group did not belong to any local association and did not receive any training. The results indicated that about 44 percent of farmers in well-organized local institutions have adapted to climate change. Thus, training and experience play a significant role in the decisions of households to choose a particular adaptation strategy. Moreover, members in farmer-based organisations (FBOs) who receive good training have a high probability of reducing the negative impacts of climate change through appropriate adaptation strategies. Thus, it is not enough to belong to a group or an association; the members must also be adequately trained.

In sum, Lin (2001, p.19) gives four explanations as to why embedded resources in social relations augment the outcome of the action. For one, the flow of information is facilitated among network members. Secondly, social ties may exert an influence on the decision of an agent. Thirdly, social ties, and their acknowledged relationship to the individual may be conceived as social credentials that reflect accessibility to resources. Finally, social networks are expected to reinforce identity and recognition. This study, therefore, establishes the key elements of social capital – bonding, bridging and linking – and ascertains how they influence the capacity of farmers to cope with and adapt to a changing climate.

5.2.2 Theoretical and conceptual framework

The theory of social capital is often discussed from either group or individual perspectives. Scholars who write from the group perspective dwell on how certain groups develop and maintain social capital as a collective asset, and how such collective assets enhance the life chances of group members (Lin, 2001, p.22). Social capital, according to Bourdieu (1986), is comprised of social obligations or connections. He defines social capital as the aggregation of potential resources which are linked to the possession of a durable network of institutionalized relationships of mutual acquaintances and recognition (Bourdieu, 1986, p.248). Putnam (1995) argues that social relations and the degree of participation show the extent of social capital in societies. Thus, association and participation promote and enhance collective norms and trust, which are central to the production and maintenance of collective wellbeing. Putnam (1995, p.664) defines social capital as "features of social life – networks, norms and trust – that enable participants to act together more effectively to pursue shared objectives". Putnam's concept of social capital comprises three components: moral obligation and norms, social values (with trust as a principal value), and social network.

From the individual perspective, Lin (2001) proposes that an individual can gain access to and use two types of resources: personal and social. Personal resources are resources possessed by an individual (e.g., ownership of a material which includes symbolic goods such as a certificate). Conversely, social resources can be accessed through an individual's social connections. These resources, according to Lin (2001), can be 'borrowed' for the intention of making a gain (e.g., a Knapsack sprayer borrowed from a friend to spray pesticides on the farm). Flap (1999) views social capital as the mobilized social resources of individuals. He defines social capital based on (i) the number of persons within the social network of individuals and who are ready to assist when the need arises, (ii) the strength of the relationship indicating readiness to help, and (iii) the resources of the personal network members. Drawing on Lin (2001), we define social capital as a relational concept and focus on the resources embedded in one's network. Moreover, we focus on how the access to and use of such resources benefits individual or group actions. The valued good in a society is termed as a resource and the actions taken by individuals to gain access to the valued good is the central theme of the theory (Lin, 2001).

We employ this theory on the ground because it will bring together several sociological concepts such as social support, integration, and social cohesion that enhance climate change adaptation at the farm level (Requena, 2003). Drawing on this theory thus helps contribute to the body of literature on the importance of social capital in promoting societal well-being.

Bourdieu (1986) argued that the volume of one's social capital largely depends on network size (i.e., the number of ties an individual has in his/her personal network). Lin (2001) adds that effective utilization of social capital also depends on the heterogeneity of the network members. This leads to two concepts of individual social capital – bonding and bridging – as identified by Granovetter (1985). Bonding social capital is the construction of intra-group ties. It is homogeneous and can be very informal (e.g., family and friends) or formal (e.g., work, union members).

Such networks exhibit distinctive forms of internal trust and are driven by values and voluntary efforts with strong ties (Granovetter, 1985). Bridging social capital, on the other hand, is a heterogeneous group. It is impersonal and horizontal, and characterised by general trust and volunteering action (Woolcock and Narayan, 2000, p.230). It is often defined as weak social ties that enable people to get ahead and gain opportunities. With bridging social capital, different groups with diverse interests are able to share and exchange information, ideas and innovation. The literature also identifies a relational concept termed linking social capital. Szreter and Woolcock (2004, p.655) define linking social capital as the "norms of respect and networks of trusting relationships between people who are interacting across explicit, formal or institutionalized power or authority gradients in society".

The fundamental difference between the three relational concepts of social capital (i.e., bonding, bridging, and linking) is social distance. According to Dufhues et al. (2011), linking social capital has two points of connection. It can be connected through bridging social capital (weak ties) or through a bonding social capital (strong ties). From this perspective, the social distances among the three relational concepts of social capital can be lower (i.e., negative) or higher (i.e., positive) or even neutral (see Figure 5.1).

Figure 5. 1: Different Forms of Individual Social Capital

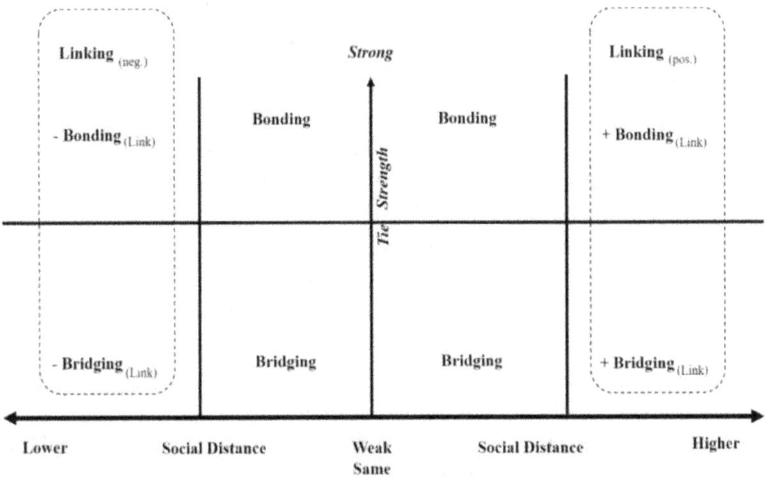

Source: Adapted from Dufhues et al., (2011)

Empirically, these connections generate four different variables of individual social capital. The generation of these variables is usually done using three modes of instruments: name generator, position generator and resource generator (Flap et al., 2000). Following Lin (2001), this study employed name and position generator instruments to establish the individual social capital.

Name Generator: The name generator instrument asks questions about the personal network of the ego (i.e., respondent). For instance, 'Whom will you seek advice from if you encounter a problem on your farm?' The name(s) generated from this question are recorded and served as the basis for the subsequent questions. Questions about each network member such as the age, sex, occupation, and relation with the ego are taken and also recorded. This process is called name interpreter (Flap et al., 2000). To fully capture the individual social capital that may enhance the flow of information on climate change adaptation strategies and to avoid bias toward a single form of social capital, ten name generator questions were employed. These questions vary in terms of skill, knowledge and resource that can potentially be exchanged among farmers and other relevant actors. In this study, to minimize the burden on the respondent (i.e., ego), the number of network members (i.e., alters) to be named was restricted to two. Exclusion of weak ties has often been one of the main weaknesses of the name generator and this is reported to cause structural distortion in ego-centered network data (Van Der Gaag, 2005). This study minimized this problem by the inclusion of the second instrument, namely, the position generator.

Position Generator: The position generator instrument utilizes a limited list of occupations to map the occupational positions of network members (Van Der Gaag et al., 2008). It is usually used to measure the weak ties of individuals. Generally, these occupational positions are considered good indicators of the resources embedded in a social network. The theoretical framework embodied in the construction of a position generator is built from the assumptions on the usefulness of the presence of alters (i.e., the number of network members), the availability of resources, and the presence of resources owned by alters (Van Der Gaag, 2005). In its pure form, a position generator does not create names, and neither does it have name interpreters. The respondents are made to answer 'yes' or 'no' as a response to whether or not they know someone from a sample of occupations that promote climate change adaptation. For example, 'Do you know any extension officers?' Responding just 'yes' or 'no', however, does not suffice for the intended social capital to be measured. To get a meaningful result from this response, the name interpreter is applied to each positive response. The study uses ten different sets of occupations based on the Standard International Occupational Prestige Scale (SIOPS) developed by Ganzeboom and Treiman (1996). Lin's theory of social resources indicates that the ability of network members to provide help and other resources is directly proportional to their social status. Sociologists have therefore established that occupational prestige in modern industrial society is the most relevant indicator of social status (Zhao, 2002).

5.2.3 Sampling and Data

This study drew on primary data and is based on both quantitative and qualitative field research. The quantitative research employed a multi-stage sampling technique to select two communities in the three agro-ecological zones of the Central Region of Ghana: the coastal, transitional and forest zones. The first stage involved the sampling of one district from each of the agro-ecological zones. The sampled districts include: Awutu Senya, Mfantseman and Assin South. This was done using simple random sampling. A similar method was used to sample two communities from each district. To ensure representativeness, 225 households were randomly selected, and questionnaires administered within the six communities. The questionnaires were administered through face-to-face interviews with respondents. The questionnaires captured information on household socio-economic and farm characteristics, climate change adaptation strategies, and the personal networks which promote climate change adaptation mechanisms. In terms of the qualitative component, three focus group discussions were held with each one involving 12 farmers. Three experts were also purposefully selected for in-depth interviews.

A plethora of approaches was used to verify the data collected. The structured questionnaire was pretested prior to the main survey. All shortfalls such as the incorrect wording of sentences, ambiguous questions, and omissions of relevant variables were factored into the final questionnaire. Also, data verification and cleaning were done to identify and correct or eliminate outliers, mismatched coding and omitted variables. Finally, we triangulated the different methods of data collection (i.e., focus group discussions, in-depth interviews, and surveys using the questionnaire) to strengthen and increase the validity of our research findings. The study utilized Stata version 13 software to analyse the quantitative data.

5.2.4 Methods of Analysis

In operationalizing individual network social capital, in this study 'personal network' is defined as the summation of all the relationships the individual possessed. To identify the resource network of the respondents, only relationships that are connected to a resource (e.g., someone who can give you US$50 – to reveal the source of credit) were measured. All measured variables for individual social capital were standardized (i.e., mean is equal to zero and standard deviation to unity)[1]. A k-means cluster analysis was employed to distinguish between strong and weak ties. Following Zhao (2002), five indicators were used to estimate the tie strength between respondents and their network members. These indicators include the type of relationship (i.e., core family, extended family, friends, and acquaintance), frequency of contact per month/year, duration of relationship in years, degree of closeness, and the level of trust. A similar tool was also employed to distinguish bonding, bridging and linking social capital. This study distinguishes linking social capital by using social distances defined as the occupational prestige of the household head and his/her network members. Social distance is precisely computed as the difference between the household head's SIOPS and the SIOPS of network members. As previously noted, social distance can be lower (i.e., negative) or higher (i.e., positive) or even neutral. Thus, the interaction of linking social capital generates four variables (i.e., clusters) using k-means cluster analysis: bonding, bridging, bonding-link, and bridging-link. The aggregation (i.e., number of ties) of these four variables (i.e., clusters) is done for each person in the social capital categories. Descriptive statistics were used to compute the response of the position generator items in the questionnaire. Qualitative data was analysed based on emerging themes and pseudonyms were used in the reporting the findings.

5.3 Results and Discussion

5.3.1 Socio-economic characteristics of the respondents

The proportions of male and female respondents in the study area was 61.5 percent and 38.5 percent, respectively. Gender diversity in rural Ghana is a socio-cultural phenomenon and it may have significant implications for welfare and ownership of resources. These implications

[1] Standardizing is a conversion process, and this allows data on different scales or derived from different distributions to be compared by bringing them to a common scale.

are highlighted along with the individual social capital and adaptation strategies in subsequent sections. Other basic characteristics of the respondents are presented in Table 5.1. The mean age of the household head is 46 years, with a standard deviation of 11 years. About 25.8 percent of respondents could not read or write, and only 9 percent had a tertiary education. The mean years of education among the respondents was 7.24 years, with a standard deviation of 5.05 years.

Table 5. 1: Basic household characteristics of survey respondents

Household Characteristics	Mean	Standard deviation
Age of household head (years)	46.29	11.77
Years of Education	7.24	5.05
Household size	5.60	2.76
Size of land cultivated (hectares)	1.45	1.33
Farming Experience (years)	16.02	10.56
Annual farm income (US$)	740.94	925.81

Household size in the study area was generally high, with an average and standard deviation of about 5.60 and 2.76 people, respectively. Studies show that 90 percent of farm holdings in Ghana are less than 2 hectares in size (MoFA, 2013). The present study lends credence to this finding as average land cultivated is 1.45 hectares, with a standard deviation of 1.33 hectares. The average amount of farming experience was 16.02 years, and the average annual farm income was US$ 740.94.

5.3.2 Individual social relations and tie strength

The name and position generator revealed that respondents (i.e., egos) have social relations with different people of different occupational prestige (usually called alters). The responses of the items used in the position generator questions are presented in Table 5.2.

Table 5. 2: Responses to the items in the position generator questions

Is anyone in your family or among your friends or acquaintance a/an:	Percentage (%)			
	Nobody	Core Family	Extended Family	Friend
National government official including a Minister[2]	51.1	1.3	2.2	5.3
Local government official	45.3	4.9	7.6	18.2
Radio, television or other announcer	23.1	9.3	9.8	24.0
Meteorologist	53.8	0.4	0.4	15.1
Agricultural veterinary officer	43.6	0.4	1.8	14.2
Agricultural extension officer	55.1	0.9	1.8	20.4
Machinery renter	12.0	13.8	21.3	36.4
Shop sales person (agro-chemical shop)	4.4	8.0	9.8	40.4
Higher education Professional (University professor, lecturer)	80.0	2.2	3.6	7.1
Secondary education teacher (SHS, JHS)	44.0	16.9	10.2	22.2
Authors, journalists or other writer	59.1	0.9	2.2	6.2
Farmer (crop production)	0.2	44.0	22.7	28.0
Animal producer (livestock and poultry production)	12.0	11.1	23.1	40.0

[2] This was defined to also capture directors in the various units of the Ministry of Food and Agriculture (excluding the extension and veterinary officers).

Some of the respondents relate to network members as core family members, extended family members, friends and acquaintances. It can be inferred that occupational prestige is scarce within the social circles of the respondents. For example, a majority of the respondents (i.e., about 80 percent) do not have any social relationships with people in the higher education profession. A majority of the respondents, however, have social relations with agro-chemical shop owners (i.e., about 96 percent). But for crop-producing farmers, they have more alters who relate to egos as friends and acquaintances than those who relate as core family and extended family. According to Lin's theory of social capital (2001), social resource (i.e., occupational prestige) in a society has a structure in the form of a pyramid which depicts control and accessibility to such resources. The peak of the pyramid (i.e., higher position) is occupied by fewer people with significant influence in terms of wealth, status and power. The pyramidal structure offers advantages and disadvantages to occupants who are close to the peak and at the bottom of the structure, respectively. To this end, those closer to the peak can more readily reach up to (i.e., contact) those at the peak to exert influence compared to those at the bottom of the structure. The responses of the position generator items in this study show this phenomenon. More than half of the respondents do not have any social relationships in each of the following occupations: higher education, journalists, extension officers, meteorologists and government officials.

The strength of ties between egos and alters is presented in Table 5.3. The standardized variables were clustered into two: weak ties and strong ties. For instance, the mean and the standard deviation values of the duration of relationships of the weak ties are 10.38 and 7.40, respectively. The minimum and the maximum years of relationships are 1 and 45, respectively. On the other hand, the duration of relationships for strong ties has mean and standard deviation values of 18.82 and 13.57, respectively. The minimum and the maximum duration of relationships are 1 and 71 years, respectively.

Table 5. 3: K-means cluster analysis of tie strength

Variable	Mean	Std. Dev.	Minimum	Maximum
Cluster 1, Weak Ties; N = 1093				
Role relation	2.111	1.080	1	4
Frequency of contact	3.274	1.206	0	4
Duration of relationship (years)	10.38	7.396	1	45
Closeness of relationship*	3.122	0.983	1	5
Level of trust	3.573	1.154	1	5
Cluster 2, Strong Ties; N = 1490				
Role relation	2.462	1.057	1	4
Frequency of contact	3.483	0.994	0	4
Duration of relationship (years)	18.82	13.57	1	71
Closeness of relationship	4.863	0.418	2	5
Level of trust	4.717	0.569	2	5

*Defined as 1=Not close at all, to 5=Very close

Strong ties are usually dominant among family members and such bonds last longer than weak ties. According to Granovetter (1985), strong ties are established among relatives (coded as 4=core family, 3=extended family), while weak ties are usually found among friends (coded with 2) and acquaintances (coded with 1). Moreover, strong ties are characterised by trustworthiness, long-lasting contacts and closeness between ego and alter. The frequency of contact determines the dormancy or the activeness of the tie relationship. The implications of strong and weak ties on the probability of information flow and behavioural change have been established. Weak ties have been revealed to diffuse information about new ideas as a result of differences in the social circle of actors (Rogers, 2003; Granovetter, 1985). A lack of weak ties can serve as a barrier to diffusion (Rogers, 2003). The positive effect of strong ties, Friedkin (1982) argues, is that it is a more efficient contributor to the probability of information flow than weak ties.

It must be stressed that the computation of strong and weak tie characteristics was based on the respondent's connectedness with alters within a defined occupational prestige that are likely to promote climate change adaptation. Each respondent had an option of generating at most 23 network members from both position and name generator items used in the study. After elimination of 'nobody' ties (i.e., respondents who do not relate to anyone from a particular occupation), the 225 respondents revealed a total of 2,583 network members (both position and name generator). Table 5.4 presents the four clusters of network members based on the differences in SIOPS, as previously explained in section 5.3. The lowest among the clusters is cluster one (i.e., 12.7 percent), which shows a very high negative SIOPS difference between the ego and the alter. The smaller the mean value in absolute terms, the more similar the ego and alter are in terms of socio-economic status. This lends credence to strong tie arguments which stipulate that the stronger the tie between actors, the more similar they are in diverse ways (Granovetter, 1985). Table 5.4 shows the clusters (i.e., clusters 1, 2 and 3) that are grouped into a 'no linking ties' category.

Table 5. 4: K-means cluster analysis of the social distance of personal network members of the household head

Variable	Mean	Std. Deviation	Min.	Max.	
Cluster 1, N=392					**No linking Ties Category**
Cluster Difference of SIOPS	-27.04	5.660	-47	-21	
Cluster 2, N=526					
Cluster Difference of SIOPS	-14.34	4.080	-20	-8	
Cluster 3, N=983					
Cluster Difference of SIOPS	-0.166	1.242	-7	4	
Cluster 4, N=682					**Linking ties**
Cluster Difference of SIOPS	12.07	6.235	6	28	

As demonstrated in section 3, linking social capital has two points of connection. The first is through bridging social capital where the connection is through a weak tie (i.e., Bridging-link). The second is bonding social capital where the connection is through a strong tie (i.e., Bonding-link). The aggregation of the number of ties for each of the linkages (i.e., connections) yields four individual social capital variables, as presented in Table 5.5. The mean and the standard deviation values of bonding social capital are 4.369 and 2.034, respectively. The minimum and maximum number of ties for bonding social capital are 1 and 10, respectively. Bridging social capital has a mean value of 3.031 and a standard deviation of 2.215, with a minimum tie of 0 and a maximum of 11.

Table 5. 5: Descriptive statistics of individual social capital variables

Variable	Mean	Std. Deviation	Min.	Max.
Bonding Social Capital	4.369	2.034	1	10
$Bonding_{link}$ Social Capital	1.742	1.720	0	8
Bridging Social Capital	3.031	2.215	0	11
$Bridging_{link}$ Social Capital	2.338	1.380	0	8

An independent sample t test was carried out to determine if the mean number of ties for each level of social capital differs by the sex of the respondents. The results of the test statistics are presented in Table 5.6.

Table 5. 6: Independent sample t test of individual social capital by gender

Variable	**Sex**	**N**	**Mean**	**Std. Dev.**	**Std. Error**	**t value***	**Ha: diff Pr |T| > |t|**
Bonding Social Capital	Female	86	3.686	0.199	1.849	-4.0991	**0.0001**
	Male	139	4.791	0.173	2.034		
$Bonding_{link}$ Social Capital	Female	86	1.407	0.156	1.450	-2.3215	**0.0212**
	Male	139	1.949	0.156	1.843		
Bridging Social Capital	Female	86	2.395	0.173	1.604	-3.4693	**0.0006**
	Male	139	3.424	0.207	2.444		
$Bridging_{link}$ Social Capital	Female	86	1.965	0.120	1.111	-3.2547	**0.0013**
	Male	139	2.568	0.125	1.480		

*Degree of freedom = 223

Taking bonding social capital, there is a significant difference in the number of ties of network members between females and males. Thus, male-headed households possess more individual social capital than female-headed households. This result gives credence to recent findings by Szreter and Woolcock (2004), which show that networks of men tend to be more formal with fewer relatives as compared to women. Moreover, women tend to delineate themselves from relationships that transcend the power differential in a society. Regarding the poor individual social capital of women, Kabeer (1996, p.19) explains that "women are generally poorer than men because they lack the range of endowments and exchange entitlements which male members of their households tend to enjoy."

5.3.3 Effect of individual social capital on climate change adaptation

Following Kuwornu et al. (2013), the dominant climate change adaptation strategies are presented in Table 5.7. The proportion of households that did not adopt any 'indigenous' strategies is approximately 28 percent. Moreover, 30 percent of respondents did not adopt any of the 'introduced' adaptation strategies. Thus, more farmers seem to be more involved in *indigenous* compared to *introduced* adaptation strategies.

Table 5. 7: Proportion of households by adaptation strategies

Adaptation Strategies	Percentage (%)
Indigenous Strategies	
Crop diversification	22.67
Livestock diversification	17.33
Soil and water conservation strategies	19.56
Diversification to non-farming activities	12.89
Introduced Adaptation Strategies	
Improved variety and breed	20.0
Soil- and plant-related strategies	25.78
Recommended agricultural practice	24.0

Prior to the enumeration of household questionnaires, focus groups discussions (FGDs) were held to identify the adaptation strategies familiar to farmers. In the FGDs, participants pointed out that they obtained information on climate change adaptation strategies mainly through friends, radio and television. The themes of information discussed during social relations were also ascertained.

The study revealed that some participants created social interaction through farmer-based organisations (FBOs). Thus, some farmers, through their participation in FBOs and other gatherings, established a plethora of relationships with people of different occupational prestige. Nevertheless, most farmers in the transitional zone belonged to FBOs, and this has created strong cohesiveness among these farmers. Some of the leaders of the FBOs have established strong relationships with the Ministry of Food and Agriculture (MoFA) and other government officials in the district. These leaders occasionally receive training from MoFA officials, and then also impart their knowledge to other farmers in the community. In some of these training sessions, participants revealed that they are taught the exact times of planting, the kind of fertilizer and insecticides to apply, and other specifics including planting distances. The community, through their relationship with MoFA officials, also received rain gauges to monitor rainfall. During the FGD, a 45-year old farmer and former local government representative explained how his personal network had helped him overcome an information barrier:

> *I get much information from my friends especially during the season of pepper production. I learned how to nurse the seedlings and the techniques used to transplant to the field from Mr. Kuma. This stage is important because there are some soil conservation measures that need to be taken. The high temperature can kill almost all your seedlings if you fail in this regard.*

This finding supports that of Dinh et al. (2012), who report that most farmers in Vietnam receive indirect information from friends. This is because extension officers only speak to the village headman who relays the information to other farmers. In establishing the relevance of individual social capital, scholars found that farmers who do not have contacts with relevant occupational prestige are excluded from certain services such as formal credit, government supplied agricultural extension services and healthcare. Thus, farmers who are excluded from public services such as extension services rely on their friends, who could be trusted family, close friend or acquaintances, for knowledge. A 52-year old farmer from the forest zone also shared his experience on how he adapted to the changing climate through personal networks:

> *As friends, we talk about farming activities anytime we meet at the durbar ground and other places… my personal experience with regards to climate change adaptation is that of soil and water conservation practice. In a conversation with a friend, he told me that he is not using fertilizer but cocoa husk and palm frond. I went to his farm and observed it myself…*

According to him, he learnt it from the extension officers. He convinced me about the relevance of organic fertilizer to the soil and the yield. I am currently practicing the same strategy at my farm.

Individual farmers who do not have access to extension services rely on their relationship with other personnel. It was discovered that some respondents use their social status to gain access to improved seeds, tractors for ploughing and fertilizer. In the coastal zone, a highly regarded community leader gained access to improved seed varieties through his social relationship which enabled him to go into commercial farming. The leader revealed that besides his large plantation of oil palm trees, he currently plants pineapples due to his colleague's influence. His colleague introduced him to the new and improved variety of pineapples and he currently owns two acres of farmland planted with pineapple. He explained that the climate has changed, and one must re-evaluate farming practice by diversifying and adapting to improved varieties. He also explained how he is preparing to go into coconut production:

I have a friend who is a soil scientist in one of the research institutes. He told me there is a new variety of coconut which takes 3 years to mature as compared to mine which take 8 years to mature. He has asked me to do the nursery and he will come and do the budding for me.

In Granovetter's (1985) language, both strong and weak ties have influenced the decisions of many respondents to adapt to the changing climate through diversifying into different crops and using improved varieties. It is difficult to predict the kind of social relationships (be it bonding, bridging or linking) between the ego and alter on a qualitative basis. However, the findings indicate that the establishment of networks has not only aided the flow of information but also changed behaviours through the adoption of certain practices. The relational concept of social capital has many positive implications for the development of a given society. For example, in rural Northern India, poor villagers utilize social relationships and groups for risk management, protection and solidarity functions (Woolcock and Narayan, 2000). In sum, individual social capital may affect the economic sustainability of farmers. As revealed in this section, networks of individuals have the potential of influencing adaptation decisions. Farmers who are well connected have a high propensity to adopt improved varieties and breeds, and therefore diversify their cropping systems and farming practices to reduce the adverse consequences of climate-related agricultural change.

5.4 Conclusions

This study investigated social capital as a relational concept and evaluated its influence on climate change adaptation strategies among smallholder farmers in the Central Region of Ghana. It shows how individual farmers make use of the opportunities of embedded social relationships, both in qualitative and quantitative realms, to adapt to the changing climate. Thus, social relationships not only facilitate the flow of information, they also influence behavioural change to manage risk.

The results also show that smallholder farmers have networked with people of different socio-economic status to promote climate change adaptation. Using position generators, the study found that most network members relate to egos as friends and acquaintances. By contrast, name generators show the opposite, as almost all alters relate to egos as family members with very few as friends. This supports Van der Gaag's (2005) view that the first name that comes to mind when asking name generator questions is that of family members or close friends. Using relevant climate-related occupational prestige, social distance between respondents and network members was constructed, and four clusters were generated using K-means cluster analysis. The number of ties of each individual social capital was aggregated and an independent t test was computed to determine if the mean difference differed by the sex of respondents. The results indicate that men possess greater individual social capital than women. This was significant across all four of the individual social capitals identified in the study.

The study also found that smallholder farmers have embraced both indigenous and introduced adaptation strategies. Indigenous adaptation strategies are perceived to have accumulated over generations of a people living in a particular environment (FAO, 2009). Thus, farmers developed their own indigenous adaptation strategies due to their experiences with the changing climate. Introduced adaptation strategies are those developed through scientific research.

The findings from the research show that the establishment of networks with relevant occupational prestige aids in the flow of information on climate change adaptation. Moreover, these connections lead to behavioural changes where smallholder farmers are influenced to adapt to the changing climate. Some respondents adapted by using improved varieties upon hearing of their adoption by family members, friends or acquaintances. Similarly, the adoption of new types of fertilizers was easily promoted through the establishment of relevant networks.

Studies on social capital come with many suggestions for government actions to build or support social capital. This study shows that individual social capital influences the adoption of both indigenous and introduced adaptation strategies. Based on this finding, it would be interesting to call for the promotion of social capital (such as bonding, linking and bridging) by government and other institutions.

This study has shown that leadership in a community has a vital role to play in promoting climate change adaptation practices. Based on the results obtained from both the position and the name generator techniques, it can be argued that the relay of climate change adaptation techniques or technologies to smallholder farmers need not only be accomplished through the usual technology transfer networks of agricultural researchers and extension agents. Rather, it is imperative to increase contact with a wide range of local actors who provide and share information and resources for agricultural production. These actors could be teachers, religious leaders, owners of agro-chemical shops, as well as local women and farmer leaders. In doing so, the transaction costs associated with obtaining the relevant information will be reduced. The danger here is that in designing such a policy, there is potential to exclude people from obtaining access to resources, and so caution is needed to ensure such policies do not increase social inequality.

Finally, leaders should be involved in training and capacity building programmes and workshops so that they can reach out to other people in their community. This will reduce the workload of agricultural extension agents while reaching more smallholder farmers within a particular geographical location. This will not only ensure widespread diffusion of climate change adaptation strategies among smallholder farmers but also help in guaranteeing the future of food security in Ghana.

References

Adger, W.N. (2001) Social Capital and Climate Change. Tyndall Centre for climate change research. Working Paper 8.

Adger, W.N. (2003) Social Capital, Collective Action, and Adaptation to Climate Change. *Economic Geography*, 79, 387–404.

Bourdieu, P. (1986) The forms of capital. In: J. G. Richardson, ed. (1986). *Handbook of Theory and Research for the Sociology of Education.* New York: Greenwood Press, 241–258.

Casey, F., and Lynne, G.D. (1999) Adoption of Water Conservation Technologies in Agriculture: The role of Expected Profit and the Public Interest. In: F. Casey, A. Schmitz, S. Swinton, & D. Zilberman, eds., (1999) *Flexible incentives for the adoption of environmental technologies in agriculture.* Norwell, MA: Kluwer Academic Publishers, 229-247.

Coleman, J. S. (1988) Social Capital in the Creation of Human Capital. *American Journal of Sociology*, 94, 94–120.

Deressa, T.T., Hassan, T., Ringler, C., Alemu, T., and Yesuf, M. (2009) Determinants of farmers' choice of adaptation methods to climate change in the Nile Basin of Ethiopia. *Global Environmental Change*, 19(2), 248–255.

Dinh, Q. H., Dufhues, T., and Buchenrieder, G. (2012) Facilitating access to rural services in Vietnam: The invisible social capital link. In: *Selected Paper prepared for presentation at the International Association of Agricultural Economists (IAAE) Triennial Conference, Foz do Iguaçu, Brazil, 18-24 August, 2012.* 18–24.

Dufhues, T. G., Buchenrider, G. Euler, D. G., and Munkung, N. (2011) Network based social capital and individual loan repayment performance. *Journal of Development Studies*, 42(8), 1199–1215.

Fafchamps, M. (2005) Social Capital. In: P. Aghion, & S.N. Durlauf, eds. (2005). *Handbook of Economic Growth* (Vol. 1). Amsterdam: Elsevier. 1639–1699.

Food and Agriculture Organisation (FAO) (2009) *Adapting to climate change, UNASYLVA 60 - 231/232*, Available at: www.fao.org/forestry/unasylva. Date accessed: 2/05/2015

Flap, H. (1999) Creation and returns of social capital: a new research program. *La Revue Tocqueville*, (XX), 5–26.

Flap, H., Snijders, T., Vllker, B., and Van Der Gaag, M. (2000) Measurement instruments for social capital of individuals. Questionnaire items as used in the 1999/2000 nationwide study in the Netherlands- *Social relations and networks in the neighbourhood and workplace: the social survey of the network of the Dutch* -(SSND). A joint project of the university of Utrecht, Groningen and Amsterdam.

Friedkin, N.E. (1982) Information Flow Through Strong and Weak Ties in an Intraorganizational Social Networks. *Social Networks*, 3, 273–285.

Ganzeboom, H., and Treiman, D. (1996) Internationally comparable measures of occupational status for the 1988 International Standard Classification of Occupations. *Social science research*, 239(25), 201–239.

Granovetter, M. (1985) Economic Action and Social Structure: The Problem of Embeddednes. *American Journal of Sociology*, 91(3), 481–510.

Institute of Statistical, Social and Economic Research (ISSER) (2014) The State of the Ghanaian Economy in 2013. Institute of Statistical, Social and Economic Research. University of Ghana, Legon. September, 2014.

Intergovernmental Panel on Climate Change (IPCC) (2007) *Climate Change 2007: Impacts, Adaptation and Vulnerability: Contribution of Working Group II to the fourth assessment report of the Intergovernmental Panel.* Genebra, Suíça. Available at: http://www.ipcc-wg2.org. Date accessed: 2/02/2015.

Kabeer, N. (1996) Agency, Well-being and Inequality: Reflections on the Gender Dimensions of Poverty, *IDS bulletin*, 27(1), 11-21.

Kuwornu, J. K. M., Al-hassan, R. M., Etwire, P. M., and Osei-owusu, Y. (2013) Adaptation Strategies of Smallholder Farmers to Climate Change and Variability : Evidence from Northern Ghana. *Information Management and Business Review*, *5*(5), 233–239.

Lin, N. (2001) *Social Capital: A Theory of Social Structure and Action*, Cambridge, Mass: Cambridge

Mandleni, B., and Anim, F.D.K. (2011) Perceptions of Cattle and Sheep Farmers on Climate Change and Adaptation in the Eastern Cape Province of South Africa. *Journal of Human Ecology*, 34(2), 107–112.

Minani, B., Rurema, D., and Lebailly, P. (2007) Rural resilience and the role of social capital among farmers in Kirundo Province, Northern Burundi. *Applied Studies in Agribusiness and Commerce-APSTRACT*, 7(2-3), 121-125.

Ministry of Food and Agriculture (MoFA) (2013) *Agriculture in Ghana: Facts and Figures*, Accra, Ghana.

Munasib, A., and Jordan, J. (2011) The effect of social capital on the choice to use sustainable agricultural practices. *Journal of Agricultural and Applied Economics*, 43(2), 213–227.

Nam, P. (2011) Social capital and private adaptation to climate change: Evidence from the Mekong River Delta in Vietnam. In: *Institute for International Economics Policy (IIEP) conference on the "Economics of Adaptation to Climate Change. George Washington University, Washington, DC.* 1–24.

Newton, K. (2001) Trust, Social Capital, Civil Society, and Democracy. *International Political Science Review*, 22 (2), 201-214.

Ngigi, M. W., Bryan, E., Claudia, R., Birner, R., and Mureithi, D. (2012) Climate Change Adaptation in Kenyan Agriculture: Could Social Capital help? *The 8th AFMA Congress Theme: "Repositioning African Agriculture by Enhancing Productivity, Market Access, Policy Dialogue and Adapting to Climate Change"*. Conference Proceeding.

Ozor, N., Madukwe, M. C., Enete, A. A., Amaechina, E. C., and Onokala, P. (2010) Barriers to Climate Change Adaptation among Farming Households of Southern Nigeria. *Journal of Agricultural Extension*, 14 (1). pp. 114–124.

Pelling, M., and High, C. (2005) Understanding adaptation: What can social capital offer assessments of adaptive capacity? *Global Environmental Change*, 15(4), 308–319.

Putnam, R.D. (1995) Tuning in, tuning out: The strange disappearance of social capital in America. *PS: Political science & politics*, 28(04), 664–683.

Requena, F. (2003) Social capital, satisfaction and quality of life in the workplace. *Social indicators research*, *61*(3), 331-360.

Rogers, E.M. (2003) *Diffusion of Innovations*, (5th Edition), New York: Free Press.

Sobel, J. (2002) Can we trust Social Capital? *Journal of Economic Literature, XL, 139–154*

Szreter, S., and Woolcock, M. (2004) Health by association? Social capital, social theory, and the political economy of public health. *International Journal of Epidemiology*, 33(4), 650–667.

Tazeze, A., Haji, J., and Assistance, M.K. (2012) Climate Change Adaptation Strategies of Smallholder Farmers : The Case of Babilie District, East Harerghe Zone of Oromia Regional State of Ethiopia. *Journal of Economics and Sustainable Development*, 3(14), 1–13.

Van der Gaag, M. (2005) Measurement of individual social capital. *Design*, 5(10), 266.

Van der Gaag, M., Snijders, T.A., and Flap, H.D. (2008) Position Generator measures and their relationship to other Social Capital measures. In: Lin, N., and Erickson, B. H., (Eds), (2000). *Social Capital: An International Research Programme*. Oxford, Oxford University Press, 27–48.

Woolcock, M., and Narayan, D. (2000) Social Capital: Implications for Development Theory, Research, and Policy. *The World Bank Research Observer*, 15(2), 225–249.

Zhao, Y. (2002) Measuring the Social Capital of Laid-off Chinese Workers. *Current Sociology*, 50(4), 555–571.

CHAPTER 6

What factors influence farmers' vulnerability to climate change and variability? Empirical evidence from smallholder women rice farmers in the Northern Region of Ghana

Suhiyini Issah Alhassan, Yaw Bonsu Osei-Asare, John K.M. Kuwornu

Abstract

The unpredictable and long-term shift in climatic conditions bestows huge effects on the livelihoods of rain-fed farmers. This effect is enormous among smallholder farmers, especially females who lack the capacity to withstand the adverse effects of climatic shocks. This study used the Livelihood Vulnerability Index (LVI) to determine smallholder women rice farmers' vulnerability to climate change and variability, and the Tobit regression model to examine the factors influencing their vulnerability. Historical data on rainfall and temperature were obtained from the Ghana Meteorological Service, while a questionnaire was used to collect primary data from 260 smallholder women rice farmers randomly selected from the Savelugu/Nanton Municipality and West Mamprusi District in the Northern Region of Ghana through a multi-stage sampling technique. The empirical results showed that most farmers interviewed were highly vulnerable to climate change and variability. It was also revealed that whereas education, age, off-farm income, participation in decision-making, formal extension contacts and farmland ownership have significant negative influence on women rice farmers' vulnerability to climatic stress; farm size, membership of a farmer-based group, hired labor, access to tractor services and noticing climate variability have a positive influence on farmers' vulnerability to climate variability. It is recommended that adult education programs such as 'night school' should be vigorously pursued to improve farmers' literacy. Also, regular engagements between farmers and agricultural extension agents are necessary to train farmers on agronomic practices and other non-farm income generation skills to help them diversify their livelihoods.

Keywords: Climate change and variability; Livelihood Vulnerability Index; smallholder women rice farmers; Tobit regression model; Vulnerability

6.1 Introduction

Agriculture's productivity in sub-Saharan Africa is adversely affected by climate change and variability. The effects of climate change and variability are particularly severe for smallholder farmers, especially female farmers (Boko et al., 2007). An obvious reason for this is that farmers in sub-Saharan Africa rely heavily on rain-fed and subsistence agriculture, and rains are increasingly erratic. Most countries in this region already struggle with the problems of poverty, food insecurity, health, and other basic social needs. The cumulative effect of climate change and variability is, therefore, a hindrance to the growth of the economies of sub-Saharan countries including Ghana, placing them at the high end of the climate change vulnerability spectrum (Nti, 2012). As stated by Al-Hassan and Poulton (2009), the perennial occurrence of floods, droughts and bushfires in the Northern Region of Ghana heightens farmers' struggles for food every year, especially among women and children.

Despite the major role played by women in rice production, and that women constitute 50.4 of the percent population of the Northern Region of Ghana (Ghana Statistical Service, (GSS), 2012), the land tenure system in this region does not favor their total ownership of land. Only one in forty-five female farmers owns land for rice cultivation in the Northern Region (Asare, 2000). This, coupled with the fact that women in the region are marginalized in certain vital decision--making including responses to climate change and variability, makes women more vulnerable and susceptible to the impacts of climate variability.

Stanturf et al. (2014) revealed that Ghana's Northern Region is more vulnerable to climate change and variability compared to the rest of the country, mainly because of the high rates of illiteracy and relatively underdeveloped infrastructure. Nti (2012) reported that farmers in the Northern Region possess the least capacity to adapt to climate change and variability and that this region is the most exposed region to climatic stressors in the country (Etwire et al., 2013a). In the Northern Region, rice is among the most vulnerable crops to climate variability, especially floods and drought (Antwi-Agyei et al., 2013). Though rice farming in the Northern Region is typically the preserve of men, female rice farmers are usually small-scale and are the most affected by climatic stressors. The effect of climate change and variability on smallholder women rice farmers in the Northern Region of Ghana is so glaring and severe that this study could not have been better undertaken anywhere else in Ghana. The study examined the vulnerability of farmers to climate change and variability, using smallholder women farmers in the Northern Region as a case study. The study objectives were to determine farmers' vulnerability to climate change and variability and to examine factors influencing the vulnerability of farmers to climate change and variability.

In the literature, there exist several studies on farmers' vulnerability and adaptation to climate change and variability. Yet, these studies are limited to male farmers, with little emphasis on women farmers in Ghana, particularly in the country's Northern Region. To the best of our knowledge, this study is the first of its kind in the region's two districts and its findings, apart from adding to the climate change vulnerability literature, will provide a regional-level ecological and gender-specific analysis of vulnerability to climate change and variability. This will form the basis for addressing the specific needs of smallholder women rice farmers in the Northern Region as a whole, and the region's two districts in particular.

6.2 Materials and Methods

6.2.1 Conceptualization of Terminologies

Climate Change and Variability: These two terms describe the variations in the values of statistics including mean, standard deviation, and frequencies of climatic variables on all time-based and spatial scales beyond single weather events (IPCC, 2001; Fussel and Klein, 2006). The interim variability is described as climate variability, while the observed long-term shift in these climatic variables is climate change. This variability may either be internal (i.e., resulting from natural internal processes within the climate system) or external (i.e., resulting from variations in natural or anthropogenic external forces).

Vulnerability: This term has no generally accepted definition due to its wide usage in different disciplines to explain particular areas of concern. The conventional meaning of the word 'vulnerability' is the capacity to be distressed. That is, the extent of injury likely to be caused to a system as a result of its exposure to a hazard, as stated by Turner et al. (2003). Vulnerability refers to the ability to forestall, manage with, resist and recuperate from the effects of natural disasters (Madhuri et al., 2014). In the views of Cutter et al. (2008) and Nelson et al. (2009), vulnerability could also be the predisposition of any group of people, location, or system to disorders determined by exposure and sensitivity to distress, including their adaptive capacity. Vulnerability, as employed in this study, refers to the degree to which smallholder women rice farmers are susceptible to, or unable to adapt to, the negative impacts of climatic stresses (i.e., mainly floods, bushfires and droughts).

Smallholder Women Farmers: Smallholder farmers are made up of farmers with landholdings of three hectares or less (Chamberlin, 2008). The average landholding of peasant farmers in Ghana's Northern Region is about two hectares (MoFA, 2011). Morton (2007) defined smallholder farming as the agricultural activities undertaken mostly by family labor in which the outputs are the main source of income for the family. However, this is not synonymous with subsistence farming in that the latter is a livelihood strategy where the main output of agricultural activity is consumed directly (Morton, 2007). Smallholder women rice farmers have been defined in this research to include farmers with a farm size of three hectares or less and who rely heavily on family labour for their farming activities.

6.2.2 Evidence of Climate Change and Variability

The intra- and inter-annual rainfall pattern in Ghana is highly erratic (World Bank, 2010). Based on historical data, rainfall trends in Ghana indicate a decreased rainfall amount between 1960 and 2006, with an average annual reduction of 2.3 millimeters. Kankam-Yeboah et al. (2011) observed a significant change in Ghana's climate, with a 1°C rise in temperature over a 30-year period. The high temperatures associated with Harmattan (i.e., a season in the West African subcontinent) which were hitherto experienced in March in northern parts of Ghana are more recently being experienced in January (GSS, 2012). The onset and end of the rainy season in the Northern Region is becoming increasingly variable each year. Initially, the first rain was received in April and the last rain came in September (Etwire et al., 2013b). However, the rainfall pattern in recent times has changed and the region now receives its first rain in late June to early July, which then usually ends abruptly in October (Amikuzuno and Donkoh, 2012).

Historical data on rainfall and temperature for Savelugu/Nanton Municipality and West Mamprusi District in the Northern Region of Ghana were sourced from the Ghana Meteorological Service. For rainfall, the analysis revealed that whereas the average rainfall in Savelugu/Nanton has decreased over the period 1986 and 2016, the case of West Mamprusi showed a different trend, with average rainfall increasing slightly over the period 1996 – 2015 (Figure 6.1). Generally, a trend of decreasing rainfall has been observed in Ghana (Fosu-Mensah et al., 2012). For the temperature pattern in the study area, data from the Ghana Meteorological Service showed a general increase in temperature over the period 1987 - 2016 (Figure 6.2) This is consistent with the findings of Amikuzonu and Donkoh (2012), Fosu-Mensah et al., (2012) and Kankam-Yeboah et al., (2011).

Figure 6. 1: Average daily rainfall by year in Savelugu/Nanton and West Mamprusi

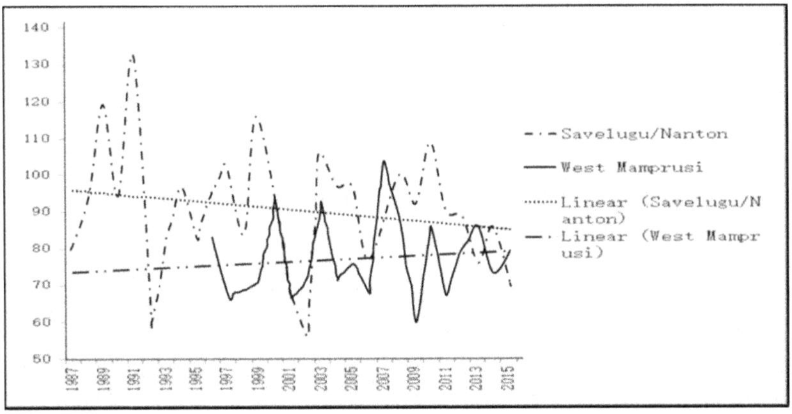

Source: Authors' computation of rainfall data from GMS (1987-2016), 2016.

Figure 6. 2: Average daily maximum and minimum temperatures by year in Savelugu/Nanton and West Mamprusi

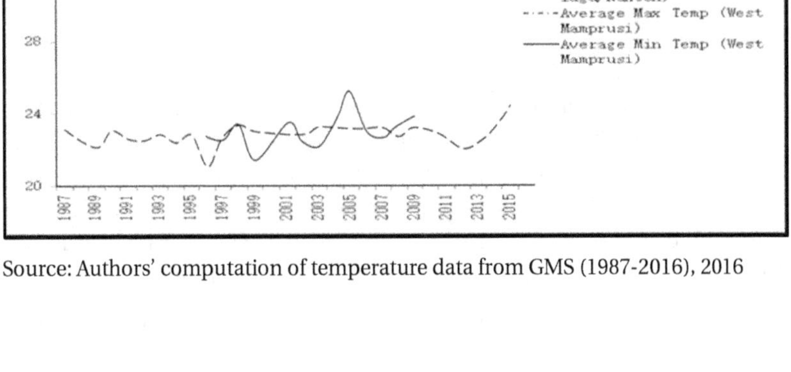

Source: Authors' computation of temperature data from GMS (1987-2016), 2016

6.2.3 Literature Review

Deressa et al. (2009) distinguished between two main approaches for measuring vulnerability to climate variability; namely, econometric and indicator techniques. The econometric approach employs regression analysis methods. The main limitation of the econometric approach is the testing of several econometric assumptions based on hypothesis, confidence intervals and standard errors, and also attributing causation without stringent assumptions. The indicator approach involves selecting components considered to be indicators of vulnerability (Deressa et al., 2009). The major limitation of the indicator approach is the likely subjectivity on the part of the researcher in selecting the components to be included in computing the index. But, the indicator approach is preferred to the econometric approach because of its practicality and ease of interpretation. This study employed the Livelihood Vulnerability Index (LVI), espoused by Hahn et al. (2009) to measure women rice farmers' vulnerability to climate change and variability. The LVI was previously employed by Adu et al. (2018) to measure smallholder maize farmers' vulnerability to climate change in the Brong-Ahafo Region of southern Ghana. The empirical results showed that farmers were more vulnerable in terms of social networks and health, with farmers in the Wenchi District being more vulnerable to climate change than farmers in the Techiman Municipality.

Nkondze et al. (2013) employed an ordered multinomial logit model to determine factors influencing households' vulnerability to climate change in Swaziland. The Household Vulnerability Index (HVI) was first computed for each household as a measure of vulnerability. The HVI is an index between 0 and 100. Using the computed HVIs and based on the FANRPAN (2011) categorisation of vulnerability, households were then classified as either lowly, moderately or highly vulnerable to climate change. These three categories were used as the dependent variable in estimating the factors influencing the vulnerability of households using an ordered multinomial logit regression model. The limitation of the multinomial logit model in estimating determinants of vulnerability is that the dependent variable is a ratio and not categorised such as the current data set, with the possibility of observing zero values (Gujarati and Porter, 2008). Tobin (1958) proposed the Tobit model to deal with the problem of zero observations by treating all farmers with observed zero LVIs as non-vulnerable, and that the probability and level of vulnerability are determined by the same sets of variables. This is known as censoring of the data (Tobin, 1958). This study employed the Tobit model to determine the factors influencing the vulnerability of women rice farmers to climate change and variability. It is appropriate where the dependent variable (i.e., vulnerability) is a continuous rather than categorical variable, and the interest is to measure the direction and intensity of the influence of each exogenous variable on the dependent variable (Tobin, 1958).

6.2.4 Study Area

The study was conducted in the Northern Region of Ghana. More precisely, it was conducted in the Savelugu/Nanton Municipality and the West Mamprusi District in the Northern Region. Figure 6.3 shows the study communities in each district. The study area is in the Guinea Savanna agro-ecological zone and a semi-arid climatic region, with its vegetation comprising mainly of grasslands interspersed with Guinea and Sudan savannah woodland. These woodlands are characterized by drought-resistant trees such as the baobab, shea nut, acacia, dawadawa, mango and neem. The average minimum and maximum temperatures of the region are 14^{0}C at night and 40^{0}C during the day, respectively (GSS, 2014). The region experiences two seasons: the dry season (i.e., November to April) and the wet season (i.e., between May and October) with an average annual rainfall of 750mm – 1050mm (GSS, 2014).

The Northern Region has a land area of 70,384 square kilometers (i.e., 31% of Ghana's land area), with a population of 2,479,461, which is 10.1% of the national population. The region has a female population of 1,249,574 (i.e. 50.4%). West Mamprusi District is the third largest district in the region with a total population of 168,011, of which 50.06% are female. Savelugu/Nanton Municipality is the fourth largest district in the region, with a population of 139,285, of which 51.51% are female.

The agricultural sector employs 73.11% of the Northern Region's population who are aged 15 years and above, with 43.06% being female. The main crops cultivated in the region include maize, rice, yam, cassava, soybeans, millet, and sorghum (GSS, 2012). About 14.1% of households in the Northern Region are headed by females, which is higher than the national average of 11% (GSS, 2012). The Northern Region is experiencing the effects of climate variability in the form of shifting rainfall patterns and more frequent and extreme floods and droughts. The impacts on rice production in the region are negative since agriculture in northern parts of Ghana is predominantly rain-fed (Yengoh et al., 2010).

Figure 6. 3: Map of Savelugu/Nanton Municipality and West Mamprusi District showing the study communities

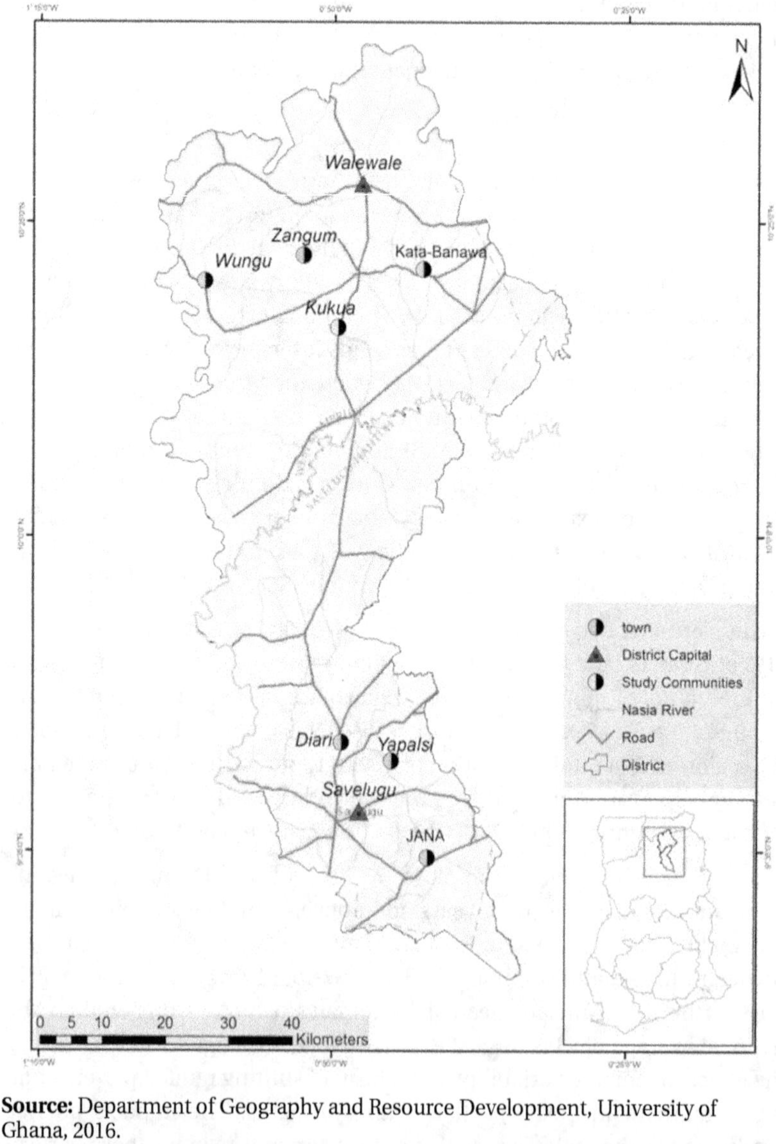

Source: Department of Geography and Resource Development, University of Ghana, 2016.

6.2.5 Sampling and Data

A multi-stage sampling technique was used. The Northern Region was purposively selected because it is in a guinea savanna ecological zone located in a semi-arid climatic region and it is Ghana's highest producer of rice. West Mamprusi District and Savelugu/Nanton Municipality have been purposively chosen among other districts because they are the top two rice producing districts in the Northern Region and they also have the most erratic rainfall patterns. Based on population and land size, 150 respondents from four communities in West Mamprusi District (i.e., Zangum, Kata-Banawa, Wungu and Kukua) and 110 respondents from three communities in the Savelugu/Nanton Municipality (i.e., Diare, Jana and Yapalsi) were randomly selected using proportionate and simple random sampling techniques. A questionnaire was then administered to the 260 smallholder women rice farmers to collect primary data on their socio-demographic characteristics, livelihood strategies, social networks, their food, health and water situations, and their access to institutional and environmental information that may influence their vulnerability to climate change and variability. Secondary data on temperature and rainfall in each district were collected from the Ghana Meteorological Service.

6.2.6 Conceptual Framework

The conceptual framework was developed by the researchers following a thorough review of literature on integrated approaches to conceptualizing vulnerability to climate change and variability. The effects of climate change and variability include floods, droughts and bushfires. The extent of damage suffered by smallholder farmers due to climatic stressors is triggered by their exposure (i.e., frequency and intensity of climatic shocks), sensitivity (i.e. depends on human and ecological conditions) and adaptive capacity (i.e. access to economic and natural resources) to climatic shocks (IPCC, 2014; Turner et al., 2003). This is the basis on which smallholder women rice farmers' vulnerability to climate change and variability was determined. According to Etwire et al. (2013b), farmers' adaptation to climate change is influenced by their socio-demographic characteristics and other institutional and environmental factors. The study examined factors influencing smallholder women rice farmers' vulnerability to climate change and variability using these three categories.

6.3 Data Analysis

6.3.1 Measuring smallholder women rice farmers' vulnerability to climate variability

This objective was achieved by estimating the LVI, as developed by Hahn et al. (2009). The Livelihood Vulnerability Framework is commonly used in assessing vulnerability to climate variability because it provides a framework that makes it possible to analyze both the essential components that constitute livelihood and the contextual factors influencing these factors. The LVI assumes equal weights for all major and sub-components (Sullivan, 2002).

Hahn et al. (2009) made use of seven major indicators to estimate the LVI. These include socio-demographic profile, livelihood strategies, social networks, health status, access to food, access to water, and events of natural disasters and climate variability. Each major indicator consists of several sub-components known as indicators. Each of these indicators is measured on a different scale and contributes positively or negatively to vulnerability. It is, therefore, necessary to standardize each indicator as an index using equation (6.1) (where the indicator contributes positively to vulnerability) or equation (6.2) (if the indicator contributes negatively to vulnerability) (Adu et al., 2018).

$$Index_{if} = \frac{S_s - S_{min}}{S_{max} - S_{min}} \tag{6.1}$$

$$Index_{if} = \frac{S_{max} - S_s}{S_{max} - S_{min}} \tag{6.2}$$

Where S_s is the observed sub-component indicator s for farmer f, and S_{min} and S_{max} are the minimum and maximum values, respectively, for the sub-component in the combined data. The sub-component indicators were then averaged using equation (6.3) to obtain the index of each major component:

$$M_{if} = \frac{\sum_{i=1}^{n} index_{if}}{n} \tag{6.3}$$

Where M_{if} is one of the seven major components [Socio-Demographic Profile (SDP), Livelihood Strategies (LS), Social Network (SN), Health (H), Food (F), Water (W), or Natural Disaster and Climate Variability (NDCV)] for farmer f; $index_{if}$ represents the sub-components, indexed by i , that make up each major component, and n is the number of sub-components in each major component. Once values for each of the seven major components for each farmer are calculated, they are also averaged to obtain the LVI for each farmer using equation (6.4):

$$LVI_f = \frac{\sum_{i=1}^{7} w_{Mi} M_{if}}{\sum_{i=1}^{7} w_{Mi}} \qquad (6.4)$$

Where w_{Mi}, the weights of each major component, is a function of the number of sub-components that make up each major component and are included to ensure that all sub-components contribute equally to the overall LVI. Based on the computed LVIs, farmers' vulnerability to climate change and variability were categorized into least ($0 \leq LVI < 0.31$), low ($0.31 \leq LVI < 0.47$), moderate ($0.47 \leq LVI < 0.52$), high ($0.52 \leq LVI < 0.61$), and extreme ($0.61 \leq LVI \leq 1.00$). This categorization is a modification of the FANRPAN (2011) categorization of farmers' vulnerability to climate change.

6.3.2 Factors influencing the vulnerability of farmers to climate variability

The study employs the Tobit model in analyzing the intensity and direction of factors influencing the vulnerability levels of smallholder women rice farmers to climate variability.

The general structural equation (6.5) of the Tobit model is specified as;

$$y^*_i = x_i \beta + \varepsilon_i \qquad (6.5)$$

Where $\varepsilon_i \sim N(0, \delta^2)$ is normally distributed with a zero mean and constant variance. y^* is a latent dependent variable, which is the vulnerability indices computed for each farmer that is observed for values greater than τ and censored otherwise. The observed y is defined by the following measurement equation (6.6), and $\boldsymbol{\beta}$ is the parameter to be estimated.

$$y_i = \begin{cases} y^* & if \quad y^* > \tau \\ \tau & if \quad y^* \leq \tau \end{cases} \qquad (6.6)$$

In the classic Tobit model, $\tau = 0$, and this implies that the data is censored at 0, as per equation 6.7.

$$y_i = \begin{cases} y^* & if \quad y^* > 0 \\ 0 & if \quad y^* \leq 0 \end{cases} \qquad (6.7)$$

The empirical model is specified in equation (6.8), as follows:

$$Y^*(LVI > 0.47) = \beta_0 + \beta_1 EDU + \beta_2 AGE + \beta_3 FLAB + \beta_4 FARMEXP + \beta_5 FBO + \beta_6 INCOME \beta_7 FARMSIZE + \beta_8 DECISION + \beta_9 EXTENSION + \beta_{10} HLAB + \beta_{11} TRACTOR + \beta_{12} FARMLAND + \beta_{13} NOTICECV + \varepsilon_i \quad (6.8)$$

Following the FANRPAN (2011) categorization of farmers' vulnerability, which has been modified in this research, farmers with an LVI ≤ 0.47 (i.e., least and low vulnerability) were considered not vulnerable to climatic stress. The Tobit regression model was therefore executed with a specified left censoring limit of 0.47. Table 6.1 presents a description of the variables considered in the Tobit regression model.

There are two basic assumptions underlying the Tobit model: homoscedasticity and normality of the disturbance term. If the error term ε_i is either heteroskedastic or non-normal, then the maximum likelihood estimates are inconsistent. Notwithstanding, it is possible to obtain consistent estimates with heteroskedastic errors if the heteroskedasticity is estimated using the maximum likelihood technique.

Table 6. 1: Description of variables used for the Tobit regression model

Variable	Description	Measurement	A Priori Expectation
Y_i	Livelihood Vulnerability index	$0 \leq LVI \leq 1$	+
Personal factors			
AGE	Age of farmer	Years	+/-
EDU	Farmer's years of education	Years	-
FLAB	Use of family labour	Dummy: 1 = Yes, 0 = Otherwise	-
FARMEXP	Farming experience of farmer	Years	-
FARMSIZE	Farm size	Hectares	-
INCOME	Off-farm income per annum	Ghana Cedis	-
Institutional factors			
FBO	Member of a farmer-based organization or group	Dummy: 1 = Yes 0 = Otherwise	-
EXTENS	Extension contacts	Average No. of visits p.a.	-
HLAB	Hired labour	Man days	-
DECISION	Participation in decision-making	Dummy:1 = Yes 0 = Otherwise	-

LAND	Ownership of farmland	Dummy: 1 = Self-owned, 0 = Otherwise	-
TRACTOR	Farmer's access to tractor services	1 = Yes 0 = Otherwise	-
Environmental factor			
NOTICECV	Noticed climate variability	1 = Noticed 0 = Otherwise	-

6.4 Results and Discussion

6.4.1 Socio-demographic characteristics of respondents

The socio-demographic profiles of respondents included family labour, farm size, off-farm income, age, farming experience and level of education of the farmer. These characteristics are presented in Table 6.2. Approximately 66% of sampled respondents reported engaging in at least one off-farm activity, with a mean income of GHS132.48. The average farm size of sampled farmers is 1.9 hectares, which is less than the Northern Region average farm size of 2 hectares. Every smallholder woman rice farmer had at least one year of experience in rice farming, with an average farming experience of 20.45 years for the sampled farmers.

Table 6. 2: Socio-demographic profile of respondents

Variable	Minimum	Maximum	Mean	Standard Deviation
Age of farmer	26	65	50.80	9.28
Farm size (hectares)	0.11	2.20	1.90	1.07
Annual off-farm income	0	1630.00	632.48	162.91
Farming experience	1	45	20.45	12.70
№ of economic activities	1	6	2.34	0.99

Experience is important in rice farming since with greater experience farmers are better able to understand and apply the appropriate agronomic practices suitable for adapting to climate-related events to reduce their levels of vulnerability. Closely related to farming experience is the age of the farmer. The age of a smallholder woman rice farmer portrays her understanding of the local weather patterns and how she has been adapting to climate stressors. The average age of the sampled farmers was 50.8 years.

A majority of the sampled respondents (i.e., 85%) had no formal education. Smallholder women rice farmers' need formal education to assist them read and comprehend concepts, especially instructional labels on agricultural inputs and technologies (Adu et al., 2018; Etwire et al., 2013 ab). Approximately 95% of respondents reported agriculture to be their first occupation. The findings revealed that rainwater and well-water are the main sources of water for domestic use, with 91.2% and 73.8% of respondents sourcing water from rain and wells, respectively.

6.4.2 Institutional factors

Use of hired labour, farmer-farmer extension contacts, formal agricultural extension contacts, access to tractor services, access to weather information, access to formal credit, membership of a farmer group and participation in decision-making are the institutional factors considered in this study. Results of these factors are presented in Table 6.3.

Table 6. 3: Institutional factors for respondents

Factor	Frequency	Percentage of Sample
Use of hired labour	173	66.5
Access to farmer-farmer extension services	138	53.1
Access to formal agricultural extension services	159	61.1
Access to tractor service	191	73.5
Access to weather information	108	41.5
Membership of a farmer-based group	162	62.3
Access to formal credit	26	10.0
Participates in decision-making	237	91.2
Borrowed from friends and relatives	70	26.9
Owned farmland	104	40.0

Approximately 66% of the sampled farmers employ the services of hired labour (i.e., by-day labourers) to supplement family labour. Almost 53% and 61% of respondents reported receiving farmer-farmer and formal agricultural extension services, respectively. Whereas farmer-farmer extension services are provided to farmers from other farmer colleagues and focus mostly on indigenous climate-related adaption strategies, formal agricultural extension services are provided to farmers by agricultural extension officers and focus mostly on research-based agronomic practices. Approximately 73% and 41% of respondents reported having access to tractor services and weather information, respectively. However, it was established that compared to male farmers, smallholder women rice farmers do not receive the services of tractors for ploughing at the time they most need it. Just as smallholder women rice farmers can increase their farm size when they are able to access tractors for ploughing, they can adopt stringent adaptation strategies when they are well-informed about the weather situation in their localities. Approximately 62% of sampled respondents belonged to at least a farmer-based organization where they share farming experiences, and also receive education on best-practice agriculture from agricultural experts. Only 10% of respondents reported accessing credit from formal financial institutions in the past twelve months. About 91% of sampled respondents reported participating in household and community-level decision-making. When smallholder women rice farmers are involved in decision-making, they are able to speak out about their problems, thereby reducing their vulnerability to climate change and variability.

Only 40% of the interviewed women rice farmers reported owning their farmlands. This implies that such farmers have more land security and access to farmland for farming activities, which can influence their decisions about whether or not to expand production as an adaptation measure to reduce vulnerability.

6.4.3 Environmental factors

Two environmental factors were considered in the study. Namely, respondents' perceptions of rainfall and temperature patterns between 1986 and 2015. In terms of the rainfall pattern, approximately 75%, 77%, and 76% of respondents in Savelugu/Nanton, West Mamprusi and the pooled data, respectively, perceived a decrease in rainfall over the three decades. In contrast, only 3.3% 10% and 6.5% of respondents in Savelugu/Nanton, West Mamprusi and in the pooled data, respectively, perceived an increase in rainfall over the period. Juxtaposing respondents' perceptions of rainfall patterns with the historical rainfall data from the Ghana Meteorological Service showed that whereas the average rainfall in Savelugu/Nanton has decreased between 1985 and 2015, the case of West Mamprusi showed an opposite trend with average rainfall increasing marginally over the same period. Generally, a decreasing trend in rainfall pattern has been observed in Ghana (Deressa et al., 2010; and Fosu-Mensah et al., 2012).

In terms of temperature patterns in the study area, a majority of the respondents in Savelugu/Nanton (78.2%), West Mamprusi (88.7%) and in the pooled data (84.2%) perceived an increase in temperatures between

1985-2015. Only a minority of the respondents (i.e., Savelugu/Nanton - 5.5%, West Mamprusi – 1.3%, and pooled data – 3.1%) perceived no change in temperatures. The perception of respondents on temperature compared to the temperature data from the Ghana Meteorological Service attests to a general trend of increased temperatures over the period. This is consistent with Deressa et al. (2010), Fosu-Mensah et al. (2012), and Kankam-Yeboah et al. (2011).

6.4.4 Farmers' vulnerability to climate variability in the Northern Region of Ghana

The computed vulnerability indices revealed that whereas farmers in West Mamprusi District are more vulnerable than farmers in Savelugu/Nanton Municipality in terms of socio-demographic profile, livelihood strategies, social networks and food; farmers in Savelugu/Nanton Municipality are more vulnerable than farmers in West Mamprusi District in terms of health, water, and natural disaster and climate change and variability. Overall, farmers from Savelugu/Nanton Municipality (0.53) are slightly more vulnerable to climate change and variability than farmers in West Mamprusi District (0.51) (Table 6.4).

The LVI computed for each farmer was used to classify farmers into five levels of vulnerability. In the combined data, farmers have a low vulnerability in terms of socio-demographic characteristics (0.44) and health (0.39). Women farmers are moderately vulnerable in terms of water, as indicated by the computed water index of 0.50. The computed indices showed that farmers in the study region are highly vulnerable in terms of livelihood strategies (0.56), social network (0.58), food (0.59), and natural disasters and climate change and variability (refer to Table 6.4). This could be attributed to inadequate intervention programs in the study region to ameliorate the effects of climate change on farmers' livelihoods. This is congruent with Adu et al. (2018), who reported that smallholder maize farmers in the Brong-Ahafo Region of Ghana were more vulnerable to social networks, livelihood strategies, and natural disasters and climate change and variability.The empirical results further revealed that in both districts and in the combined data, a majority of the sampled farmers are highly vulnerable to climate data and variability (Table 6.5). However, farmers in Savelugu/Nanton Municipality (67.3%) are more vulnerable than farmers in West Mamprusi District (51.4%). This is evident by the computed LVIs of 0.53 and 0.51 for Savelugu/Nanton and West Mamprusi, respectively.

Table 6. 4: Computed Major Components Indices and LVIs

Major Component of Vulnerability	Savelugu/Nanton	West Mamprusi	Pooled Data
Socio-demographic profile	0.43	0.45	0.44
Livelihood strategies	0.56	0.57	0.56
Social network	0.51	0.65	0.58
Health	0.47	0.32	0.39
Food	0.57	0.61	0.59
Water	0.58	0.43	0.50
Natural disaster and climate change and variability	0.57	0.52	0.55
LVI	0.53	0.51	0.52

Table 6. 5: Classified levels of vulnerability of women rice farmers in study districts

Level of Vulnerability	Savelugu/Nanton		West Mamprusi		Pooled data	
	Freq.	%	Freq.	%	Freq.	%
Least	0	0.0	0	0.0	0	0.0
Low	0	0.0	35	23.3	35	13.5
Moderate	36	32.7	38	25.3	74	28.5
High	74	67.3	77	51.4	151	58.0
Extreme	0	0.0	0	0.0	0	0.0

6.4.5 Determinants of women rice farmers' vulnerability in the Northern Region of Ghana

The goodness of fit of the Tobit regression model was established by using the log likelihood ratio of the chi square (x^2=140.01) as the basis of testing for normality. This was found to be significant at 1%, indicating that the error term is normally distributed, and variation in the vulnerability levels of farmers to climatic stresses is significantly explained by variations in the independent variables considered in the model. The robust regression of the Tobit model was selected and the robust standard error command in the Stata software was also enabled to ensure homoscedasticity.

The Tobit regression results showed that in the pooled data, whereas education, age, off-farm income, participation in decision-making, formal extension contacts and farmland ownership have a significant negative influence on women rice farmers' vulnerability to climatic stress; farm size, membership of a farmer-based group, hired labour, access to tractor services and noticing climate variability were found to have a positive influence on farmers' vulnerability to climate variability.

In Savelugu/Nanton Municipality, family labour and off-farm income were found to have a significant inverse relationship with the vulnerability of women rice farmers to climatic stress, while farm size, farming experience, formal extension contacts, hired labour and access to tractor services positively influence farmers' vulnerability to climate variability. In West Mamprusi District, the vulnerability of smallholder women rice farmers is significantly negatively influenced by off-farm income, participation in decision-making and formal extension services. On the other hand, farm size, membership of a farmer-based group and hired labour have a significant positive influence on farmers' vulnerability to climate change. This is presented in Table 6.6.

Table 6. 6: Tobit regression results of the determinants of women rice farmers' vulnerability to climate variability

Independent / Explanatory variables		Savelugu/Nanton		West Mamprusi		Pooled Data	
		Coefficient	P-value	Coefficient	P-value	Coefficient	P-value
Constant		0.4906***	0.000	0.5341***	0.000	0.4987***	0.000
Personal Factors	Farmer's education	-0.0006	0.402	0.0007	0.269	-0.0019***	0.000
	Age of farmer	-0.0003	0.113	-0.0002	0.427	-0.0003*	0.090
	Family labour	-0.0001*	0.095	-0.2340	0.877	-0.0001	0.213
	Farming experience	0.0006***	0.001	0.0002	0.111	0.0002	0.139
	Off-farm income	-0.0002***	0.000	-0.0001**	0.045	-0.0001***	0.000
	Farm size	0.0054***	0.002	0.0051*	0.096	0.0029***	0.000
Institutional Factors	Member of FBO	0.0064	0.267	0.0232***	0.001	0.0089**	0.018
	Decision-making	-0.0027	0.650	-0.0169**	0.018	-0.0090**	0.011
	Extension contacts	0.0020*	0.077	-0.0026*	0.099	-0.0015*	0.054
	Hired labour	0.0002***	0.000	0.0001***	0.000	0.0001***	0.000
	Access to tractor services	0.0140*	0.100	-0.0049	0.620	0.0102*	0.075
	Farm land ownership	-0.0076	0.356	-0.0012	0.861	-0.0094*	0.087
Environmental Factor	Noticed climate variability	0.0119	0.108	-0.0036	0.521	0.0111**	0.025
Sigma		0.0233		0.0297		0.0265	
Number of observations		110		150		260	
F (13,97)		131.23		122.16		140.01	
Prob > F		0.0000		0.0000		0.000	
Pseudo R2		0.3831		0.1088		0.1840	
Log-likelihood		-154.8279		-210.8240		-450.5055	
Left-censored observations (LVI≤0.47)		18		15		33	

Note: *, **, and *** means statistical significant at 10%, 5% and 1%, respectively.

Age of a farmer negatively and significantly contributes to the vulnerability of the farmer to climate-related events such as floods, drought and bushfires. This is probably because elderly female rice farmers are acquainted with certain adaptation strategies to avert the effects of climatic shocks, rendering them less vulnerable than younger female rice farmers. The finding of this study is contrary to that of FANRPAN (2011) but is consistent with Bayard et al. (2007).

As expected, a farmer's years of education has a significant negative influence on her vulnerability to climatic stressors. Literate female rice farmers are able to read and easily understand issues relating to climate variability adaptation as well as instructional labels on farming inputs. This finding is congruent with the findings of Ayuya et al. (2012) and Nkondze et al. (2013). Off-farm income negatively influences smallholder women rice farmers' vulnerability to climate variability. This suggests that farmers who engage in other income-generating activities outside the farm are less vulnerable than those who depend solely on the farm as their source of livelihood. The reason for this may be that farmers engaged in other income-generation activities are not as hard hit by the effects of low crop yields resulting from erratic rainfalls since they can augment their income from the other sources. Almost 34% of the sampled farmers reported not engaging in any off-farm activity, suggesting their high levels of vulnerability. This finding is supported by Bawakyillenuo et al. (2016), who reported off-farm activities to be a major adaptation strategy for abating the effects of climate variability by farmers in the rural northern savannah zone of Ghana.

Farm size was found to have a positive relationship with vulnerability to climate variability. The larger the farm size, the higher the vulnerability of a farmer to climatic stress. For effective management, almost 52% of sampled women rice farmers in the region reported that they had reduced their farm size. The average farm size of sampled farmers is 1.90 hectares, which is less than the Northern Region average farm size of 2 hectares. This finding is supported by Uddin et al. (2014), who reported a significant negative relationship between farm size and farmers' adoption of adaptation strategies in Bangladesh. Though family labour is not significant in the pooled data, it significantly and inversely influences women rice farmers' vulnerability to climatic stress in the Savelugu/Nanton Municipality. This agrees with Quisumbing (2007) and Peterman et al. (2010), who reported a positive family labour influence on the yield of staple crops. Thus, women rice farmers who rely on family labour for their farming activities (i.e., 85%) are less vulnerable to the impacts of climate variability. The reason is that women farmers who supplement hired labour with family labour are in a better position to adapt to climate variability than farmers who do not.

In contrast to expectations, being a member of farmer-based organizations renders a farmer more vulnerable to climatic stress. This may be partly explained by the fact that being a member of a farmer-based group will mean a farmer is spending more time at meetings and other group activities, which may not necessarily be productive for farming activities. Most farmers interviewed (i.e., 62.3 %) are members of at least one farmer-based group. This finding is consistent with Pulhin et al. (2006), who found that the number of organizations joined by a farmer does not necessarily ensure decreased vulnerability of communities in the Pantabangan-Carranglan Watershed in the Philippines to climate variability and extremes. Participation in decision-making and the number of formal extension services received by a woman rice farmer have a significant negative contribution to her vulnerability. This is not different from the *a priori* expectation and indicates that smallholder women rice farmers' participation in decision-making decreases their vulnerability to climatic stresses, in that their challenges of responding are unearthed and the appropriate measures put in place to combat the vulnerability. The number of extension contacts received by a smallholder woman rice farmer has a significant inverse effect on her vulnerability to climatic stress. This is consistent with the *a priori* expectation. More than 61% of sampled farmers reported receiving support from a formal extension service in the past twelve months. The more frequently agricultural extension officers visit and train smallholder women rice farmers on agronomic practices, the lower their vulnerability to climatic shocks. This finding agrees with that of Fatuase et al. (2015), who reported a positive relationship between access to extension agents by arable crop farmers in Owo Local Government Area of Ondo State in Nigeria and their adoption of adaptation measures.

Hired labour has a significant positive influence on the vulnerability of smallholder women rice farmers to climatic shocks. Farmers depending on hired labour (i.e., 66% of the sampled farmers) for their farming activities are more vulnerable. Hired labour is not a reliable labour source since most farmers who are often hired are usually busy working on their farms at the time that smallholder women rice farmers need their labour. Although access to tractor services has been found to have a significant influence on smallholder woman rice farmers' vulnerability to climate variability, it deviates from the *a priori* expectation, as the result indicates that it positively contributes to farmers' vulnerability. Farmland ownership met its *a priori* expectation, with a significant negative contribution to farmers' vulnerability. Land is a major resource for agricultural activities and women rice farmers who own their farmlands are able to adopt long-term adaptation strategies such as tree planting and installation of irrigation equipment without any hesitation or fear of losing their farmlands. This finding agrees with Pulhin et al. (2006), who found that farmers who do not own land are likely to be more vulnerable to climate variability and extremes in the Pantabangan-Carranglan Watershed in the Philippines. These authors further noted that climate variability has a significantly positive influence on a farmers' vulnerability to climatic stresses. This implies that farmers who perceived variations in the weather elements such as rainfall and temperature are more vulnerable to climate variability. The possible reason could be that female farmers perceiving unpredictability of climatic conditions and being risk-averse, will prefer to not adopt adaptation strategies which will require additional expenditure and the effectiveness of which is not guaranteed. This finding is consistent with the findings of Mandleni and Anim (2011).

6.5 Conclusions

It is conclusive from this study's empirical findings that most female rice farmers in the Northern Region of Ghana are highly vulnerable to climate variability, with farmers in Savelugu/Nanton Municipality being more vulnerable to climatic stresses than those in West Mamprusi District. In Savelugu/Nanton Municipality, family labour and off-farm income have a significant negative influence on farmers' vulnerability to climate variability; whereas farm size, farming experience, formal extension contacts, hired labour and access to tractor services have a significant positive effect on farmers' vulnerability to climate variability. Similarly, in West Mamprusi District, off-farm income, participation in decision-making and formal extension services had a significant negative effect on famers' vulnerability to climate variability; while farm size, membership of a farmer-based group and hired labour have a significant positive influence on the farmers' vulnerability. Overall, farmers' education, age, off-farm income, participation in decision-making, formal extension contacts and farmland ownership have a significant negative influence on women rice farmers vulnerability to climate variability; while farm size, membership of a farmer-based group, hired labour, access to tractor services and noticing climate variability have significant positive effect on farmers vulnerability to climate variability.

Based on these conclusions, this study recommends that agricultural extension officials of Ghana's Ministry of Food and Agriculture should intensify engagements with smallholder women rice farmers by educating them on effective agronomic practices, while also encouraging farmer-to-farmer extension services to promote the adoption of agronomic practices for adapting to climate change and variability. Also, the 'Night School System' initiated by School for Life, an NGO based in the Northern Region of Ghana, should be re-introduced with the aim of improving adult literacy among smallholder women rice farmers. This will strengthen the adaptive capacities of the farmers and therefore help to reduce their vulnerability to climatic stressors. Furthermore, to foster higher adoption of adaptation strategies to reduce vulnerability to climate change and variability, smallholder women rice farmers should be particularly encouraged by government, research institutions and other agriculture-based NGOs to participate in all stages of adaptation projects or initiatives. Finally, interventional projects of government and NGOs should concentrate on facilitating the ability of women rice farmers to diversify their livelihood sources by providing them with mentorship training on off-farm income-generation ventures, and up-scaling their participation in the agribusiness value-chain to processing.

Acknowledgements

The authors of this study remain indebted to the A. G. Leventis Foundation Scholarship Scheme, College of Basic and Applied Science, University of Ghana, Legon for partly funding the study.

References

Adu, D. T., Kuwornu, J. K. M., Anim-Somuah, H, and Sasaki, N. (2018) Application of livelihood vulnerability index in assessing smallholder maize farming households' vulnerability to climate change in Brong-Ahafo region of Ghana, *Kasetsart Journal of Social Sciences*, 39(1), 22-32.

Al-Hassan, R., and Poulton, C. (2009) Agriculture and Social Protection in Ghana. *Future Agriculture.*

Amikuzuno, J., and Donkoh, S. A. (2012) Climate variability and yield of major staple food crops in Northern Ghana. *African Crop Science Journal*, 20(2), 349-360.

Antwi-Agyei, P., Dougill, A. J., Fraser, E. D., and Stringer, L. C. (2013) Characterising the nature of household vulnerabilityto climate variability: Empirical evidence from two regions of Ghana. *Environment, Development and Sustainability*, 15(4), 903-926.

Asare, A. (2000) *Operational guidelines on community forest (management) committees.* ITTO/FSD Collaborative Off-Reserve Forest Management Project.

Ayuya, O. I., Kenneth, W. S., and Eric, G. O. (2012) Multinomial Logit Analysis of Small-Scale Farmers' Choice of Organic Soil Management in Bungoma Country, Kenya. *Current Research Journal of Social Sciences*, 4(4), 314-322.

Bawakyillenuo, S., Yaro, J. A., and Teye, J. (2016) Exploring the automous adaptation strategies to climate change and climate variability in selected villages in rural northern savannah zone of Ghana. *Local Envoronment,* 21(3), 361-382.

Bayard, B., Jolly, C.M., and Shannon, D.A. (2007) The economics of adoption and management of alley cropping in Haiti. *Journal of environmental management, 84*(1), 62-70.

Boko, M. A., Niang, A., Nyong, C., Vogel, M., Githeko, M., Mednay, et al. (2007) *In climate change 2007: Impact, Adaptation, and Vulnerability. Contribution of Working Group II to the Fourth Assessment Report of the Intergovernmental Panel on Climate Change, ed. M.L. Parry,J.P. Canziani, J.P.Palutikof, P.J. Van der Linden & C.E.Hanson*. Cambrige: Cambridge University Press.

Chamberlin, J. (2008) *It's Small World After All: Definding Smallholder Agriculture in Ghana.* Intl Food Policy Res Inst.

Cutter, S. L., Barnes, L., Berry, M., Burton, C., Evans, E., Tate, E., et al. (2008) A place-based model for understanding community resilience to natural disasters. *Global Climate Change,* 18, 598-606.

Deressa, T. T., Hassan, R. M., and Ringler, C. (2010) Perception of and adaptation to climate change by farmers in the Nile basin of Ethopia. *The Journal of Agricultural Science,* 149(01), 23-31.

Deressa, T. T., Hassan, R. M., Ringler, C., Alemu, T., and Yesuf, M. (2009) Determinants of farmers' choice of adaptation methods to climate change in the Nile Basin of Ethopia. *Global Environmental Change,* 19(2), 248-255.

Etwire, P. M., Al-Hassan, R. M., Kuwornu J, K. M., and Osei-Owusu, Y. (2013b) Smallholder farmers' adoption of technologies for adaptation to climate change in Northern Ghana. *Journal of Agricultural Extension and Rural Development,* 5(6), 121-129.

Etwire, P. M., Al-Hassan, R. M., Kuwornu, J. K., and Osei-Owusu, Y. (2013a) Application of the Livelihood Vulnerability Index in Assessing Vulnerability to Climate Change and Variability In Northern Ghana. *Journal of Environment and Earth Science,* 3(2), 157-170.

Fatuase, A. I., Aborisade, A. S., and Omisope, E. T. (2015) Determinants of Adaptation Measures to Climate Change by Arable Crop Farmers in Owo Local Government Area of Ondo State, Nigeria. *World Rural Observation,* 7 (1), 49 - 57.

Food Agruculrure and Natural Resources Policy Analysis (FANRPAN) (2011) Measuring Vulnerability - Challenges and Opportunities. *FANRPAN,* 2(11).

Fosu-Mensah, B., MacCarthy, D., Vlek, P., and Safo, E. (2012) Simulating impact of seasonal climatic variation on the response of maize (Zea mays L.) to inorganic fertilizer in semi-humid Ghana. *Nutrient Cycling in Agroecosystems,* 94(2-3), 255-271.

Fussel, H. M., and Klein, T. R. (2006) Climate change vulnerability assessment: an evolution of conceptual thinking. *Climate Change, 75,* 301-329.

Ghana Statistical Service, GSS (2012) *2010 Population and Housing Census.* Analytical Report, Northern Region.

Ghana Statistical Service, GSS (2014) *Ghana Living Standards Survey Round 6 (GLSS 6).* Poverty Profile in Ghana (2005-2013), Accra, Ghana.

Gujarati, D. N., and Porter, D. C. (2008) *Basic Econometrics* (5th ed.). New York, USA: McGraw-Hill/Irwin.

Hahn, M. B., Riederer, A. M., and Foster, S. O. (2009) The Livelihood Vulnerability Index: A pracgmatic approach to assessing risks from climate variability and change - A case study in Mozambique. *Global Climate Chnage, 88,* 74-88.

IPCC (2001) *Climate Change 2001: Synthesis Report.* Contribution of Working Groups I, II and III to the Third Assessment Report of the Intergovernmental Panel on Climate Change, Cambridge University Press, Cambridge.

IPCC (2014) *Climate Change: Impact, Adaptation and Vulnerability. Contributions of Working Groups I, II nad III to the Fourth Assessment Report.* Cambridge University Press, Cambridge, UK.

Kankam-Yeboah, K., Adjei, K. A., Ren, L., Appiah-Adjei, E. K., and Agyapong, A. A. (2011) Validation of TRMM Data in the Black Volta Basin of Ghana. *Journal of Hydrologic Engineering,* 17(5), 647-654.

Madhuri, S. R., Divase, G. T., Devkar, T. B., and Tekade, A. R. (2014) Solubility and Bioavailability Enhancement of Poorly Aqueous Soluble Atorvastatin: In Vitro, and In Vivo Studies. *2014.* Hindawi Publishing Corporation.

Mandleni, B. K., and Anim, F. D. (2011) Climate Change Awareness and Decision on Adaptation Measures by Livestock Farmers. *85th Annual Conference of the Agricultural Economics Society, Warwick University,* (p. 26).

Ministry of Food and Agriculture, MoFA (2011) *Agriculture in Ghana: Facts and Figures.* Accra, Ghana: MoFA, Accra.

Morton, J. F. (2007) The impact of climate change on smallholder and subsistence agriculture. *Proceedings of the National Academy of Sciences,* 104(50), 19680-19685.

Nelson, R., Kokic, P., Crimp, S., Martin, P., Meinke, H., Howden, S. M., et al. (2009) The vulnerability of Australian rural communities to climate variability and change: Part II - Integrating impacts with adaptive capacity. *Environmental Science and Policy,* 13(2010), 18-27.

Nkondze, M. S., Masuku, M. B., and Manyatsi, A. (2013) Factors Affecting Households Vulnerability to Climate Change in Swaziland: A Case of Mpolonjeni Area Development Programme (ADP). *Journal of Agricultural Science,* 5(10), 108 -122.

Nti, F. K. (2012) Climate Change Vulnerability and Coping Mechanisms among Farming Communities in Northern Ghana. *A Thesis Submitted in Partial fulfillment for the award of MSc Degree, Kansas State University, Manhattan, Kansas.*

Peterman, A., Quisumbing, A., Behrman, J., and Nkonya, E. (2010) Uderstanding gender differences in agricultural productivity in Uganda and Nigeria: Citeseer.

Pulhin, J. M., Peras, R. J., Cruz, R. O., Lasco, R. D., Pulhin, F. B., and Tapia, M. A. (2006) Vulnerability of Communities to Climate Varaibility and Extremes: Pantabangan-Carranglan Watershed in the Philippines. *AIACC Working Papers (44).*

Quisumbing, A. R. (2007) Investment, bequest, and public policy: International transfers and the escape from poverty. *Chronic Poverty Research Center Working Papers (98).*

Stanturf, J. A., Palik, B. J., Williams, M. I., Dumroese, R. K., and Madsen, P. (2014) Forest Restoration Paradigms. *Journal of Sustainable Forestry,* 33, 161-194.

Sullivan, C. (2002) Calculating a Water Poverty Index. *World Development,* 30 (7), 1195 - 1210.

Tobin, J. (1958) *Estimation of Relationships for Limited Dependent Variables.* Econometrica (Econometric Society).

Turner, B. L., Kasperson, R. E., Matson, P. A., McCarthy, J. J., Corell, R. W., Christensen, L., et al. (2003) Aframework for vulnerability analysis in sustainability science. 100(14), 8074-8079.

Uddin, M. N., Bokelmann, W., and Entsminger, J. S. (2014) Factors Affeccting Farmers' Adaptation Strategies to Environmental Degradation and Climate Effects: A Farm Level Study in Bangladesh. *Climate,* 2, 223-241.

World Bank. (2010). *Economics of Adaptation to Climate Change: Social synthesis Report.* 1818 H Street NW, Washington DC, USA.

Yengoh, G. T., Armah, F. A., Onumah, E. E., and Odej, J. O. (2010) Trends in agriculturally-relevant rainfall characteristics for small-scale agriculture in Northern Ghana. *Journal of Agricultural Science,* 2(3), 3 - 16.

CHAPTER 7

Impacts of participation in climate change projects on Ghanaian maize farmers' yields, income and resilience to climate shocks

Armah Ralph Nii Armah, Ramatu M. Al-Hassan,
John K. M. Kuwornu

Abstract

The impact of climate change and variability on Ghana's agriculture, especially in northern parts of the country, is now recognized. In response, the Government of Ghana and development partners, including civil society organizations, have developed projects to support farmer adaptation and resilience to climate change. This paper identifies farmers' perceptions of climate change and assesses the impact of participation in climate change-related projects on farmers' maize yields, farm and off-farm income, and their resilience to climatic shocks. The study uses cross-sectional data from 241 smallholder farmers in Ghana's Northern Region, Afram Plains and Accra Plains. The propensity score matching (PSM) method is used to evaluate the impact of participation in climate change-related projects. The PSM results show that there is an insignificant difference between participants and non-participants in terms of the outcome variables. This notwithstanding, participation in climate change-related projects has the potential to make a positive impact, the limitation, however, is with the scope and depth of the projects.

Keywords: Accra Plains; Afram Plains; Agriculture; Climate; Ghana; Northern Region; Propensity Score Matching

7.1 Introduction

One of the greatest challenges confronting most countries in sub-Saharan Africa is how to increase agricultural production in the face of harsh environmental conditions. Agriculture, the main source of income for about 90% of the rural population in sub-Saharan Africa, has not been a sufficient vehicle for solving the region's food insecurity. This is a result of low agricultural productivity, due in part to climate change and variability (Boko et al., 2007; Owusu et al., 2010). The effects of climate change are mostly indiscernible from the effects of other risks because the indirect consequences of climate change have the greatest impact on the poor (Athula and Scarborough, 2011).

Ghana's agricultural sector contributes about 26% of the country's GDP and plays an important role in job creation, employing about 44.7% of the country's total labour force (Ghana Statistical Service (GSS), 2014). Agricultural production in Ghana is mainly smallholder in nature, although it has great prospects for growth given the country's vast arable lands. On average, only 58% of Ghana's arable land is cropped annually. Agricultural productivity in Ghana is challenged by climate change, which may lead to floods, drying up of water bodies, and droughts as well as their ripple effects on land and soil degradation.

Ghana has benefitted from numerous climate change-related projects over the last decade, with technical and financial support from international donors or research organizations, the Government of Ghana or its mandated institutions, non-governmental organizations (NGOs) and civil society (Würtenberger et al., 2011; Al-Hassan et al., 2013b). The initiatives promoted through these projects focus on building farmers' capacity, creating awareness about climate change and its effects, and promoting climate change adaptation or coping strategies (Al-Hassan et al., 2013b). These activities mostly aim to either build farmers' resilience or mitigate the impacts of climate change and variability. However, there have not been any evaluations of the effectiveness of the projects in achieving their objectives.

The main objective of the paper is to assess the impact of farmers' participation in climate change-related projects on their resilience to climatic shocks and their income (i.e., on-farm and off-farm) and maize yields. The paper also assesses farmers' perceptions about climate change.

The remainder of this paper is organized as follows: the research methodology is presented in the next section, followed by the results and discussion, and the conclusion and recommendations. The scope of this paper is limited to the Alliance for a Green Revolution in Africa (AGRA) bread-basket zones of the Northern Region, Afram Plains and Accra Plains in Ghana.

7.2 Materials and Methods

7.2.1 Study Area, Sampling and Data

A multi-stage sampling technique was used to select three zones, six districts and 241 farmers from 24 communities for the study. The Northern Region, the Afram Plains and the Accra Plains were purposively sampled since they constitute AGRA's bread-basket areas in Ghana. One district was randomly sampled from each of the Afram and Accra Plains, whiles four districts were randomly selected from the Northern Region due to its relatively large size. In each of the selected districts, four communities and ten households per community were randomly selected.

The data for this paper was collected using a structured questionnaire administered to smallholder farmers. The information gathered from the questionnaire included farmers' demographic characteristics, assets, household income sources, and farm characteristics. Farmers' perceptions of climate change, their participation in climate change-related projects, and their recovery from climate and other environmental shocks were also obtained from the questionnaire.

7.2.2 Theoretical and Analytical Framework

There are many reasons why agricultural project initiatives or technologies might improve farm household wellbeing (Mendola, 2007). However, whether or not an initiative is effective in meeting a project's objectives forms an integral part of project evaluation. A well-designed intervention (or treatment) is typically based on theory or research evidence that articulates how its core mechanisms will work to achieve its goals and produce the desired outcomes (Heinrich et al., 2010).

There is an extensive literature on the relevance and use of impact evaluation models. Many studies (e.g., Baker, 2000; Blundell and Dias, 2000; Duflo et al., 2008; Ravallion, 2008; Heinrich et al., 2010) have provided useful overviews of these models. Dehejia and Wabba, (1999) and Khandker et al. (2003) also outline approaches for impact evaluation and the data requirements for each approach. According to Becerril and Abdulai (2010) and Heinrich et al. (2010), the main challenge of evaluation is the construction of the counterfactual outcome, which measures what would have happened to participants in the absence of the intervention (or treatment). This allows for addressing the question of whether improved wellbeing of project participants compared to non-participants is a result of their participation.

Heinrich et al. (2010) have discussed a primer for applying the propensity score matching (PSM) approach, whereas Khandker and Faruqee (2003) have opined that the model works under the conditional independence and common support assumptions. The basis for these assumptions is that all factors influencing participation are observable, and individuals with similar characteristics have an equal chance of belonging to the participant or non-participant group (Awunyo-Vitor et al., 2013).

The propensity score is a conditional probability of participation given pre-participation characteristics. Awunyo-Vitor et al. (2013) used the PSM model to assess the impact of formal financial market participation on farm size and expenditure, whereas Al-Hassan et al. (2013a) used it to evaluate the effect of participating in an information communication technology (ICT)-based agricultural market information project on farm households in northern Ghana. Owusu et al. (2010) also used the PSM model to evaluate the impact of off-farm work on food security among farm households in northern Ghana. Many other authors (e.g., Mendola, 2007; Faltermeier and Abdulai, 2008; Becerril and Abdulai, 2010; Wainaina et al., 2012) have used the PSM model to analyse the impacts of technology adoption, contract farming, water conservation, and intensification technologies among others.

This paper uses the PSM method to assess the impact of participation in climate change-related projects, where the impact or outcome variables are farmers' maize yields, income and perceived resilience to shocks. To assess the impact, it is necessary to assess what the situation would have been had farmers not participated in climate change-related projects; i.e., the counterfactual situation. Ideally, experimental data would provide information on the counterfactual situation that would address issues of causal inference (Mendola, 2007). The basis for the use of PSM is that participation and non-participation decisions may be selective and dependent on pre-participation variables. The PSM model thus helps to address the bias introduced through the selection of participants for a project.

The relationship used to examine the impact of participation on outcome variables (R_i) is a linear function, with a vector of explanatory variables (X_i) and a participation dummy variable (V_i). The regression equation is specified as:

$$R_i = X_i'\beta + \delta V_i + \varepsilon_i, \tag{7.1}$$

Where R_i represents the outcome variables (i.e., resilience to drought, flood, and livestock deaths; farm and off-farm income; and maize yield) and ε_i is a normal random disturbance term. The variable V_i is a 0 or 1 dummy variable for participation in climate change-related projects; V_i=1 if the farmer participated and V_i =0 otherwise; whereas the vector X_i denotes household and farm-level characteristics.

To successfully evaluate the impact of participation on the outcome variables, there needs to be a specification of the average treatment effect (Becerril and Abdulai, 2010; Heinrich et al., 2010). Rosenbaum and Rubin (1983) define the average treatment effect (D_i) in a counterfactual framework as:

$$D_i = R_i^A - R_i^N \qquad (7.2)$$

Where R_i^A and R_i^N denote the outcome variables of participants and non-participant farmers, respectively. In estimating the impact of equation (7.2), a problem arises because either R_i^A or R_i^N is normally not observable, and both are not observed for each household.

The PSM, which is the probability of assignment to treatment conditional on pre-treatment variables, is given by:

$$p(X) = \Pr[D = 1] = E[D|X]: p(X) = F\{h(X_i)\} \qquad (7.3)$$

Where *F{.}* can be a normal or logistic cumulative distribution, p is the probability, and X is a vector of pre-treatment characteristics.

If all the information relevant to farmers' participation in climate change-related projects and the outcome variables are observable, then the propensity score will yield valid matches for the estimation of impact. This is termed the average treatment effect of the treated (ATT) and is a measure of the impact of participation. The ATT is predicted as follows:

$$ATT = \{(E(R_A|D = 1. E(P(X)) - E(R_N|D = 1. E(P(X))\} \qquad (7.4)$$

Where $E(P(X))$ denotes the expectation with respect to the distribution of the propensity score in the entire population.

The first step in the PSM model involves the estimation of a binary choice model of participation. In this paper, a logit model equation (7.5) is used to estimate the probability of participation in climate change-related projects, as follows:

$$Y_i = \beta_0 + \beta_i X_i + \mu_i \qquad (7.5)$$

Where Y_i is the participation variable (Yi = 1 if participant and Yi = 0 if non-participant); β represents the coefficients; X_i denotes the farmer and household characteristics; and μ_i is the disturbance term.

The models are estimated simultaneously with the twelfth version of the STATA software. The variables used for the Logit regression model are presented in Table 7.1.

Table 7. 1: Variables used for the Logit regression model

Variable	Variable description
Sex	Sex of respondent (1=female, 0=otherwise)
Education	Education of respondent (1=at least basic education, 0= otherwise)
Household size	Total number of people in the household
Farm size	Total farm size (hectares)
If north	Location variable (1=if Northern Region, 0=otherwise)
Mobile	Mobile phone ownership (1=owner, 0=otherwise)
Agricultural extension	Access to agricultural extension services (1=Yes, 0=otherwise)
Landowner	Land ownership (1=owner, 0=otherwise)
Previously participated in an agricultural project	Beneficiary of past agricultural project or program (1=participant, 0=otherwise)

7.3 Results and Discussion

7.3.1 Socioeconomic characteristics of respondents

Tables 7.2 and 7.3 present the socioeconomic characteristics of participants and non-participants. These characteristics help to identify possible sources of bias in the selection of project participants. There are significant differences between participants and non-participants in terms of their location (i.e., north or south), having contact with an agricultural extension agent, being a landowner, and having previously participated in an agricultural project. More participants than non-participants have access to extension, and more participants than non-participants have participated in an agricultural project. The dominance of participants from the north is because of the larger proportion of the sample from the Northern Region.

Table 7. 2: Description of dichotomous variables

Variable	Percentage
Sex:	
Male	84.2
Female	15.8
Education:	
At least basic education	46.9
Otherwise	53.1
Off-farm income:	
Yes	59.7
No	40.3
Mobile phone owner:	
Yes	89.6
No	10.4
Northern Region (location variable)	
Yes	66.4
No	33.6
Access to extension services:	
Yes	61.0
No	39.0
Land ownership	
Owner or family land	27.8
Lease, rent, sharecrop or contract	72.2
Participant in agric. program since 2000:	
Yes	55.6
No	44.4

Table 7. 3: Characteristics of participants and non-participants in climate change-related projects

Variables	Participants (N=109)		Non-participants (N=132)		P-value
	Mean	Standard deviation	Mean	Standard deviation	
Age of respondent (years)	45.91	15.25	45.27	16.29	0.757
Respondent's farming experience (years)	21.37	14.21	21.79	14.60	0.822
Total farm size of household (hectares)	3.13	3.09	4.88	2.60	0.488
Household size (number)	7.42	3.28	7.18	2.93	0.549

***, **, and * denote 1%, 5%, and 10% significance levels, respectively.

The outcome variables were farmer resilience to climatic shocks (i.e., flood, drought) and death of livestock, farm income, non-farm income and maize yield. Resilience to shock was measured as the extent to which farmers reported they had recovered (i.e., percentage recovery from the pre-shock situation) from a particular shock which occurred in the last 10 years. Incomes and maize yields were average values over the last 5 years. Maize was selected as a target crop because it is the dominant cereal in the

Northern Region of Ghana. Livestock are an important asset for households in the Northern Region of Ghana and death of livestock is often related to climate variables such as drought and disease outbreaks.

Table 7.4 suggests that farmers' recovery from shocks is rather slow; the average rate of recovery is highest for drought, and least for death of livestock. This shows that it is more difficult for farmers to rebuild their livestock assets than to recover from drought; recovery from drought can happen quickly following a good rain that favors crop production. Project participants recover faster from drought than non-participants, and participants also have higher maize yields. However, non-participants have higher farm and non-farm incomes. This suggests that selection or participation in climate change-related projects is probably targeted at low performing farmers who are likely to be more vulnerable to climate variability. The maize yield of participants is higher, although not statistically significant.

Table 7. 4: Descriptive statistics of outcome variables for participants and non-participants

Outcome variables	Participants (N=109)		Non-participants (N=132)		P-value
	Mean	Standard deviation	Mean	Standard deviation	
Drought recovery (percentage)	41.61	33.74	25.39	30.73	0.000***
Flood recovery (percentage)	14.68	28.55	18.83	31.62	0.291
Death of livestock recovery (percentage)	7.06	19.84	6.14	19.02	0.712
Off-farm income (Ghana Cedi)	815.69	1779.9	1218.03	2928.01	0.211
Farm income (Ghana Cedi)	1812.90	3429.31	4501.96	15483.72	0.077*
Maize yield (100kg bag)	8.12	6.68	6.17	6.42	0.022**

***, **, and * denote 1%, 5%, and 10% significance levels, respectively.

7.3.2 Farmers' perceptions of climate change

The main indicators of climate change in Ghana are changes in temperature and rainfall. The World Bank Group reported in 2010 that although there will be fluctuations in both annual temperatures and precipitation, the trend for temperature over the period 2010 to 2050 indicates warming throughout Ghana (World Bank, 2010). During the data collection, farmers generally defined climate change as changes in weather conditions particularly rainfall and temperature. Figures 7.1 and 7.2 summarize the farmers' responses about whether or not there have been changes in both temperature and rainfall over the past decade (i.e., 2004 – 2014).

Figure 7. 1: Farmers' Perceptions of Temperature Changes

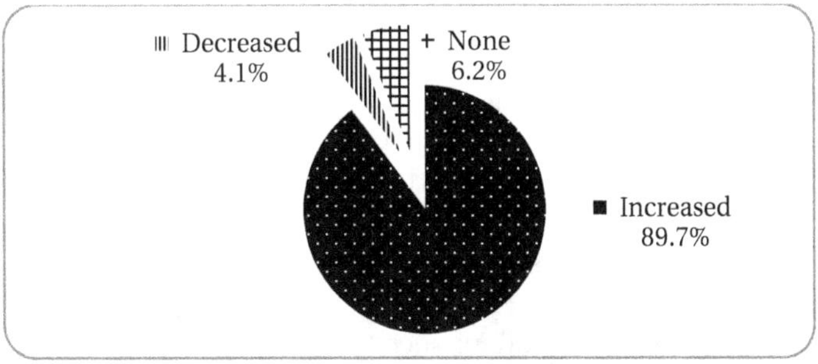

A majority of the respondents (i.e., about 90%) perceived temperature to have increased since 2000, while 4% said there has been a decrease in temperature. About 6% of the respondents claimed there has been no change in temperature (Figure 7.1).

Figure 7. 2: Farmers' Perceptions of Rainfall Changes

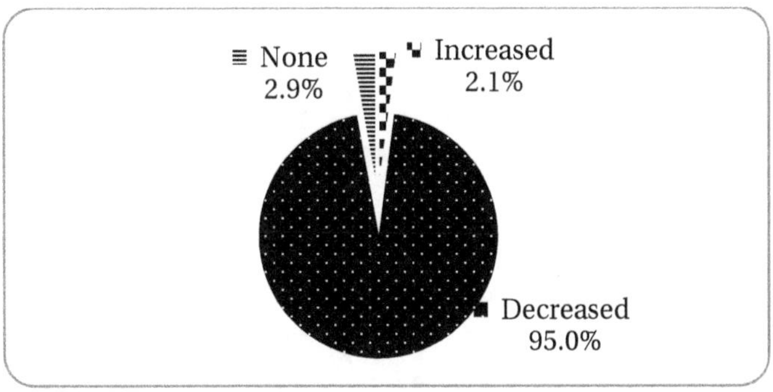

Ninety-five percent (95%) of the respondents perceived that the frequency and amount of rainfall had decreased. Only approximately 2% of the respondents agreed that rainfall had increased, whereas about 3% had not observed any changes (Figure 7.2).

The farmers' perceptions about changes in temperature and rainfall agree with statements from the farmers that: "*the weather has gotten hotter, rainfall has been lower than expected over the years*" as well as "*the weather becomes more variable from year to year*".

7.3.3 Practices adopted to respond to climate variability

Farmers adopt numerous strategies to respond to the changing climate. These are categorized as soil and plant health strategies, improved varieties and breeds, land management, and others introduced by projects. The details of the technologies adopted are presented in Table 7.5. The table shows that adoption is highly skewed in favour of soil and plant health, as well as land and crop management practices. This reflects the importance of declining soil fertility and an increased prevalence of pests and diseases in the study area (Ellis-Jones et al., 2012). The use of improved varieties is low, despite research efforts focused on variety improvements. This is consistent with the findings of Ellis-Jones et al. (2012) that farmers in the Northern Region ranked pests, diseases and weeds highest among all constraints, followed by a lack of improved seed and declining soil fertility. Adoption of long-term practices for land improvements and conservation, such as tree planting and creation of fire-belts is rather low.

Table 7. 5: Farmers' responses to changing climatic conditions

Farmer practice	Participants (N= 102) % of respondents	Non-Participants (N= 126) % of respondents
Soil and plant health – short-term		
- Inorganic fertilisers - Compost - Herbicide - Insecticides/pesticides	70	72
Improved varieties or breeds		
- High yielding - Early maturing - Drought tolerant - Improved breeds	35	27
Land & crop management		
- Harrowing before planting - Planting at recommended time - Planting in rows	56	60
Long-term strategies		
- Tree planting - Irrigation - Establish fire-belt - Reduce farm size	8	3

Note: Sum of numbers in columns exceeds 100% because farmers used combinations of the practices.

7.3.4 Factors influencing farmers' participation in climate change-related projects

Table 7.6 presents the maximum likelihood estimates from the Logit model. Among the independent variables considered, the location variable (i.e., Northern Region), mobile phone ownership, contact with agricultural extension, and previous participation in agricultural programs significantly influenced the probability of participation in climate change-related projects.

Table 7. 6: Results of the Logit regression model of the factors influencing participation in climate change-related projects

Covariates	Odd Ratio	Marginal effects
Sex	-0.0138	0.0214
Education	-0.3836	-0.1093
Household size	-0.0224	-0.0056
Farm size	-0.0013	-0.0005
If Northern Region	0.9567***	0.2842
Mobile phone	0.6226*	0.2207
Agric. extension	0.9368***	0.3201
Land ownership	0.1726	0.0573
Past agricultural program	1.4922***	0.5267
Constant	-1.3465	

***, **, and * denote 1%, 5%, and 10% significant levels, respectively.

Number of observations= 241; LR chi^2=121.27; Prob >chi^2 =0.0000; Pseudo R^2=0.3685

Farmers in the Northern Region are more likely to participate in climate change-related projects. This is because more climate change-related projects have been implemented in the Northern Region compared to the other AGRA bread-basket areas. Farmers who own mobile phones are also 22% more likely to be participants than those who do not, although the level of significance is 10%. Mobile phones form part of the media for the transfer of information to farmers, including that on upcoming projects. Mobile phone ownership could, therefore, form the basis of a farmer being a beneficiary of such projects. This may then present the problem of selection bias.

The coefficient of farmers' access to agricultural extension services is positive and significant at 1%. The marginal effect of access to agricultural extension shows that the variable increases the likelihood of participation in projects by about 32%. This is because farmers with access to agricultural extension may be privy to information about upcoming projects, through their interaction with extension agents.

Furthermore, farmers who have participated in past agricultural programs are about 52% more likely to participate in climate change-related projects. Benefiting from a previous project may induce a farmer's willingness to participate in subsequent ones. Such farmers ascribe higher levels of expectation to introduced projects, given they have benefited from other projects.

7.3.5 Impact of participation in climate change-related projects on the outcome variables

The effect of participation in climate change-related projects on resilience to climatic shocks, income and yield was estimated with the nearest neighbor algorithm, and the results are presented in Table 7.7. The results show that there is no significant difference between participants and non-participants in terms of resilience to shocks (i.e., drought, flood, death of livestock), maize yield, and farm and non-farm income. This could be due to certain project characteristics.

According to Al-Hassan et al. (2013b), many projects have very short durations (i.e., they are implemented over a period of 5 years or less). Given that project benefits flow over fairly long periods, the short duration of most of the projects poses a danger of disappointment to participants who tend to have high expectations of such interventions (Al-Hassan et al., 2013b). The short duration of projects also limits the time for assessing responses, learning lessons and adjusting implementation measures. These challenges also exist for projects with limited geographical coverage.

Another possible reason for the similarity in outcome variables between participants and non-participants is the low degree of sustainability of projects. Some farmers participated in projects only because there was something being offered, and once the project ended they stopped carrying out the activities promoted by the project.

Table 7. 7: Average treatment effect on treated (i.e., effect of participation)

Outcome variable	ATT	Critical level of hidden bias (r)	Number treated	Number control
Drought recovery	1.17 (0.12)	4.95 – 5.00	109	132
Flood recovery	-6.88 (-0.80)	1.00 – 1.05	109	132
Death of livestock recovery	-3.30 (-0.48)	1.20 – 1.25	109	132
Farm income	-4285.85 (-0.65)	3.40 – 3.45	109	132
Off-farm income	-161.08 (-0.25)	1.80 – 1.85	109	132
Maize yield	0.17 (0.13)		109	132

The *t*-statistics are in parenthesis.

7.4 Conclusions

This paper assessed the impact of farmers' participation in climate change-related projects on their resilience to climatic shocks as well as income (i.e., on-farm and off-farm) and maize yield. The result from the Logit model indicates that farmers in the Northern Region and those who own mobile phones are more likely to participate in climate change-related projects. Furthermore, access to agricultural extension services and farmers' participation in previous agricultural projects increase the probability of participating in climate change-related projects. The PSM model results show no significant difference between participants and non-participants in relation to resilience to shocks (i.e., drought, flood, death of livestock), maize yield, and on-farm and off-farm income.

The climate change-related projects promote adaptation strategies such as irrigation, tree planting, use of inoculum as a soil management practice, and capacity building. Therefore, farmers who apply the strategies should be expected to improve their resilience. The lack of impact from the PSM analysis probably points to project unsustainability arising from limited geographical coverage and the short duration of the projects. Also, projects which build the capacity of farmers but do not enhance access to inputs are not likely to achieve expected results.

Farmers need rapid results and are more inclined to adopt short-term practices. Any promotion of conservation strategies for agriculture to build long-term resilience of farmers or contribute to the mitigation of climate change will require complementary incentives to partly cover the costs of the investment or help farmers recover part of the opportunity cost of the land before the investments begin to yield results. Although farmers are aware of the changing environment and its impact on their livelihoods, they need to be constantly informed and educated about the causes of the increasing environmental challenges. Such education could be through the regular extension system, but there is also a need for more intensive and extensive awareness-raising campaigns.

Although this study has identified the different practices farmers adopt to adapt to or cope with climate variability, it was not able to isolate and assess the impacts of individual practices on yields, incomes and resilience. Farmers commonly combine practices (e.g., harrowing land, planting an improved seed and applying fertilizer). The impact of participation in climate change-related projects was therefore assessed instead of individual practices.

References

Al-Hassan, R.M., Egyir, I.S., and Abakah, J. (2013a) Farm household level impacts of information communication technology (ICT)-based agricultural market information in Ghana. *Journal of Development and Agricultural Economics*, 5(4), 161-167.

Al-Hassan, R.M., Armah, R.N.A., and Etwire, P.M., (2013b) Climate Change Related Projects Implemented in AGRA Bread Basket Areas in Ghana. Paper prepared under the Environment Policy Action Node, Ghana.

Athula, S., and Scarborough, H. (2011) Coping with Climatic Variability by Rain-fed Farmers in Dry Zone, Sri Lanka: Towards Understanding Adaptation to Climate Change. Australian Agricultural and Resource Economics Society (AARES), 55th Annual National Conference 8-11 February 2011, Melbourne, Victoria.

Awunyo-Vitor, D., Al-Hassan, R.M., and Sarpong D.B. (2013) Impact of Formal Financial Market Participation on Farm Size and Expenditure on Variable Farm Inputs: The Case of Maize Farmers in Ghana. ISRN Economics, Vol 2013, Article No. 32969. http://dx.doi.org/10.1155/2013/329674.

Baker, J.L. (2000) Evaluating the Impact of Development Projects on Poverty: A Handbook for Practitioners. The International Bank for Reconstruction and Development/The World Bank.

Becerril, J., and Abdulai, A. (2010) The impact of improved maize varieties on poverty in Mexico: A propensity score marching approach, *World Development*, 38(7), 1024–1035.

Blundell, R., and Dias, M.C. (2000) Evaluation Methods for Non-Experimental Data, *Fiscal Studies*, 21(4), 427–468.

Boko, M., Niang, I., Nyong, A., Vogel, C., Githeko, A., Medany, M., Osman-Elasha, B., Tabo, R., and Yanda, P. (2007) Africa. *Climate Change 2007: Impacts, Adaptation and Vulnerability. Contribution of Working Group II to the Fourth Assessment Report of the Intergovernmental Panel on Climate Change*, M.L. Parry, O.F. Canziani, J.P. Palutikof, P.J. van der Linden and C.E. Hanson, Eds., Cambridge University Press, Cambridge UK, 433-467.

Dehejia, R.H., and Wabba, S. (1999) Causal Effects in Non-experimental Studies: Reevaluating the Evaluation of Training Programs, *Journal of the American Statistical Association*, 94(448),1053-1062.

Duflo, E., Glennerster, R., and Kremer, M. (2008) “Using randomization in development economics research” Handbook of Development Economics, 4, 25-39.

Ellis-Jones J., Larbi, A., Hoeschle-Zeledon, I., Dugje, I.Y., Teli, I.A., Buah, S.S.J., Kanton, R.A.L., Kombiok, J.M., Kamara, A.Y., and Gyamfi, I. (2012) Sustainable intensification of cereal-based farming systems in Ghana's Guinea savanna: constraints and opportunities identified with local communities. IITA report. IITA, Ibadan, Nigeria.

Faltermeier, L., and Abdulai, A. (2008) The adoption of water conservation and intensification technologies and farm income: A propensity Score Analysis for rice farmers in Northern Ghana”. Paper presented at the American Agricultural Association, Ornaldo.

Ghana Statistical Service (2014) Ghana Living Standards Survey Round 6 (GLSS 6): Labour Force Report. August, 2014.

Heinrich, C., Maffioli, A., and Vázquez, G. (2010) A Primer for Applying Propensity-Score Matching: Impact-Evaluation Guideline. Technical Notes. No. IDB-TN-161, August.

Khandker, R. S., and Faruqee, R. R. (2003) “The Impact of Farm Credit in Pakistan”. *Agricultural Economics*, 28, 197-213.

Mendola, M., (2007) “Agricultural technology adoption and poverty reduction: A propensity-score matching analysis for rural Bangladesh”. *Food Policy 32* (2007) 372-393.

Owusu, V., Abdulai, A., and Abdul-Rahman, S. (2010) “Non-farm Work and Food Security Among Farm Households in Northern Ghana”. *Food Policy 36* (2011) 108-118.

Ravallion, M. (2008) “Bailing Out the World's Poorest”. *Challenge*. 52(2), 55-80.

Wainaina, W., Okello, J., and Nzuma, J. (2012) “Impact of contract farming on smallholder poultry farmers' income in Kenya”, The International Association of Agricultural Economists Conference, Foz do Iguacu, 18-24 August.

Würtenberger, L., Bunzeck, I. G., and Tilburg, X.V. (2011) Initiatives related to climate change in Ghana: Towards coordinating efforts. Energy research Centre of the Netherlands, ECN-E--11-010. Retrieved October 20, 2011 from http://www.ecn.nl/docs/library/report/2011/e11010.pdf

Rosenbaum, P. R., and Rubin, D.R. (1983) The central role of the propensity score in observational studies for causal effects. *Biometrika* 70: 41-55.

The World Bank Group (2010) “Economics of Adaptation to Climate Change-Ghana Country Study.” http://climatechange.worldbank.org/sites/default/files/documents/EACC_Ghana.pdf

CHAPTER 8

Vulnerability of smallholder maize farming households to climate variability in the Eastern Region of Ghana

Mustapha Abubakar Sadiq, Ramatu M. Al-Hassan,
John K. M. Kuwornu

Abstract

Climate change and variability is one of the great challenges facing farming households in Ghana due to an over-reliance on rainfall for agricultural production. This study applied the Livelihood Vulnerability Index (LVI) to measure the vulnerability of smallholder maize farming households to climate variability in Ayensuano District and Yilo Krobo Municipality in the Eastern Region of Ghana. Both primary and secondary data were collected on the following household variables: socio-demographic profile, livelihood strategies, social networks, health, food, water, and natural disaster and climate variability. The data were collected from 75 respondents in each district, giving a total sample size of 150. Excel (2013) was employed in analyzing the data. The results of the LVI indicate that households in Ayensuano District are more vulnerable to the impacts of climate variability than those in Yilo Krobo Municipality. Weather stations should be installed in the two districts to make information about any impending natural disasters available to farmers.

Keywords: Ayensuano; climate change and variability; Yilo Krobo; vulnerability

8.1 Introduction

Scholze et al. (2006) assert that the lives of people across the entire world are progressively being endangered by the adverse effects of climate variability and change. Climate variability involves annual climatic fluctuations (i.e., short-term) above or below a long-term average value (IPCC, 2014). Droughts and floods, which are the main effect of climate variation, have immediate negative impacts on food production. The World Bank (2010) posits that climate variability and change have been experienced in Ghana, with three major physical impacts, namely change in rainfall, rise in sea level and temperature change. Further, the Environmental Protection Agency of Ghana predicted expected mean temperature increases of 0.6 °C, 2.0 °C and 3.9 °C over 1961-2000 mean value by the years 2020, 2050 and 2080, respectively (EPA, 2009). The increase in temperatures has resulted in low maize yields due to a reduction in the growing period and an increase in the evapotranspiration rate (Daze, 2007).

Vulnerability, according to the IPCC (2007), is a function of the character, magnitude and rate of climate variation to which a system is exposed, along with its sensitivity and adaptive capacity. Impacts of climate variation do not affect all people and communities at the same time, or to the same degree, because people are exposed to different rates of variation. The factors influencing smallholder maize farming households' vulnerability to climate variability may include the gender and age of the household head, their formal education, total farm size, farming experience, household size, hired labour, family labour, membership of a social group, access to formal credit, and their temperature and rainfall perceptions (Tiwari et al., 2008; Al-Hassan et al., 2013). Climatic shocks are expected to disproportionately affect the poor, young, elderly, sick, and otherwise marginalized populations (Kasperson and Kasperson, 2001).

Agriculture in Ghana is dominated by smallholder farmers, whose farming is characterized by a reliance on rain-fed conditions, and limited use of improved seeds, fertilizer and mechanization. This results in productivity levels well below their potential. Although these smallholders have considerable experience in dealing with climate variability, and they possess the local knowledge to cope during difficult periods, the unprecedented and sustained levels of variability associated with long-term climate change are outside the realm of what traditional coping strategies are able to manage (Pettengell, 2010). Agricultural activities are rural-based in Ghana's Eastern Region and about 69% of the households in the region are engaged in agricultural activities. Maize plays an important role in food security for many poor households in Ghana (MoFA, 2011), with a per capita consumption of over 100 kg. Maize also serves as a cash crop for these households (FAO, 2008).

The Eastern Region of Ghana is the country's second largest producer of maize. About 19% of the maize grown in Ghana is grown in this region (GSS, 2012). However, climate variability is expected to affect food production (MoFA, 2011). Maize needs a regular supply of water and suffers badly in times of drought. Tweneboah (2000) posited that soil moisture deficiency that causes wilting for 1-2 days during tasselling can reduce yields by up to 20%, and 6-8 days of wilting at this stage can reduce yields by 50%. These yield reductions cannot be made up by the later availability of soil moisture, either by precipitation or irrigation.

Models and information about climate variability are available at macro levels, but little work has been done at the micro level (Etwire et al., 2013). Without the appropriate policies and/or adaptive strategies in place at a household level, smallholder farmers will find it extremely difficult to practice sustainable agriculture in environmentally unpredictable climatic conditions (Nelson et al., 2007). This study aids in better understanding smallholder farmer households' vulnerability to climate variability and highlights potential areas for intervention to support maize production by these farmers.

8.2 Materials and Methods

8.2.1 Measuring Vulnerability

Vulnerability, in general, can be measured using two approaches, namely the indicator approach and the econometric approach. The indicator approach involves the selection of variables that a researcher considers contribute to climate-related vulnerability, while the econometric technique involves the application of economic, mathematical and statistical methods in analyzing climate vulnerability. Both techniques have setbacks. The indicator approach has some level of subjectivity in selecting the variables that may account for vulnerability, whereas the econometric approach has the possibility of violating an econometric assumption.

There are several methods of measuring vulnerability using the indicator approach. The United Nations General Assembly (1997) used a Sustainable Livelihoods Approach (SLA) to design development programs at the community level by looking at five types of household assets—natural, social, financial, physical, and human capital. This approach has proven useful as it addresses the issues of sensitivity and adaptive capacity to climate change and variability, but it fails to integrate climate exposures (Hahn et al., 2009). Due to the limitation of the SLA, Hahn et al. (2009) employed two techniques to assess households' vulnerability levels to climate change and variability in Mozambique. The first technique expresses the Livelihood Vulnerability Index (LVI) as a composite index comprised of seven major components (i.e. socio-demographic profile, livelihood strategies, social networks, health, access to food, access to water, and natural disasters and climate change), while the second technique aggregates the seven components into the Intergovernmental Panel on Climate Change's (IPCC) three contributing factors to vulnerability (i.e., exposure, sensitivity, and adaptive capacity). Etwire et al. (2013) employed the LVI technique developed by Hahn et al. (2009) to assess households' vulnerability to climate change and variability in northern Ghana.

This study employed the two approaches used by Hahn et al. (2009) in assessing the levels of Mozambique households' vulnerability to climate change and variability.

8.2.2 Livelihood Vulnerability Framework

The Livelihood Vulnerability Framework is particularly relevant to understand households' vulnerability to climate variability because it provides a framework for analyzing both the key components that make up livelihoods and the contextual factors that influence the components.

The livelihood vulnerability will be determined for all the districts selected for the study, taking into account the IPCC definition of vulnerability to climatic impacts developed by Hahn et al. (2009). This definition makes use of seven major components, namely socio-demographic profile, livelihood strategies, social networks, health, access to food, access to water, and natural disasters and climate change. Each component is made up of several sub-components, each of which is measured on a different scale. It is, therefore, necessary to standardize each sub-component as an index using equation (8.1).

$$index_{sa} = \frac{S_d - S_{min}}{S_{max} - S_{min}} \quad (8.1)$$

Where S_d is the observed sub-component indicator for the household, and S_{min} and S_{max} are the minimum and maximum values, respectively. After each sub-component is standardized, the sub-component indicators are averaged using equation (8.2) to obtain the index for each major component:

$$M_d = \frac{\sum_{i=1}^{n} index_{s_{di}}}{n} \quad (8.2)$$

Where M_d is one of the seven major components [Socio-Demographic Profile (SDP), Livelihood Strategies (LS), Social Network (SN), Health (H), Food (F), Water (W), or Natural Disaster and Climate Variability (NDCV)] for a household; $index_{sdi}$ represents the sub-components, indexed by i, that make up each major component, and n is the number of sub-components in each major component. Once values for each of the seven major components for a household are calculated, they are averaged using Equation (8.3) to obtain the district-level LVI:

$$LVI_d = \frac{\sum_{i=1}^{7} W_{Mi} M_{di}}{\sum_{i=1}^{7} W_{Mi}} \quad (8.3)$$

This can also be expressed as

$$LVI_d = \frac{W_{SDP}\, SDP_d + W_{LS}\, LS_d + W_{SN}\, SN_d + W_F F_d + W_H H_d + W_W W_d + W_{NDCV}\, NDCV_d}{W_{SDP} + W_{LS} + W_{SN} + W_F + W_H + W_W + W_{NDCV}}$$

(8.4)

The weights of each major component, WMi, were determined by the number of sub-components of each major component. This thereby ensures that the overall LVI is the result of equal contributions of each sub-component. In this study, the LVI is scaled from 0 (i.e., minimum level of vulnerability) to 1 (i.e., the maximum level of vulnerability). The summary of the LVI is also presented in a spider diagram.

Descriptions of the major components and sub-components of the LVI developed for the study area are presented in Table 8.1.

Table 8. 1: Components of the LVI

Major Components	Sub-components	Explanation of the Components and Measurement
SOCIO-DEMOGRAPHIC PROFILE	Percentage of female-headed households.	Percentage of female-headed households. If a male head is away from home > 3 months per year, the female is counted as the head of the household.
	Percentage of households where head of the household had not attended school	Percentage of households where the head of the household reported that they had not attended school.
	Dependency level.	Percentage of population below 15 and over 65 years of age.
	Percentage of households with orphans.	Percentage of households that have at least 1 orphan living in their home.
LIVELIHOOD STRATEGIES	Percentage of households with family members working outside the community.	Percentage of households that reported at least one family member who worked outside the community to earn a wage.
	Percentage of households solely dependent on agriculture as source of income.	Percentage of households that reported agriculture as their only source of income.
	Average number of households who have the burden of a loan.	Percentage of households who took a loan in the past 12 months.
HEALTH	Percentage of households with family member suffering from chronic illness.	Percentage of households that reported at least one member of their family suffering from chronic illness.
	Average time to health facility (minutes).	Average time it takes the households to get to the nearest health facility.
	Average number of households receiving proper facilities for child delivery and immunization.	Percentage of households receiving proper facilities for child delivery and immunization.
	Households that do not have toilet facilities.	Percentage of households that do not have toilet facilities.
	Average Malaria Exposure *Prevention Index (range: 0–12).	Months reported exposure to malaria *Owning at least one bed net indicator (have bed net = 0.5, no bed net = 1) (e.g., Respondent reported

Major Components	Sub-components	Explanation of the Components and Measurement
		malaria is a problem from January–March and they do not own a bed net = 3 *1=3).
SOCIAL NETWORK	Average Receive: Give ratio (range: 0–15).	Ratio of (the number of types of help received by a household in the past one year) to (the number of types of help given by a household to someone else in the past one year).
	Average Borrow: Lend Money ratio (range: 0.5–2).	Ratio of a household borrowing money in the past 12 months to a household lending money in the past month, e.g., If a household borrowed money but did not lend money, the ratio = 2:1 or 2 and if they lent money but did not borrow any, the ratio = 1:2 or 0.5.
	Percentage of households that did not apply to their local government for assistance in the past 12 months.	Percentage of households that reported they did not receive any assistance from their local government in the past 12 months.
FOOD	Average number of households who struggle to find food.	Percentage of households who struggle to obtain food for their family
	Percentage of households using only forest-based energy for cooking purposes.	Percentage of households using only forest-based energy for cooking purposes.
	Average Crop Diversity Index.	Percentage of households who grow 1 additional crop.
	Percentage of households that do not save crops.	Percentage of households that do not save crops.
	Percentage of households that do not save seeds.	Percentage of households that do not save seeds.
WATER	Percentage of households reporting conflicts over water.	Percentage of households that heard about water conflicts in their community.
	Percentage of households that utilise natural water sources.	Percentage of households that reported they collected water from wells, rivers or lakes as their primary water source.
	Percentage of households that do not have a consistent water supply.	Percentage of households that reported they cannot avail themselves a constant supply of water.

Major Components	Sub-components	Explanation of the Components and Measurement
	Average time to water source (minutes).	Average time it takes the households to travel to their primary water source.
	Inverse of the average number of liters of water stored per household (range: >0–1).	The inverse of (the average number of liters of water stored by each household +1).
NATURAL DISASTERS AND CLIMATE VARIABILITY	Percentage of households that reported the occurrence of floods and droughts in the past 10 years.	Percentage of households that reported the occurrence of floods and droughts in the past 10 years.
	Percentage of household members with an injury or death as a result of the most severe natural disaster in the past 10 years.	Percentage of households that reported that their household members suffered from injury or death.
	Percent of households that did not receive a warning about the pending natural disaster	Percentage of households that did not receive a warning about the most severe flood and drought event in the past 10 years.
	Mean standard deviation of the daily average maximum temperature by month.	Standard deviation of the average daily maximum temperature by month between 2006 and 2015 was averaged for each province.
	Mean standard deviation of the daily average minimum temperature by month.	Standard deviation of the average daily minimum temperature by month between 2006 and 2015 was averaged for each province.
	Mean standard deviation of the daily average rainfall by month.	Standard deviation of the average daily rainfall by month between 2006 and 2015 was averaged for each province.

From equations (8.1) - (8.3), Hahn et al. (2009) calculated a new variable, LVI-IPCC. This takes into consideration the IPCC definition of vulnerability. The LVI-IPCC diverges from the LVI when the major components are combined. Rather than merge the major components into the LVI in equation (8.3), the major components are first combined according to three categories, namely exposure, adaptive capacity and sensitivity (Table 8.2), using the following equation (8.5):

$$CF_d = \frac{\sum_{i=1}^{n} W_{Mi} M_{di}}{\sum_{i=1}^{n} W_{Mi}} \tag{8.5}$$

Where CF_d is an IPCC-defined contributing factor (exposure, sensitivity or adaptive capacity) for the district, M_{di} are the major components for the district indexed by *i*, W_{Mi} is the weight of each major component, and *n* is the number of major components in each contributing factor.

Table 8. 2: IPCC contributing factors to vulnerability

Major Components	Sub-components
Exposure	Natural disasters and climate variability
Sensitivity	Health Food Water
Adaptive capacity	Socio-demographic profile Livelihood strategies Social networks

Source: Hahn *et al.* (2009)

Once exposure, adaptive capacity and sensitivity are calculated, the three contributing factors are combined using equation 8.6;

$$LVI-IPCC_d = (e_d - a_d) * S_d \tag{8.6}$$

The LVI-IPCC is on a scale of -1 (i.e., least vulnerable) to 1 (i.e., most vulnerable), and the result is presented in a vulnerability triangle. For the purpose of illustration, a detailed example of calculating the LVI for one of the districts in the current study area is presented in Appendix I.

8.2.3 Sampling and Data

Primary data at the household level in the two selected districts were solicited using a questionnaire, semi-structured interviews and focus group discussions. A questionnaire covering 34 key variables (including household composition and resources, institutional setting, perceptions about changes in temperature) was administered at the household level and applied to calculate the LVI. Secondary data on rainfall and temperature were obtained from the National Meteorological Department of Ghana. The reference period selected for the data on climate events was from 2006 to 2015. Relevant literature from secondary sources such as journal articles, publications and the internet were reviewed to support or refute arguments and conclusions about the subject matter.

A multi-stage sampling technique was employed to sample households. The Eastern Region was selected for purposive sampling due to its high level of maize output for the whole of Ghana. The two districts, Yilo Krobo and Ayensuano, were also purposively selected because of their prevalence of agrarian households with small farm holdings. Also, to date, there has been little research in the two districts. Simple random sampling (i.e., picking from a list) was used to select five communities from each district, making a total of ten communities sampled for the research. Fifteen household heads were selected randomly from each community, giving a sample size of 150. The questionnaire and interviews were conducted in January 2016.

Both a Spider Diagram and a Vulnerability Triangle were used to visually present the impacts of the various contributing factors to the households' vulnerability.

8.3 Results and Discussion

The LVI values for all 30 variables were computed for both Yilo Krobo Municipality and Ayensuano District. The assessment of the individual major components along with their respective sub-components of the LVI value indicates the level of vulnerability of various blocks within each district. The overall LVI value of Ayensuano was found to be higher (0.390) than Yilo Krobo (0.363). Thus, Ayensuano is more vulnerable to climate variability than Yilo Krobo.

Ayensuano has a higher proportion of female-headed households (39%) than Yilo Krobo (15%). This makes Ayensuano more vulnerable in terms of gender, as females have been found to be more vulnerable to climate variability than their male counterparts (Hahn et al. 2009). This may be attributed not only to the limited physical capacity of women but also to their overwhelming family burdens. The number of household heads without a formal education is higher in Ayensuano (0.067) than in Yilo Krobo (0.013), but both the dependency ratio and percentage of orphans in households are higher in Yilo Krobo than in Ayensuano. However, after aggregating the sub-components, Ayensuano is more vulnerable in terms of the socio-demographic profile (0.362) than Yilo Krobo (0.327).

Households in Ayensuano are more dependent on rain-fed agriculture as a source of income. For this reason, any shock in climate variability will have greater impacts on farming productivity of the respondents in Ayensuano District. Also, the burden of loans held by households tends to make them more vulnerable to climate events when the crops fail to thrive as a result of unfavourable climatic conditions. The LVI value from the burden of loans is higher in Ayensuano (0.400) than in Yilo Krobo (0.293). A summary of the major LVI components is presented in Table 8.3.

Table 8. 3: Summary of the major LVI components

Major components	Yilo	Ayensuano	t-test	Prob.	Pooled
Socio-demographic Profile	0.327	0.362	-12.972***	0.000	0.344
Livelihood Strategies	0.413	0.502	-85.548***	0.000	0.458
Health	0.209	0.252	-16.267***	0.000	0.231
Social Network	0.387	0.431	-11.390***	0.000	0.409
Food	0.202	0.227	-54.259***	0.000	0.215
Water	0.519	0.402	4.912***	0.000	0.460
Natural disaster and climate variability	0.459	0.528	-55.760***	0.000	0.493
OVERALL VULNERABILITY	0.363	0.390	11.8533***	0.000	0.377

The major component of health is made up of 4 sub-components. The LVI value for health is generally low (i.e., 0.209 for Yilo Krobo and 0.252 for Ayensuano). This could be a result of good health facilities being accessible to all in the study area. Ayensuano recorded more households that reported members suffering from chronic illness (0.187). Ayensuano again tends to be more vulnerable in terms of availability of toilet facilities at the residence. The Average Malaria Exposure and Prevention Index was low for both districts (i.e., 0.167 for Yilo Krobo and 0.163 for Ayensuano). This could be attributed to the Ghana Ministry of Health's sensitization campaigns and free distribution of mosquito nets to residents in the study communities.

The fourth component is the household Social Network. It consists of three sub-components computed for both districts. Good social relationships among members of a community mostly reduce the impact of climatic stresses on individual households. In both districts, no respondent applied to their local government for household assistance during the past twelve months. This makes the LVI value of the Social Network component extremely high (1.00). This is consistent with the findings of Etwire et al. (2013), who reported that households with less access to emergency cash loans/aid are likely to be more vulnerable. Access to emergency loans/aid in times of need is important assets for the households' livelihood. Ayensuano is more vulnerable in terms of the LVI value for the average borrow to lend ratio (0.133) than Yilo Krobo (0.040).

The overall vulnerability of Ayensuano is higher (0.431) than that of Yilo Krobo (0.387).

Climate variability may affect agriculture and food security by altering the spatial distribution of rainfall and the availability of water. It may heighten uncertainties throughout the food chain, from farm to fork and yield to trade dynamics (Hanjra and Qureshi, 2010). This makes the food component of the LVI an important component in the analysis. It is computed using five key sub-components and the indices show Ayensuano to be more vulnerable in terms of food security than Yilo Krobo. The LVI value for households who use only forest-based energy for cooking purposes is the same for both districts (0.973). This value means that the use of petroleum products for cooking has not gained much popularity in the study area. Households in Yilo Krobo do not struggle to find food, but about 6% of the respondents in Ayensuano do struggle to find food for home consumption. Approximately 99% of respondents grow one additional crop to maize, hence the LVI value for the average crop diversity index is low (i.e., 0.011 for Yilo Krobo and 0.014 for Ayensuano). Ayensuano is more vulnerable in terms of the food component of the LVI (0.040) than Yilo Krobo (0.010).

The sixth major component of the LVI is water and it consists of five sub-components. After aggregating the sub-components, the indices show that Yilo Krobo is more vulnerable in terms of water (0.202) than Ayensuano (0.227). This is because the majority of respondents in Yilo Krobo (i.e., 89%) reported having an inconsistent water supply due to inadequate water sources such as boreholes and wells. Ayensuano showed a lower vulnerability value of 0.219 in terms of the average time it takes to travel to water sources than Yilo, which showed a value of 0.379. This result is supported by the findings of Hahn et al. (2009), who reported a lower vulnerability score (0.099) for respondents in Mabote District in Mozambique because of their consistent and easy access to potable water. The average volume of water stored by households in the study area is 306 litres per day, with Yilo Krobo households storing about 340 litres and Ayensuano households storing only 272 litres. In both districts, women and children are mostly the household members who fetch the water. This is very common in Ghanaian culture. Yilo Krobo showed a higher vulnerability value in terms of conflict over water (0.320), with the value for Ayensuano being substantially lower (0.240).

The last component of the LVI is natural disasters and climate variability. Six components were aggregated for this major component. The result shows that Ayensuano was more vulnerable (0.773) than Yilo Krobo (0.253) in terms of reported cases of floods and droughts. Approximately 7% of households in Ayensuano contained a member who had sustained an injury over the past ten years as a result of an occurrence of flood or drought, but no household in Yilo Krobo reported members having sustained an injury over the same period. Adaptation to climate variability is an ongoing process of planning and adjustment that is required to respond to an evolving context and risks. To do this effectively requires an availability of appropriate, timely and locally-relevant climate information such as short-range weather forecasts, seasonal forecasts and early warnings for pending climate-related hazards. Publicity about pending climate disasters was low in Yilo Krobo, where only around 5% of the respondents reported having received warnings. This means Yilo Krobo was more vulnerable (0.947) than Ayensuano (0.813) in terms of information dissemination about pending natural disasters. This is consistent with the findings of Etwire et al. (2013) and Hahn et al. (2009). The LVI value for Ayensuano is lower than that of Yilo Krobo for both the mean standard deviation of average maximum temperature recorded over the past ten years and the mean standard deviation of average rainfall recorded over the past ten years. Thus, Yilo Krobo is more vulnerable in mean standard deviation of average maximum temperature (0.684) and mean standard deviation of average rainfall (0557) than that of Ayensuano (i.e., 0.663 for mean standard deviation of average maximum temperature and 0.504 for that of average rainfall).

After aggregating the seven major components of the LVI, Ayensuano was found to be more vulnerable (0.390) than Yilo Krobo (0.363). For the pooled data, the LVI value was 0.377. The LVI values for the major vulnerability components are presented in Figure 8.1.

Figure 8. 1: Vulnerability Spider Diagram

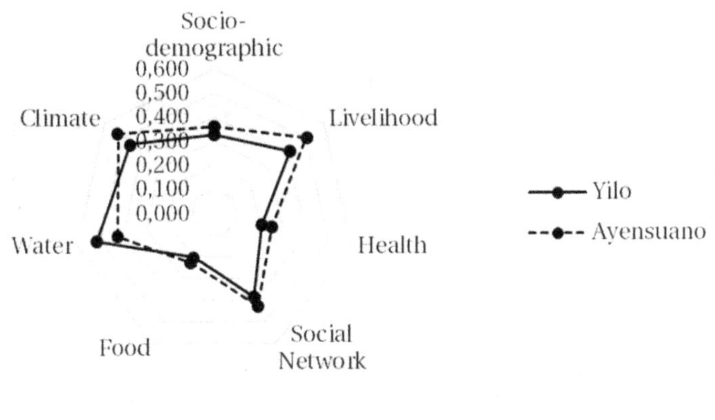

The spider diagram provides information on household characteristics that contribute most to their vulnerability to climate variability in the study area. The closer the major component is to the centre of the diagram, the least vulnerable is that component. For both study districts, the contribution of the food component to the LVI is the least amongst the other components, with the natural disaster and climate variability component having the greatest impact on the households' vulnerability level.

A two-tailed t-test was conducted to test the statistical difference in the mean LVI of the two districts. The difference in the mean LVI between the two districts is significant at 1% (Table 8.4).

Table 8. 4: Results of a Two-Tailed T-Test for Differences in Mean LVI Between the Two Districts

Variable	Count	Mean	Std. Dev.	Std. Err. of Mean	t-test value	Prob	df
Ayensuano	75	0.3648	0.0153	0.0028	11.8533***	0.000	148
Yilo Krobo	75	0.3347	0.0158	0.0018			
Pooled	150	0.3500	0.0216	0.0018			

The results of the LVI-IPCC show Ayensuano to again be more vulnerable (0.031) than Yilo Krobo (0.028). The LVI-IPCC values range from -1 (i.e., least vulnerable) and +1 (i.e., highly vulnerable). This result is depicted in Table 8.5, showing the sub-components of the major LVI components.

Table 8. 5: LVI-IPCC Contributing Factors

Major components	Sub-components	Yilo Krobo	Ayensuano
Exposure	Natural disasters and climate variability	0.459	0.528
	Sub-LVI	0.459	0.528
Sensitivity	Health	0.209	0.252
	Food	0.202	0.227
	Water	0.519	0.426
	Sub-LVI	0.317	0.305
Adaptive capacity	Socio-demographic profile	0.327	0.362
	Livelihood strategies	0.413	0.502
	Social networks	0.387	0.431
	Sub-LVI	0.371	0.425
Overall LVI-IPCC		0.028	0.031

Yilo Krobo is least exposed to climate variability (i.e., LVI value of 0.459) than Ayensuano (0.528), but in terms of sensitivity, Yilo Krobo is more vulnerable (0.317) than Ayensuano (0.239). Ayensuano is found to be less adaptive to the shocks of climate variability: the LVI value for Ayensuano is 0.425 while that of Yilo Krobo is 0.371. Figure 8.2 summarizes the LVI-IPCC contributing factors.

Figure 8. 2: Vulnerability Triangle of LVI-IPCC for Ayensuano and Yilo Krobo

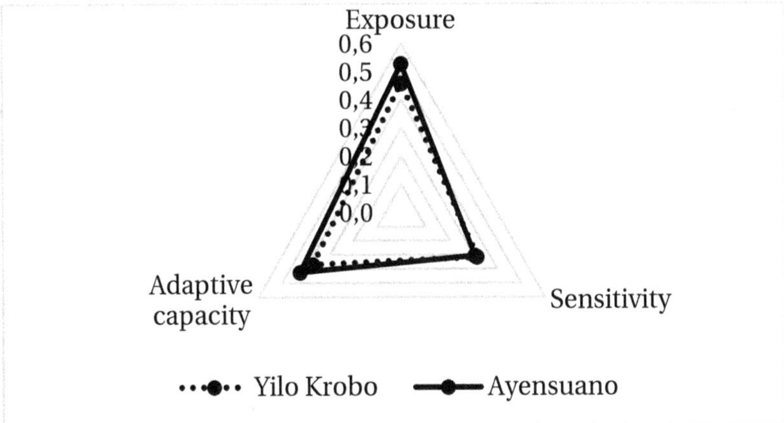

A two-tailed t-test was used to assess the statistical significance of the difference in the mean LVI of the districts, and the hypothesis is rejected at the 1% level of significance (Table 8.6).

Variable	Count	Mean	Std. Dev.	Std. Err. of Mean	t-test value	Prob.	df
Ayensuano	75	0.0278	0.0105	0.0012			
Yilo Krobo	75	0.0210	0.0026	0.0003	5.4348***	0.000	148
Pooled	150	0.0244	0.0083	0.0007			

8.4 Conclusions

Households in Ayensuano were found to be more vulnerable to the impacts of climate variability than those in Yilo Krobo, with Livelihood Strategies, Social Network and Natural Disaster and Climate Variability components being the most vulnerable components. This implies that households in Ayensuano are sensitive to modest variations in climatic conditions and their ability to adapt is severely constrained, thereby putting the vulnerable households at risk of becoming food insecure or malnourished.

The overall vulnerability index of households in Yilo Krobo is reduced as a result of their perceptions of temperatures and access to credit. Formal extension visits and farming experience of the household heads in Ayensuano makes these households less vulnerable since they are equipped with knowledge and skills relating to new and improved agricultural practices. The pooled data shows that access to credit, formal extension contacts and temperature perceptions are factors that reduce the vulnerability of households in the study area.

Urgent intervention is necessary in Ayensuano to enhance the financial and social capitals of the district's households. Additionally, interventions are required to enhance their human, physical and social capitals. The District Assembly should assist to establish a community financial co-operative to promote household savings and investment in the district. This has the potential to reduce the financial vulnerability of households and help to build their social networks. Water supply should be improved in Yilo Krobo to reduce the vulnerability of households to climate variability. This can be achieved through the extension of pipe-borne water projects to more deprived communities, and the construction of potable water sources such as boreholes and wells.

Weather stations should be installed in all settlements of the country which are significantly different in their geographical and socio-economic conditions. This is because climate-related research in Ghana suffers from a lack of weather data for specific sites. Scientifically-recorded weather data can help provide site-specific information about climate variability. This will aid in the dissemination of information on any impending natural disasters to farmers.

For a long-term solution to the over-reliance on rainfall for agriculture in all wards of the study regions, it is suggested that dams should be constructed at suitable sites to provide irrigation services to the farmers. This would help to support year-round farming and the raising of financial standings of farmers in the region.

References

Al-Hassan, R., Kuwornu, J. K. M., Etwire P.M., and Osei-Owusu, Y. (2013) Determinants of choice. *British Journal of Environmental Climate Change*, 3(2), 172-187.

Dazé, A. (2007) Climate change and poverty in Ghana. Unpublished report by CARE International Ghana.

Environmental Protection Agency, EPA (2009) National action programme to combat drought and desertification. Accra: *Environmental Protection Agency*, Ghana.

Etwire, P.M., Al-Hassan, R.M., Kuwornu, J.K.M., and Osei-Owusu, Y. (2013) Application of the Livelihood Vulnerability Index in Assessing Vulnerability to Climate Change and Variability in Northern Ghana. *Journal on Environmental Earth Science 3(2):157-170.*

FAO Statistical Databases (2008) FAOSTAT: Agriculture Data. Available online: http://faostat.fao.org.

Ghana Statistical Service, GSS. (2012) 2010 Population and Housing Census. Analytical Report, Northern Region.

Hahn, M. B., Riederer, A. M., and Foster, S. O. (2009) The Livelihood Vulnerability Index: A pragmatic approach to assessing risks from climate variability and change—A case study in Mozambique. *Global Environmental Change,* 19(1), 74-88.

Hanjra, M. A., and Qureshi, M.E. (2010) Global water crisis and future food security in an era of climate change. *Food Policy,* 35, 365–377.

IPCC (2007) Climate Change 2007: Synthesis Report. *Contribution of Working Groups I. II and III to the Fourth Assessment Report of the Intergovernmental Panel on Climate Change. Cambridge University Press,* Cambridge, UK.

IPCC (2014) Climate Change 2007: Impacts, Adaptation, and Vulnerability, Contribution of Working Group II to the Fourth Assessment Report. *Cambridge University Press,* Cambridge, UK.

Tiwari, K.R., Sitaula, B.K., Nyborg, I.P., and Paudel, G.S. (2008) Determinants of Farmers' adoption of improved soil conservation technology in a Middle Mountain watershed of Central Nepal. *Environmental Management,* 42(2), 210-222.

Kasperson, J.X., and Kasperson, R.E. (2001) International workshop on vulnerability and global environmental change. SEI Risk and Vulnerability Programme Report 2001-01, *Stockholm Environment Institute,* Stockholm, Sweden.

Ministry of Food and Agriculture, MoFA (2011) Agriculture in Ghana: Facts and Figures (2010). *Accra, Ghana. p. 53.*

Nelson, D. R., Adger, W. N., and Brown, K. (2007) Adaptation to environmental change: contributions of a resilience framework. *Annual review of Environment and Resources.*

Pettengell, C. (2010) Climate Change Adaptation: Enabling people living in poverty to adapt. *Oxfam Policy and Practice*: Climate Change and Resilience.

Scholze, M., Knorr, W., Arnell, N. W., and Prentice, I. C. (2006) A climate change risk analysis for world ecosystems. *Proceedings of the National Academy of Sciences.*

Tweneboah, C. K (2000) Modern Agriculture in the Tropics, Food crops. *Co-wood Publishers.*

World Bank (2010) Economic of Adaptation to Climate Change: *Social Synthesis Report, 1818* H Street NW Washington DC, USA.

CHAPTER 9

A multinomial logit analysis of farmers' adoption of climate variability adaptation strategies: The case of smallholder women rice farmers in the Northern Region of Ghana

Suhiyini Issah Alhassan, John K.M. Kuwornu, Yaw Bonsu Osei-Asare

Abstract

There are several Indigenous and contemporary technologies available to farmers for adaptation to climate variability. This study identified women rice farmers' adaptation strategies to climate variability using descriptive statistics, and determined the factors influencing their adoption of adaptation strategies (i.e., Indigenous and research-based) using a multinomial logit model. A multi-stage sampling technique was used to select 260 women rice farmers from Savelugu/Nanton Municipality and West Mamprusi District in the Northern Region of Ghana. Primary data was solicited from these farmers using a questionnaire. The empirical results revealed that soil-related strategies (i.e., loosening of soil, creating bunds or drainage channels, and organic fertilizer) are the predominant Indigenous adaptation strategies adopted by smallholder women rice farmers. On the other hand, the use of improved rice varieties (i.e., high-yielding varieties, early-maturing varieties and drought/flood resistant varieties) is the most-used research-based adaptation strategy adopted by women rice farmers in their bid to adapt to climate variability. The choice of Indigenous adaptation strategies among women rice farmers is significantly determined by their ownership of farmland, farmer-farmer extension contact, the distance to market, access to informal credit from friends and relatives, farming experience, farm size, the number of economic activities engaged in by a farmer, and family labour. The factors influencing the adoption of research-based adaptation strategies are the age of a farmer, participation in decision-making, farming experience, family labour, access to tractor services, access to formal credit, noticing decreased rainfall and increased temperatures, membership of a farmer-based organization, access to weather information and farmers' location. To foster adoption of adaptation strategies to climate variability, this study recommends that the government of Ghana, through its planting for jobs and food program should assist women farmers to form farming support groups. In addition, the program should provide the farmers with training on modern agronomic practices, give them input credit and access to tractor services, connect them with appropriate markets, and make available timely information on expected seasonal weather patterns for informed farming decisions.

Keywords: Climate variability; Indigenous and research-based adaptation strategies; multinomial logit analysis; smallholder women rice farmers.

9.1 Introduction

The impact of climate variability on people and ecosystems is increasing worldwide, with the possibility of impeding efforts towards sustainable growth and development (World Bank, 2008). Recent temperature increases, erratic rainfalls leading to droughts/floods and water scarcity are all evidence of climate variability (Adger et al., 2003; IPCC, 2007, 2014). Based on climate models, by the year 2100, mean daily temperatures are expected to increase by between 2.5°C - 3.2°C and total annual rainfall is projected to decrease by between 9-27 percent across various geographical locations (Minia, 2004).

The effects of climate variability differ based on agro-ecological regions, within countries in such regions, and due to socio-economic clusters such as gender differentials. Regions which rely heavily on rain-fed and subsistence agriculture are noted for their high climate-sensitivity in relation to systems of agricultural production and such regions are thus more vulnerable to climate variability (Harvey et al., 2014).

About 80 percent of Ghanaian rice production is produced by smallholder farmers in lowland areas which are rain-fed (MoFA, 2009). The national average yield of rice has remained as low as 2.5 tons/hectare/year

(MoFA, 2011) but was reported by the Crop Research Institute (CRI), Savannah Agricultural Research Institute (SARI), and International Food Policy Research Institute (IFPRI) to be only 2.2 tons/hectare/year in 2013. Although Ghana has the potential to achieve 6-8 tons/hectare/year, the survey attributed the low yield to climate variability and low adoption of climate variability adaptation strategies by smallholder farmers (MoFA, 2011). The highest rice producing region in Ghana is the Northern Region, which produces about 37 percent of the national rice output (MoFA, 2014).

To avert the menace of decreasing rice yields, SARI, the University for Development Studies, Non-Governmental Organisations (NGOs) and other research-based institutions in the Northern Region of Ghana have developed numerous technologies to aid adaptation to climate variability by all farmers in the region. However, these technologies do not seem to be carefully documented in the Ghanaian adaptation literature. The technical and economic prospects of these innovations have not been fully explored by farmers. This is due to their low awareness of the innovations, which has subsequently resulted in low rates of adoption by Ghanaian farmers (Etwire et al., 2013a).

Generally, compared to their male compatriots, the agricultural activities of female rice farmers are limited by a lack of the needed resources. This difference reduces efficient agricultural investments by female farmers, which constrains their investments aimed at enhancing their resilience to climate variability (FAO, 2011). In the Northern Region, just as in many parts of northern Ghana, relative to male-headed households, female-headed households have less ownership of and access to land and other resources to support production (Blackden and Wodon, 2006; Doss and Morris, 2000; Koru and Holden, 2008). Yet, females constitute about 50 percent of the region's population. The limited access to land and resources is partly due to cultural practices in the region. According to Asante and Amuakwa-Mensah (2014), rice and rooted crops (especially cassava) production are likely to reduce due to climate variability. These undesirable effects of climate variability worsen the plight of the poor, vulnerable and marginalised groups, who are mostly women and children. Targeting female farmers among other factors is vital in stimulating the adoption of newly-introduced climate-related technologies (Etwire et al., 2013a).

In the adaptation literature, much is known about how farmers adapt to climate change and variability. But very little has been researched on the climate variability adaptation strategies of women, especially in the Northern Region of Ghana where most women farmers operate on a small-scale, are under-resourced, and often suffer cultural discrimination.

Therefore, the focus of this research was to understand how smallholder women rice farmers in the West Mamprusi District and the Savelugu/Nanton Municipality adapt to the current climatic variability (by adopting Indigenous and research-based adaptation strategies), and to determine the factors influencing their ability to adopt the available climate variability adaptation strategies.

9.2 Materials and Methods

9.2.1 Evidence of Climate Variability

The Word Bank (2010), after assessing the intra- and inter-annual rainfall pattern in Ghana, concluded that Ghana's annual rainfall is highly erratic. Available historical data on rainfall indicates that the high amount of rainfall experienced in Ghana in the 1960s significantly declined between the late 1970s and early 1980s. Thus, Ghana experienced a decreasing rainfall trend between 1960 and 2006, with an average annual reduction of 2.3 millimeters. Kankam-Yeboah et al. (2011) also observed a significant change in Ghana's climate, with a $1^{\circ}C$ rise in temperature over a 30-year period. The high temperatures, which were hitherto experienced in March in northern Ghana, are more recently being experienced in January. The onset and offset of the rainy season continue to vary each year. Initially, April was noted for the start of the rainy season in the Northern Region, while September always marked the end of the rainy season (Etwire et al., 2013b). But the timing of the rainy season has changed in recent times. The region now receives its first rain in late June to early July and mostly experiences heavy rainfalls between September and October. Floods are commonly associated with the heavy rains, and these floods take lives and destroy farm produce and properties. The rains then usually end abruptly, resulting in drought.

Two climatic variables were considered in the study, namely, rainfall and temperature patterns between 1987 and 2015. In terms of the rainfall pattern, historical data from the Ghana Meteorological Service is presented in Figures 9.1 and 9.2. From this data, it is evident that the average rainfall in Savelugu/Nanton has decreased between 1987 and 2015, while in West Mamprusi there has been a marginal increase in average rainfall between 1996-2015. Generally, a trend of decreasing rainfall has been observed in Ghana (Deressa et al., 2009; Fosu-Mensah et al., 2012). Regarding the temperature pattern in the study area, data from the Ghana Meteorological Service showed a general increase in temperature over the period 1987 – 2015. This is consistent with Fosu-Mensah et al. (2012) and Kankam-Yeboah et al. (2011).

Figure 9. 1: Average daily rainfall by year in Savelugu/Nanton and West Mamprusi (1987-2015)

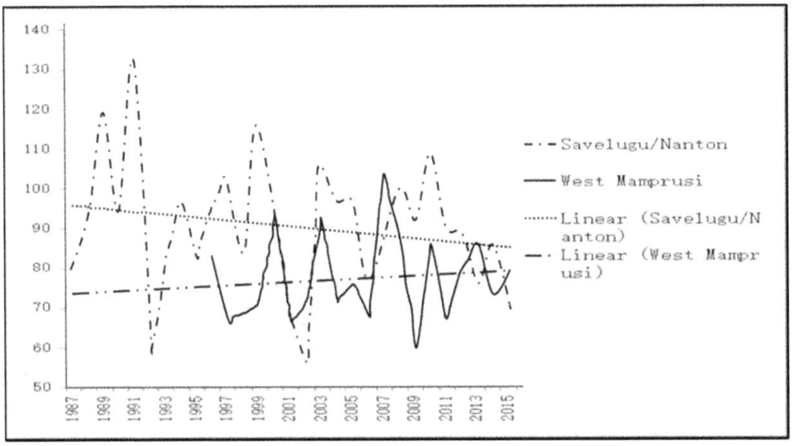

Source: Authors' computation of rainfall data from GMS (1987-2015), 2016.

Figure 9. 2: Average daily maximum and minimum temperatures by year (1987-2015)

Source: Authors' computation of temperature data from GMS (1987-2015), 2016.

To determine the effects of decreased rainfall and increased temperatures on the yield of rice, data on rice outputs from 2009 – 2014 were collected from MoFA's Statistics, Research and Information Directorate (SRID). The data indicates that the average yield of rice in Savelugu/Nanton Municipality and West Mamprusi District (and all of the Northern Region) had decreased over the period between 2009 – 2014, as presented in Figure 9.3. Amikuzuno and Donkoh (2012) attributed the low yield of major staple crops in northern Ghana under rain-fed conditions, including rice, to climate variability. This conclusion was drawn after establishing a causal relationship between yield and climate variability in the Northern Region of Ghana using co-integration analysis.

Figure 9. 3: Average Annual Yield of Rice under Rain-fed Conditions (2009 – 2014)

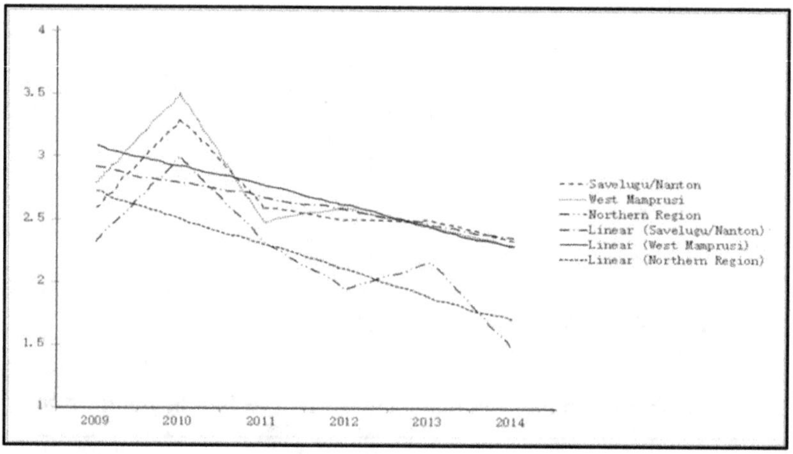

Source: Authors' computation of data from SRID, MoFA, 2016.

9.2.2 Study Area

The study was conducted in the Savelugu/Nanton Municipality and the West Mamprusi District in the Northern Region of Ghana. Figure 9.4 shows the study communities in each district. The region is in the Guinea Savanna Agro-ecological Zone, which is a semi-arid climatic region. The vegetation comprises mainly of grassland areas, interspersed with Guinea and Sudan savannah woodland characterised by drought-resistant trees such as the baobab, shea nut, acacia, dawadawa, mango and neem (GSS, 2012). The average minimum and maximum temperatures of the region are 14^0C at night and 40^0C during the day, respectively (GSS, 2012). The region experiences two seasons: the dry season (i.e., between November to April) and the wet season (i.e., between May and October), with an average annual rainfall of 750mm – 1050mm (GSS, 2014).

The Northern Region's agricultural sector employs 73.11 percent of the region's population of 15 years and above. Around 43 percent of these agricultural workers are female. The main crops cultivated in the region include maize, rice, yam, cassava, soy beans, millet and sorghum (GSS, 2012). About 14 percent of households in the Northern Region are headed by females, which is higher than the national average of 11 percent (GSS,

2012). West Mamprusi District is the third largest district in the region, with a total population of 168,011, of which 50.06 percent are female. Savelugu/Nanton Municipality is the fourth largest district in the region, with a population of 139,285, of which 51.51 percent are female.

Figure 9. 4: Map of Savelugu/Nanton Municipality and West Mamprusi District showing the Study Communities

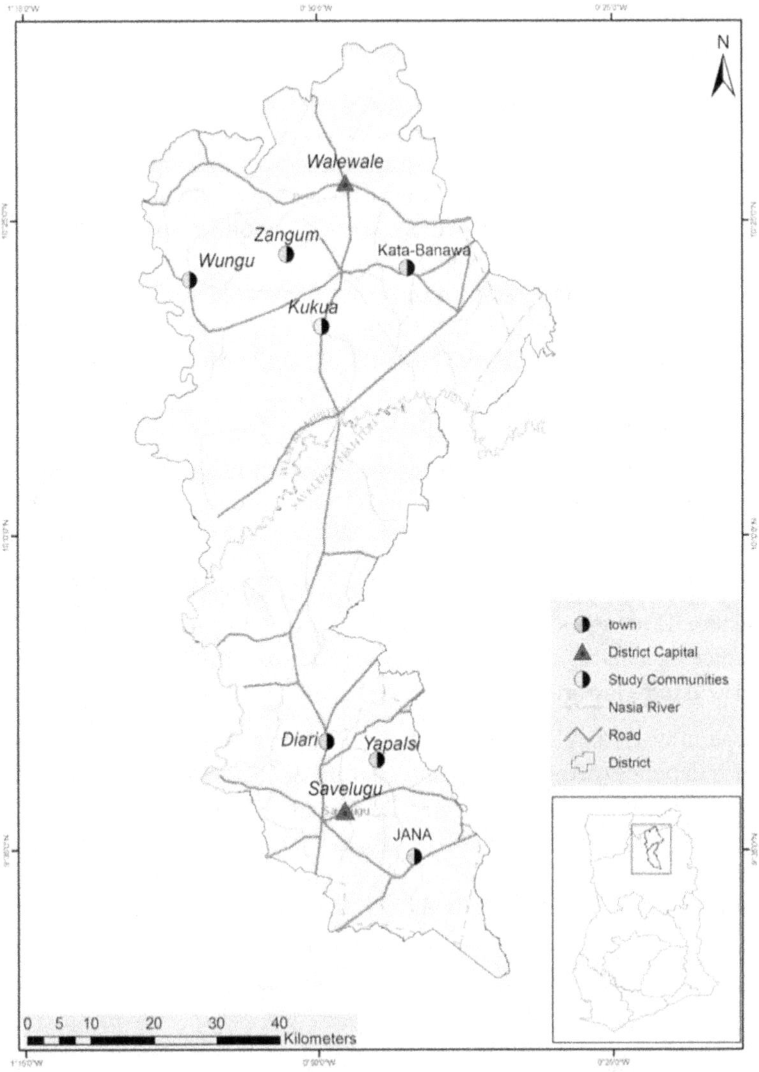

Source: Department of Geography and resource Development, University of Ghana, 2016.

9.2.3 Sampling and Data

A multi-stage sampling technique was used. The Northern Region was purposively selected because it is in a Guinea Savanna Ecological Zone located in a semi-arid climatic region, and it is the Ghana's largest producer of rice. West Mamprusi District and Savelugu/Nanton Municipality were purposively chosen among other districts because they are the top two rice producing districts in the Northern Region (SRID, MoFA, 2015), and they also have the most erratic rainfall patterns (GSS, 2012) in the region.

Based on population and land size, four communities in West Mamprusi District and three communities in the Savelugu/Nanton Municipality were randomly selected. Simple random sampling was used to proportionately select a total of two hundred and sixty (260) smallholder women rice farmers from the four communities. A questionnaire was administered to the 260 smallholder women rice farmers to gather primary data. Secondary data on temperature and rainfall in each district were collected from the Ghana Meteorological Service weather stations, while data on rice outputs were sourced from MoFA's SRID (SRID, MoFA, 2015).

9.2.4 Theoretical Framework

The theoretical framework adopted for this study is based on the random utility model, as specified by Greene (2003). A common formulation of the model is given in equation (9.1):

$$U = X\beta + \varepsilon \tag{9.1}$$

Where U denotes a farmer's decision on an adaptation strategy, X denotes explanatory variables, β denotes the parameter to be estimated, and ε denotes the error term associated with the estimation.

Assuming Y_{ia} and Y_{ib} are farmer i's decision on two adaptation choices, which depends on the utilities derived from them, they can be represented as U_i^a and U_i^b. For example, U_i^a and U_i^b could be the utilities derived by farmer i from crop-related strategies and soil-related strategies, respectively. The choice of farmer i between the two adaptation strategies reveals which one provides a higher utility; however, the farmer's utility is latent. Hence, the observed indicator is equal to one (1) if $U_i^a > U_i^b$, and zero (0) if $U_i^a \leq U_i^b$. This is commonly specified as:

$$U_i^a = X_i\beta_{ia} + \ell_{ia} \tag{9.2}$$

$$U_i^b = X_i\beta_{ib} + \ell_{ib} \tag{9.3}$$

Given the assumption that a farmer prefers and practices the adaptation strategy with the greatest net returns, the observable discrete choice strategy can be related to the latent continuous dependent variable, as specified in equation (9.4):

$$Y_{ia} = \begin{cases} 1 & if \quad U_i^a > U_i^b \\ 0 & Otherwise \end{cases} \tag{9.4}$$

Thus, Y_{ia} is a dichotomous dependent variable and takes the value 1, if farmer *i* adopts strategy *a* over others; and 0, if otherwise (Greene, 2003). The probability that farmer *i* will choose strategy *a* over other strategies is given by equation (9.5):

$$P(Y_{ia} = 1|X) = P(U_i^a > U_i^b) \tag{9.5}$$

Substituting equations (9.2) and (9.3) into equation (9.5) results in equation (9.6):

$$\begin{aligned} P(Y_{ia} = 1|X) &= P(X_i\beta_{ia} + \ell_{ia} - X_i\beta_{ib} - \ell_{ib} > 0|X) \\ &= P[(\beta_{ia} - \beta_{ib})X_i + \ell_{ia} - \ell_{ib} > 0|X] \\ &= P(\beta^* X_i + \ell^* > 0|X) = F(\beta^* X_i) \end{aligned} \tag{9.6}$$

Where β^* is a vector of parameters to be estimated, ℓ^* is a disturbance term, and $F(\beta^*X_i)$ is a cumulative distribution of the disturbance term evaluated at β^*X_i.

Following Greene (2003), depending on the distribution of the random disturbance term, the linear probability, logit or probit are the appropriate qualitative choice models for such a regression. Given that the identified adaptation strategies are more than two, the Multinomial Logit (MNL) and Multinomial Probit (MNP) models are the most applicable econometric models. In the literature, the MNL is the most widely used model in estimating the influence of explanatory variables on a dependent variable with multiple but unordered categories of choices (Gbetibouo, 2009).

9.2.5 Method of Analysis

Given that the dependent variable is not specified in any order of importance or magnitude, this study employed the unordered MNL in modeling women rice farmers' adaptation behaviors in response to climate variability. The choice of this model is based there being more than two adaptation strategies identified and that they are not ordinal in nature. The adaptation strategies were categorized into two sets: Indigenous adaptation strategies and research-based adaptation strategies. A farmer is said to have adopted one of the categories of either Indigenous or research-based strategies if she had employed at least one of the specific strategies in the main category within the 2016 farming season.

For the Indigenous adaptation strategies, smallholder women rice farmers have the option of choosing among four main strategies in adapting to climate-related events (i.e., mainly bushfires, floods and droughts). These main Indigenous climate-related adaptation strategies are broadly categorized into four, namely Crop-Related Strategies (CRS), Soil-Related Strategies (SRS), Cultural Practices-Related Strategies (CUPRS) and Other-Indigenous Strategies (OIS) (Ewtire et al., 2013b).

The probability of a farmer's choice of crop-related strategies over the other strategies, for instance, is assumed to be a function of socioeconomic, institutional and environmental characteristics. This is presented in equation (9.7).

$$Pr\,o\,b(Y_i = 1) = \frac{\ell^{\beta_j X_i}}{1+\sum_{k=0}^{3} \ell^{\beta_k X_i}}, \qquad j = 0,1, \ldots\ldots,3. \qquad (9.7)$$

Where B_j is a vector of coefficients on each of the exogenous variables, X.

Assuming $\beta_0 = 0$, equation (9.7) will be normalized to remove indeterminacy and the probability estimated as follows:

$$Pr\,o\,b(Y_i = 1|X) = \frac{\ell^{\beta_j X_i}}{1+\sum_{k=0}^{3} \ell^{\beta_k X_i}}\, j = 0, 2, \ldots\ldots, J, \beta_0 = 0. \qquad (9.8)$$

Estimating Equation (9.8) using the maximum likelihood method yields the log-odds ratio presented in Equation (9.9):

$$\ln\left(\frac{P_{ij}}{P_{ik}}\right) = X_i(\beta_j - \beta_k) = X_i\beta_j, \text{ if } k = 0. \qquad (9.9)$$

The choice of any Indigenous adaptation strategy to climate variability is, therefore, the log-odds in relation to the base alternative. According to Greene (2003), the coefficients of the Multinomial Logit are difficult to interpret, and associating B_j with the jth outcome is tempting and misleading. Instead, the marginal effects are usually derived to explain the effects of the exogenous variables on the endogenous variable in terms of probabilities, as presented in Equation (9.10).

$$\frac{\delta P_j}{\delta X_i} = P_j\left[B_j - \sum_{k=0}^{j} P_k B_j\right] = P_j\left(B_j - \overline{B}\right) \qquad (9.10)$$

The marginal effects measure the expected change in the likelihood of choice of a particular climate-related strategy with respect to a unit change in an exogenous variable (Greene, 2003).

The empirical multinomial model for the Indigenous adaptation strategies is specified in equation (9.11).

$$Y_i = \beta_0 + \beta_1 Age + \beta_2 FFE + \beta_3 Mkt + \beta_4 Borrow + \beta_5 Employment + \beta_6 FarmExp + \beta_7 FLab + \beta_8 Edu + \beta_9 FarmSize + \beta_{10} Land + \beta_{11} Location + \mu_i \qquad (9.11)$$

Where Y_i is the independent variables, β_is are parameters to be estimated, and μ_i is the error term.

The independent variables for this study are the four broad categories of Indigenous adaptation strategies. The soil-related strategies category was used as the base outcome. The parameters for the model were estimated using the maximum likelihood method. The likelihood ratio (LR) statistic was used to test for the overall significance of the model, while the Z statistic was used to test for the significance of each of the explanatory variables.

The Indigenous climate-related adaptation strategies' multinomial logit model was tested for two major econometric problems associated with these models, namely, the Independence of Irrelevant Alternatives (IIA) assumption and multicollinearity. Greene (2003) stated that the IIA assumption follows from the assumption that disturbances are independent and homoscedastic. The IIA assumption suggests that the probability of a smallholder rice farmer's choice of any Indigenous climate-related adaptation strategy is not influenced by the exclusion or inclusion of any other alternative strategy. The Hausman test was used to test for the violation or otherwise of the IIA assumption. The statistic has a limiting chi-squared distribution with K degrees of freedom and tests the null hypothesis that the odds are not dependent from irrelevant alternatives (Greene, 2003). Farmers' adoption of research-based adaptation strategies was also analysed using the multinomial logit model. However, with the research-based adaptation model, the dependent variable was specified as a set of research-based climate-related adaptation strategies, with few modifications to the explanatory variables. The empirical model for the research-based adaptation multinomial logit model is specified in equation (9.12):

$$Y_i = \beta_0 + \beta_1 Age + \beta_2 Edu + \beta_3 Decision + \beta_4 Credit + \beta_5 Land + \beta_6 Tractor + \beta_7 FLab + \beta_8 Rain + \beta_9 Temp + \beta_{10} Extension + \beta_{11} FBO + \beta_{12} Info + \beta_{13} Farmsize + \beta_{14} HLab + \beta_{15} Location + \mu_i \qquad (9.12)$$

The categories of the research-based adaptation strategies used as the dependent variable in equation (9.12) are Improved Varieties Strategies (IVB), Soil and Plant Health Strategies (SPH), Husbandry-Related Strategies (HRS), and Other research-based Adaptation Strategies (ORS). However, some farmers reported not adopting any research-based adaptation strategies in response to climate variability. No Adoption of Research-based adaptation Strategy (NRS) was therefore included as a fifth category because some farmers felt better off by not adopting any of the adaptation strategies.

Table 9.1 presents a description of the independent variables considered in the multinomial logit models.

Table 9. 1: Description of Independent Variables for the Multinomial Logit Models

Variable	Description	Measurement
Education	Farmer's years of education	Years
Hired labour	Farmer's use of hired labour	Man-days
Family labour	Farmer's use of family labour	Dummy: 1=yes, 0=otherwise
Farming experience	Farming experience of farmer	Years
Income	Off-farm income	Ghana Cedis
FBO	Member of a farmer-based group	Dummy: 1=yes, 0=otherwise
Decision	Voice in decisions of household and community	Dummy: 1=yes, 0=otherwise
Farmer-farmer extens. contact	Extension services received by a farmer from another farmer	Average number of visits per annum
Formal extens. contact	Extension service received by a farmer from a MoFA extension officer	Average number of visits per annum
Tractor	Access to tractor services	Dummy: 1=yes, 0=otherwise
Farmland	Ownership of farmland	Dummy: 1=yes, 0=otherwise
Rain	Perception of rainfall	Dummy:1=decrease, 0=otherwise
Age	Age of farmer	Years
Temperature	Perception of temperature	Dummy: 1=increase, 0=otherwise
Credit	Access to formal credit	Dummy: 1=yes, 0=otherwise
Borrow	Borrow from friends and relatives	Dummy: 1=yes, 0=otherwise
Weather	Access to weather information	Dummy: 1=yes, 0=otherwise

Employment	Number of economic activities engaged in by a woman rice farmer	Number
Location	The geographical location of the farmer	Dummy: 1=Savelugu/Nanton, 0=Otherwise
Market	Distance to market	Kilometers

Temperature Perception: Farmers' perception of temperature can be that it has increased, decreased, remained unchanged or not certain. Generally, warm (i.e., high) or cold (i.e., low) temperatures affect the moisture level of the soil which in turn affects the growth of plants. The influence of farmers' perception of temperature on their adoption of adaptation strategies is varied in the literature. Farmers perceiving increased temperatures are hypothesized to adopt adaptation strategies to reduce their vulnerabilities compared to those who have not perceived a temperature increase.

Rainfall Perception: Agriculture in the study area is predominantly rain-fed, and rainfall is increasingly erratic. Farmers may perceive that rainfall has increased, decreased, remained unchanged or be uncertain of the rainfall patterns over the period. A pattern of increased rainfall may result in floods, while a pattern of decreased rainfall may result in drought. While farmers perceiving a pattern of decreased rainfall are expected to have a high probability of adopting drought-related measures, farmers perceiving a pattern of increased rainfall are expected to adopt more of the flood-related adaptation strategies. However, this depends on the specific location of farmers, since the rainfall amount may not be evenly distributed across geographical locations.

Noticing climate variability: Perceiving climate variability heightens the likelihood of farmers adopting adaptation measures in response to climatic stresses. Farmers' perception of their climate variability can have a significantly negative influence on their adoption of climate-related adaptation strategies, thereby increasing her vulnerability to climate stress (Hassan and Nhemachena, 2008).

Agricultural Extension Services: The most reliable source of training and information to farmers on new technologies aimed at adapting to climate variability is agricultural extension services. These services are either formal (i.e., extension agent-farmer) or informal (i.e., farmer-farmer). Extension services contribute to efficient technology adoption decisions of farmers (Gbetibouo, 2009). The expected influence of access to extension services on farmers' adoption decisions regarding adaptation strategies is positive.

Access to credit: Farming is becoming very expensive and farmers need to complement their owned capital with borrowed capital (i.e., formal and informal) to meet the increasing investment and transaction costs associated with adaptation technologies. In this study, access to credit is measured as a dummy, with 1 denoting a farmer receiving credit within the 2016 farming season, and 0 for otherwise. Access to affordable credit increases the financial capacities of farmers to adopt adaptation strategies. This study hypothesized a negative relationship between farmers' access to credit and their overall vulnerability to climatic stresses.

Distance to market: The availability and accessibility of markets is a panacea to farmers' complaints of no demand for their products, especially in periods of bumper harvests. Thus, the distance farmers must travel to sell their farm produce is inversely related to their willingness to adopt strategies that will result in higher yields. According to Hassan and Nhemachena (2008), farmers with ready access to markets have a greater probability of adapting to climate vulnerability.

Farm size: According to Daberkow and McBride (2003, cited in Gbetibouo, 2009), due to the high initial investment cost and risks associated with certain technologies, large-scale farm enterprises are more likely to adopt such technologies given their strong financial resources relative to small-scale farm enterprises. Also, the larger the farm size, the larger the proportion of land allocated for the cultivation of other crops, and the greater the chances of adopting adaptation strategies to avert the effects of climatic shocks (Feder et al., 1985).

Educational level of farmer: There are mixed reports in the literature regarding the influence of education on the adoption of adaptation strategies among farmers and their vulnerability to climate variability. Whereas Deressa et al. (2010) and Maddison (2006) report that an increase in farmers' years of education increases the probability of their adoption of adaptation strategies, Mandleni and Anim (2011) found that education has no significant influence on farmers' adaptation to climate variability.

Age of farmer: The influence of age on farmers' decisions to adopt an adaptation strategy is not straightforward in the literature (Gbetibouo, 2009). The effect of age on adoption depends on the location and the type of innovation (Gbetibouo, 2009). Some older farmers may be more conservative, risk-averse, and lack the strength and flexibility to easily adopt new adaptation strategies compared to younger farmers. However, it could also be the case that given their vast farming experience, older farmers are more able to assess, understand and adopt new technologies to adapt to climate variability than younger farmers. The expected influence of age on adoption is, therefore, a matter of empirical evidence.

This study has hypothesized age to have either a positive or negative influence on adoption of adaptation strategies and hence, the vulnerability of farmers to climate variability.

Farming Experience: Farmers with many years of experience in rice farming are hypothesized to adopt adaptation strategies, thereby reducing their level of vulnerability to climate variability. It is believed that an experienced farmer who has lived in a particular locality over a long period of time is more acquainted with the climatic conditions in the area. Compared to a lesser experienced farmer, they may, therefore, more easily adapt to the changing environment by adopting either or both Indigenous and new research-based technologies (Deressa et al., 2009).

Farmer-based Group or Organization: A farmer-based organization is a group of farmers who come together purposefully to share farming ideas and to assist members in fostering good farming practices. Such groups function as platforms for information exchange, learning and the sharing of new innovations and perceived challenges relating to agricultural and non-agricultural activities (Fafchamps, 1992). It was hypothesized that farmers who belong to farmer-based groups are more likely to adopt adaptation strategies and thus will be less vulnerable to climatic stresses than those who are not members of such a group.

Family Labour: Women rice farmers who have a majority of their household members assisting them in their day-to-day farming activities including the adoption of adaptation strategies makes them less vulnerable to climate variability compared to those without assistance from family members. Peterman et al. (2010) and Quisumbing (2007) reported positive relationships between family labour and crop productivity. In this study, family labour was measured as a dummy, with 1 denoting farming with family labour assistance and 0 denoting otherwise.

Hired Labour: Farmers who are able to employ hired labour to supplement family labour in their farming activities are more likely to adopt adaptation strategies, making them less vulnerable to climatic stresses. Most literature has reported a positive relationship between hired labour and productivity (e.g., Adebayo and Adewuni, 2008; Ajibefun and Abdul kadri, 2004; Mazvimavi et al., 2014).

Off-farm income: Response to climate variability through the adoption of adaptation strategies requires some level of financial commitment (Deressa et al., 2009). Thus, farmers who have other sources of income beyond the farm more easily adapt to climatic stresses than farmers who depend on their farms as their sole source of livelihood. Off-farm income is hypothesized to have a negative influence on farmers' vulnerability to climate variability, but a mixed influence on farmers' adoption of adaptation strategies.

Number of economic activities engaged by a farmer: The number of economic activities undertaken by a farmer is a proxy for how diversified the farmer is in terms of livelihood strategies, and thus has a direct effect on their vulnerability or resilience to climate variability. FANRPAN (2011) found that smallholder farmers who are engaged in several sources of livelihood easily adapt to climate extremes.

Location of farmer: Adoption of technologies is location specific (Gbetibouo, 2009). Location varies in terms of vegetation and institutional arrangements, which serve as key determinants of the adoption of adaptation strategies by farmers to abate the effects of climate variability. Bawakyillenuo et al. (2016) reported that farmers in Savelugu/Nanton Municipality employed more crop-related adaptation strategies because of the interior savannah woodlands vegetation.

Land ownership: Ownership of land is a panacea to farmers' farm investments including investments in adaptation and good crop and livestock management practices. Adaptation strategies have higher chances of being adopted when farmers have the security of land ownership. The adoption of innovations which require huge investments is thus stimulated by land tenure security (Gbetibouo, 2009). This study hypothesized land ownership to have a positive influence on farmers' adoption of adaptation strategies and a corresponding negative influence on their vulnerability to climatic stresses.

Access to weather information: Farmers with information on the likely dates for the onset and end of rainfall for a given season are better placed to adjust to the effects of climatic stress than those who do not. Access to weather information through television, radio and other communication channels increases farmers' awareness of adaptation measures such as growing varieties of crops suited to the forecast weather conditions, changing planting dates, and using irrigation. Access to weather information was hypothesized to have a direct relationship with farmers' adoption decisions and hence, an adverse influence on their vulnerability to climate variability (Mbakahya and Ndiema, 2015).

9.3 Results and Discussion

9.3.1 Adaptation strategies to climate variability

Ten (10) Indigenous strategies for adaptation to climatic stresses were identified to be widely practiced by smallholder women rice farmers in the

Northern Region of Ghana. These strategies were grouped into four major categories, namely, crop-related strategies, soil-related strategies, cultural practices-related strategies and other Indigenous strategies. The results revealed that most women rice farmers (i.e., 55 percent) adopted soil-related strategies (i.e., loosening of soil, creating bunds or drainage channels, and use of organic fertilizer) in adapting to climate variability. About 15 percent of the respondents mostly employed crop-related strategies (i.e., mixed cropping and Indigenous varieties), while about 16.5 percent adopted cultural practice-related strategies (i.e., regular or early weeding and land rotation). The remaining 13.5 percent of respondents reported adopting other Indigenous strategies such as early or late planting to coincide with the timing of the rain. These results are presented in Table 9.2.

Table 9. 2: Categories of Indigenous adaptation strategies used by smallholder women rice farmers in the Northern Region of Ghana

Main Indigenous Adaptation Strategy Category	Frequency	Percentage of sample using strategy (%)
Crop-Related Strategies	39	15.0
Soil-Related Strategies	143	55.0
Cultural Practice-Related Strategies	43	16.5
Other Indigenous Strategies	35	13.5
Total	260	100.0

Also, the study identified sixteen (16) research-based adaptation strategies that were used by smallholder women rice farmers. These specific research-based adaptation strategies were categorized into four groups, namely, improved variety strategies, soil and plant health-related strategies, husbandry practices and other research-based adaptation strategies. Some farmers, however, reported not using any of these strategies. A fifth category known as no adoption of research-based strategies was therefore added as an option or alternative for farmers who did not adopt any research-based strategy.

About 19 percent of respondents reported adopting the improved variety strategy (i.e., high-yielding, early-maturing and drought/flood resistant varieties). About 31 percent of farmers interviewed reported to have employed soil and plant health-related strategies (i.e., inorganic fertilizer, compost, herbicides, insecticides and ploughing rice straw back into the soil). About 23 percent and 13 percent of farmers interviewed reported employing husbandry practices (i.e., harrowing, earth bunds, planting during recommended periods, dibbling and timely harvesting of produce) and other research-based strategies (i.e., fire breaks, reduced farm size, irrigation and planting of trees), respectively. The remaining 12.7 percent of respondents did not adopt any research-based strategy in adapting to the effects of climate variability. These results are presented in Table 9.3.

Table 9. 3: Categories of research-based adaptation strategies used by women rice farmers in the Northern Region of Ghana

Main Research-based Adaptation Strategy Category	Frequency	Percentage of sample Using strategy
Improve Variety Strategies	51	19.6
Soil and Plant Health-related strategies	81	31.2
Husbandry practices strategies	60	23.0
Other research-related strategies	35	13.5
No Adoption of Research Based Strategy	33	12.7
Total	260	100.0

9.3.2 Determinants of the adoption of climate variability adaptation strategies

9.3.2.1 Factors influencing the adoption of adaptation strategies

The Wald chi square values of 79.63 and 141.69 for the Indigenous and research-based adaptation strategy models for the pooled samples, respectively, are both statistically significant at 1 percent. Results of the Hausman test could not reject the null hypotheses that the probability of a farmer's choice of any Indigenous adaptation strategy is not dependent on the inclusion or exclusion of another Indigenous adaptation strategy, and that the probability of a farmer's choice of any research-based adaptation strategy is not dependent on the availability or otherwise of an alternative research-based adaptation strategy. Thus, the independence from an irrelevant alternative assumption was not violated by both multinomial logit models, thereby justifying its application in determining factors influencing the choice of Indigenous adaptation strategies and adoption of research-based adaptation strategies among smallholder women rice farmers in the Northern Region.

Since the estimated coefficients only provide the direction of change without the magnitude or probabilities, the marginal effects were derived and are presented in Tables 9.4 and 9.5. For the Indigenous adaptation model, the soil-related strategy was used as the base strategy, while no adoption strategy was used as the base strategy for the research-based adaptation model.

9.3.2.2 Factors influencing the choice of Indigenous adaptation strategies

Farmland ownership, farmer-farmer extension contacts, distance to markets, access to credit from friends and relatives, and the vegetation of the farmers' location are the factors that significantly influence the choice of crop-related strategies vis-à-vis soil-related strategies by smallholder women rice farmers. Yet, farming experience, farm size, number of economic activities and use of family labour significantly contribute to a farmer's choice of cultural practice-related strategies relative to soil-related strategies. Finally, farmland ownership, distance to market, access to credit from friends and relatives, number of economic activities, and age of a farmer are the significant determinants of the choice of other Indigenous-related strategies by smallholder women rice farmers over soil-related strategies. The choice of any Indigenous adaptation strategy by a smallholder women rice farmer is not significantly influenced by her years of education. The reason for this could be that farmers need farming rather than formal education to enable them to make informed choices about traditional adaptation strategies. This confirms the findings of Mandleni and Anim (2011), that education has no significant influence on the choice of climate change and variability adaptation strategy. Results of the factors influencing the choice of Indigenous adaptation are presented in Table 9.4.

Owing farmland increases the probability of smallholder women rice farmers' choice of crop-related and other Indigenous-related strategies by 4.7 percent and 6.8 percent, respectively, relative to soil-related strategies. When a smallholder woman rice farmer receives farmer-farmer extension contact, it reduces the probability of her choice of crop-related strategies vis-à-vis soil-related strategies by almost 3 percent. The probable reason for this is that farmers tend to encourage the use of adaptation strategies which are capable of enriching the nutrient status and porosity of the soil for improved yields rather than relying on Indigenous varieties and the cultivation of many crops. This finding contradicts Al-Hassan et al. (2013), who reported a higher likelihood of farmers' choice of crops and other Indigenous-related adaptation strategies vis-a-vis soil-related strategies if the recommendation comes from another farmer.

Table 9. 4: Factors influencing the choice of Indigenous adaptation strategies by women rice farmers in the Northern Region of Ghana

Independent Variable	Crop-Related Strategies		Cultural Practices-related strategies		Other Indigenous-Related strategies	
	Marginal effect	P value	Marginal effect	P value	Marginal effect	P value
Farmland ownership	0.0467*	0.085	0.0375	0.223	0.0682*	0.078
Farmer-farmer extension contact	-0.0279***	0.006	-0.0048	0.512	0.0079	0.745
Distance to market	-0.0193***	0.004	0.0047	0.734	-0.0261**	0.025
Credit from friends and family	-0.1068**	0.020	-0.0685	0.208	0.0737*	0.072
Farming Experience	-0.0016	0.271	0.0028*	0.085	-0.0003	0.951
Farm size	-0.0039	0.943	0.0251*	0.096	0.0031	0.605
№ of economic activities	-0.0032	0.770	0.0402*	0.046	-0.0434***	0.006
Family labour	-0.0423	0.162	-0.2289***	0.002	-0.0243	0.218
Education	-0.0033	0.563	-0.0011	0.884	0.0050	0.362
Age of farmer	0.0001	0.864	0.0004	0.946	-0.0029*	0.093
Location	0.1445***	0.001	0.0377	0.189	-0.0040	0.626
Base Outcome: Soil-Related Strategies						
Number of Observations					260	
Wald Chi^2 (33)					79.63	
Prob > Chi^2					0.0000	
Pseudo R2					0.1613	
Log pseudo likelihood					-250.20084	

Note: *, ** and *** denotes statistical significance at 10%, 5% and 1%, respectively.

The probability that a smallholder woman rice farmer will choose crop-related and other Indigenous-related strategies rather than soil-related strategies reduces by about 2 percent and 2.6 percent, respectively, for any additional kilometer farther away from the market. Access to market is not only for the sale of farm produce and purchases of farm inputs, but it also serves as a socialization avenue. At the market, farmers regularly meet to interact and share farming experiences and information about farming practices such as when rains are expected, which crops can be cultivated on the same farmlands, and the availability of high-yield Indigenous rice varieties. This finding confirms the results of Etwire et al. (2013b) that the presence of a market in a community increases the probability of farmers adopting crop-related strategies over soil-related strategies. This finding also agrees with Hassan and Nhemachena (2008).

A smallholder woman rice farmer's access to informal credit from friends and relatives reduces the probability of choosing crop-related strategies over soil-related strategies by 10.7 percent but increases the probability of choosing other Indigenous-related strategies relative to soil-related strategies by 7.4 percent. The reason for this could be that women rice farmers who are able to access informal credit can easily hire labour for early planting or planting immediately after the rain starts, rather than applying manure, loosening the soil or ridging which are more costly and require more labour.

An additional year of farming experience on the part of a smallholder woman rice farmer increases the probability of choosing cultural practice-related strategies over soil-related strategies by a marginal 0.3 percent. Highly experienced farmers are acquainted with the methods of farming that will ensure higher yields, even with climate variability. Experienced women rice farmers thus prefer regular or early weeding and land rotation, which are more effective and less expensive ways of adapting to the effects of climatic stress compared to soil-related strategies.

Expanding the size of a rice farm by one hectare increases a woman rice farmer's probability of choosing cultural practice-related strategies compared to soil-related strategies by about 2.5 percent.

For any additional economic activity engaged in by a woman rice farmer, the probability of choosing cultural practice-related strategies rather than soil-related strategies increases by almost 4 percent. However, this also results in a more than 4 percent decrease in the probability of choosing other Indigenous-related strategies compared to soil-related strategies. This supports the finding of FANRPAN (2011).

A smallholder woman rice farmer is almost 23 percent less likely to choose cultural practice-related strategies vis-a-vis soil-related strategies if she depends on family labour for farming activities. The probable reason for this could be that other family members will already be engaged in their own farming activities when she needs them to assist with weeding her farm. Instead, ridging and manure application, which are less labour intensive, may be preferred strategies.

Any additional year in the age of a woman rice farmer reduces the likelihood of her choice of other Indigenous-related adaptation strategies relative to soil-related strategies by a marginal 0.3 percent. This may be explained by the fact that elderly women rice farmers may have a weak physical condition and depend on other family members (mostly children) for their farming activities. These family members may not be available in time to assist in early planting or when a woman farmer expects rains.

Being located in Savelugu/Nanton Municipality, which is in a Sudan Savannah Agro-ecological Zone, increases the chances of a woman rice farmer's choice of crop-related strategies vis-à-vis soil-related strategies by 14.5 percent. The possible reason for this is to adapt to climate variability by choosing crop varieties which suit the ecology rather than loosening the soil, applying manure or creating bunds.

9.3.2.3 Factors influencing the adoption of research-based adaptation strategies

The research findings revealed that the age of a farmer, access to tractor services, noticing a decrease in rainfall, and being a member of a farmer-based organisation or group are the factors which significantly influence the adoption of improved variety strategies by smallholder women rice farmers over a 'no adoption' strategy. Meanwhile, the adoption of soil and plant health-related strategies by smallholder women rice farmers vis-à-vis a 'no adoption' strategy is significantly dependent on the age of the farmer, farmer's participation in decision-making, farm size, farming experience, access to tractor services, being a member of a farmer-based organisation, and having access to weather information. Age of a farmer, family labour, noticing a decrease in rainfall and an increase in temperatures, and access to formal credit are the main factors inducing the adoption of recommended agricultural practices by smallholder women rice farmers compared to a 'no adoption' strategy. And family labour, access to tractor services, access to weather information and agroecology are the factors which significantly contribute to the adoption of other research-based adaptation strategies by smallholder women rice farmers relative to a 'no adoption' strategy. Farmers' years of education, hired labour, farmland ownership and formal agricultural extension contact have no significant influence on the adoption of research-based adaptation strategies by smallholder women rice farmers in the Northern Region. These results are presented in Table 9.5. This study discusses only the independent factors that have a significant influence on the adoption of research-based adaptation strategies by smallholder women rice farmers.

The probability that a woman rice farmer will adopt improved variety-related strategies, soil and plant health-related strategies, and recommended agricultural practices relative to 'no adoption' each reduce only marginally (i.e., by almost 0.3 percent) for every year added to her age. The reason for this is that elderly female farmers lack the physical strength required to implement some of the adaptation strategies and they may not be able to employ adaptation strategies that require more physical exertion. This finding is contrary to the findings of Deressa et al. (2010) but consistent with the findings of Etwire et al. (2013b), who reported a negative relationship between farmers' age and their adoption of introduced adaptation strategies compared to 'no adoption' among farmers in northern Ghana. It is also supported by the findings of Hassan and Nhemachena (2008).

Women's participation in decision-making increases the probability of their adoption of recommended agricultural practices vis-à-vis 'no adoption' by almost 12 percent. When women are allowed to participate in climate variability adaptation projects implemented in their communities by the government and other agriculture-based NGOs, they will prefer to plant at the appropriate time of the season, harvest their produce as recommended and also construct earth bunds on their farms rather than not employing any adaptation strategy.

An additional year of farming experience gained by a smallholder women rice farmer decreases their probability of adopting soil and plant health-related strategies relative to 'no adoption' of any research-based adaptation strategy by a marginal 0.5 percent. This can be attributed to the fact that farmers become used to certain ways of doing things as they continue to cultivate rice over a long period of time and may become resistant to adopting new innovations. This finding is consistent with the findings of Awotide et al. (2012).

Table 9. 5: Factors influencing the adoption of research-based adaptation strategies by women rice farmers in the Northern Region of Ghana

Independent Variable	Improved variety strategies		Soil and plant health-related strategies		Husbandry strategies		Other research-based adaptation strategies	
	Marginal effect	P value	Marginal effect	P value	Marginal effect	P value	Marginal effect	P value
Age of farmer	-0.0032***	0.004	-0.0028***	0.007	-0.0028***	0.003	-0.0004	0.121
Years of education	0.0082	0.326	0.0002	0.914	-0.0050	0.736	-0.0021	0.834
Participates in decision-making	0.0287	0.457	0.1206***	0.004	-0.0709	0.748	-0.0289	0.831
Farmland ownership	0.0452	0.251	0.0004	0.632	0.0987	0.132	-0.0733	0.371
Farming Experience	0.0023	0.976	-0.0048***	0.005	-0.0008	0.278	0.0006	0.329
Hired labour	0.0001	0.663	-0.0001	0.846	-0.0001	0.833	0.0002	0.316
Farm size	0.0216	0.871	0.2181*	0.098	0.1325	0.303	0.0562	0.682
Family labour	0.0001	0.472	-0.0001	0.341	-0.0003*	0.072	-0.0001**	0.031
Access to tractor services	-0.1613***	0.001	-0.0580**	0.015	0.0400	0.174	-0.0556**	0.019
Access to formal credit	-0.1197	0.706	-0.0575	0.768	0.2134*	0.098	0.0861	0.160
Noticed decreased rainfall	-0.0514**	0.031	-0.0121	0.176	-0.1304***	0.004	-0.0016	0.208
Noticed increased temperature	-0.0380	0.229	0.0311	0.826	-0.1938**	0.046	0.0450	0.881
Formal extension contacts	-0.0067	0.655	0.0058	0.865	0.0025	0.685	-0.0127	0.237
Member of a farmer-based group	0.2212***	0.002	0.0886**	0.019	-0.1705	0.431	-0.0366	0.991
Access to weather information	-0.0304	0.541	0.6637**	0.036	0.0492	0.130	0.0405*	0.100
Location	-0.0410	0.710	-0.0678	0.578	0.0361	0.205	0.1901***	0.001
Base Category	No Adoption Strategy							
Number of Observations	260							
Wald Chi^2 (60)	141.69							
Prob > Chi^2	0.0000							
Pseudo R2	0.1655							
Log pseudo likelihood	-336.98423							

Note: *, ** and *** denotes statistical significance at 10%, 5% and 1%, respectively.

Women rice farmers who depend on family labour for their farming activities are slightly unlikely to adopt husbandry practices and other research-based adaptation strategies relative to 'no adoption' of any research-based adaptation. In the focus group discussions, it was revealed that family members are sometimes not readily available to assist in planting or harvesting of produce and thus, cannot be relied upon. Women farmers, therefore, complement family labour with hired labour. This finding agrees with Uddin et al. (2014), who reported a negative relationship between family labour and farmers' adoption of strategies to adapt to the environmental effects of climate change in Bangladesh.

Farmers who have timely access to tractor services are 16.1 percent, 5.8 percent and 5.5 percent less likely to adopt improved variety, soil and plant health-related and other research-based adaptation strategies, respectively, vis-à-vis not adopting any research-based adaptation strategy. Most farmers interviewed asserted that they have limited financial resources for farming activities such as ploughing, harrowing, hiring of labour and purchasing of farm inputs (i.e., seeds, inorganic fertilizer, herbicides). When they are able to access the services of a tractor to plough their farmland, they are normally unable to adopt any other research-based adaptation strategy that requires expenditure.

The likelihood of a smallholder woman rice farmer adopting soil and plant health-related strategies increases by 21.8 percent if the farm size is expanded by one hectare. Larger farms have higher costs, and for farmers to achieve value for money, they need to ensure that the soil is fertile and the plants are healthy by applying inorganic fertilizer, compost, ploughing back rice straw after harvest, and spraying herbicides and insecticides on their rice plants. This finding is consistent with Gbetibouo (2009), who opined that large-scale farmers are more likely to adopt soil and plant health-related strategies since they have the capital and resources to easily invest in technologies which require high capital investments.

A smallholder woman rice farmer who has access to credit from formal financial institutions is about 12 percent unlikely to adopt husbandry practices strategies relative to not adopting any research-based strategy. It was discovered from women rice farmers during the focus group discussions that farming is very risky, and when given formal credit, they will prefer to invest in rice processing and other non-farm activities rather than hiring labour to plant or harvest farm produce at recommended times, paying for the harrowing on their farm or creating earth bunds as strategies for adapting to climatic stresses. This finding agrees with Obayelu et al. (2014), who found that smallholder farmers in Ekiti State of

Nigeria prefer to diversify to non-farm activities relative to adopting soil and conservation measures when given credit.

The probability of adopting an improved variety strategy and husbandry practices rather than not adopting any research-based adaptation strategy decrease by 5 percent and 13 percent, respectively, if a farmer has noticed a decrease in rainfall. This is contrary to the a priori expectation. The reason for this may be that with limited resources and noticing a decrease in rainfall, women will be unwilling to buy improved rice varieties (which are normally priced higher) or pay for a harrowing service as adaptation measures since they are not certain of their effectiveness. They may prefer to diversify rather than just relying on the farm produce.

The probability of adopting husbandry practices over not adopting any research-based adaptation strategies decreases by 19.4 percent if a woman rice farmer notices an increase in temperature. This was not the a priori expectation of this research. The finding is, however, consistent with the finding of Etwire et al. (2013b), who observed that farmers in northern Ghana are less likely to adopt improved varieties and other recommended agricultural practices if they noticed unpredictable or high temperatures.

A female farmer who is a member of at least one farmer-based group or organisation is 22.1 percent and 8.8 percent more likely to adopt improved variety and soil and plant health-related strategies, respectively, relative to not adopting any research-based adaptation strategy. Organised women farmers in the form of interest groups receive more assistance from governmental organisations (e.g., MoFA) and other agriculture-based NGOs. Such assistance includes introductions of new crop varieties, training on modern agronomic practices and access to credit. This no doubt increases the likelihood of the farmers' adoption of strategies such as early maturing, drought/flood tolerant and high-yielding varieties, ploughing back rice straw after harvest, and the use of compost, inorganic fertilizer and herbicides.

A smallholder woman rice farmer is 66 percent and 4 percent more likely to adopt soil and plant health-related strategies and other research-based adaptation strategies, respectively, than not adopting any research-based adaptation strategies in the Northern Region. Information on the climatic conditions such as patterns relating to the start and end times of seasonal rainfall will assist women farmers to make informed decisions about reducing farm size, irrigating their farms, and applying inorganic fertilizer, compost or herbicide on their farms. Obayelu et al. (2014) also found that farmers with access to information on climatic conditions are more likely to use improved varieties, reduce farm size, and diversify to other non-farm activities.

By location, a women rice farmer in the Savelugu/Nanton Municipality is 19 percent more likely to adopt other research-based adaptation strategies vis-à-vis not adopting any research-based adaptation strategy than a farmer in the West Mamprusi District. The reason for this may be that women farmers in Savelugu/Nanton Municipality, due to their proximity to Tamale (the Northern Region capital), have greater access to improved technologies and are more concerned with wild fires, reducing their farms to more manageable sizes, and using the few water bodies available to water their food crops. This finding agrees with Mensah-Bonsu et al. (2011) and Morris et al. (2012). It is also consistent with Al-Hassan et al. (2013), who reported a positive relationship between farm location and decisions to adopt Indigenous adaptation strategies among farmers in northern Ghana.

9.4 Conclusions and Recommendations

It is conclusive from the findings of this study that:

- Whereas soil-related strategies (i.e., loosening of soil, creating bunds or drainage channels, and using organic fertilizer) are the predominant Indigenous adaptation strategies; improved variety strategies (i.e., high-yielding varieties, early-maturing varieties and drought/flood-resistant varieties) are the most used research-based adaptation strategies by women rice farmers in the Northern Region of Ghana.
- The choice of Indigenous adaptation strategies among women rice farmers is significantly determined by ownership of farmland, farmer-farmer extension contact, distance to market, access to informal credit from friends and relatives, farming experience, farm size, number of economic activities engaged in by a farmer, and family labour.
- The factors influencing the adoption of research-based adaptation strategies are the age of a farmer, participation in decision-making, farming experience, family labour, access to tractor services, access to formal credit, noticing decreased rainfall and increased temperatures, membership of a farmer-based organization, access to weather information, and farmers' location.

This study provides the following recommendations:

- Community markets should be established in the various communities by the region's Local Assemblies in collaboration with community leaders to ensure easy access to markets by smallholder women rice farmers to sell their farm produce and purchase other farm inputs. This will abate the high transportation cost farmers incur in travelling to the town markets where prices for their farm produce are normally low.

- To foster higher adoption of research-based adaptation strategies in response to climate variability, smallholder women rice farmers should be particularly encouraged by the government, research institutions and other agriculture-based NGOs to participate in all stages of adaptation projects or initiatives.
- Through the Youth in Agriculture Policy, the government should make the agricultural sector more attractive to the youth in the Northern Region by intensifying youth training on modern agricultural technologies, agribusiness value-adding, accessing start-up capital and linking farmers to the appropriate markets.
- The Ghana Meteorological Service in both study districts should be better equipped for accurate weather forecasting, and through the National Disaster Management Organisation (NADMO) and the Information Service Department, provide smallholder women rice farmers with reliable and community-specific information on seasonal rainfall patterns to enable them to make informed farming decisions to adapt to climate variability.

Acknowledgments

This study was partly funded by the A. G. Leventis Foundation Scholarship Scheme, College of Basic and Applied Science, University of Ghana, Legon.

References

Adebayo, F. A., and Adewuni, M. O. (2008) Profitability and technical efficiency of sweet potato production in Nigeria. *Journal of Rural Development,* 31(3), 105-120.

Adger, W. N., Huq, S., Brown, K., Conway, D., and Hulme, M. (2003) Adaptation to climate change in the developing world. *Progress in Development Studies,* 3(3), 179-195.

Ajibefun, I. A., and Abdul Kadri, A. O. (2004) Impact of size of farm operation on resource use efficiency in small scale farming: Evidence from Southwestern Nigeria. *Journal of Food, Agriculture and Environment,* 2, 359 - 364.

Al-Hassan, R. M., Kuwornu, J. K.M., and Etwire, P. M. (2013) Determinants of Choice of Indigenous Climate Related Strategies by Smallholder Farmers in Northern Ghana. *British Journal of Environment & Climate change,* 3(2), 172-187.

Amikuzuno, J., and Donkoh, S. A. (2012) Climate variability and yield of major staple food crops in Northern Ghana. *African Crop Science Journal,* 20(2), 349-360.

Asante, F. A., and Amuakwa-Mensah, F. (2014) Climate change and variability in Ghana: Stocktaking. *Climate , 2015(3),* 78-99.

Awotide, B. A., Diagne, A., and Omonona, B. T. (2012) Impact of Agricultural Technology Adoption onSustainable Rice Productivity and Rural Farmers' Welfare in Nigeria: A Local Average Treatment Effect (LATE) Technique. *A Paper Prepared for Presentation at the African Economic Conference.* October 20 - November 2, 2012. Kigali, Ruwanda.

Bawakyillenuo, S., Yaro, J. A., and Teye, J. (2016) Exploring the automous adaptation strategies to climate change and climate variability in selected villages in rural northern savannah zone of Ghana. *Local Envoronment,* 21(3), 361-382.

Blackden, M., and Wodon, Q. (2006) Gender Time Use, and Poverty in Sub-Saharan Africa. *World Bank Working Paper, 73.*

Deressa, T. T., Hassan, R. M., and Ringler, C. (2010) Perception of and adaptation to climate change by farmers in the Nile basin of Ethopia. *The Journal of Agricultural Science,* 149(01), 23-31.

Deressa, T. T., Hassan, R. M., Ringler, C., Alemu, T., and Yesuf, M. (2009) Determinants of farmers' choice of adaptation methods to climate change in the Nile Basin of Ethopia. *Global Environmental Change,* 19(2), 248-255.

Doss, C. R., and Morris, M. L. (2000) How does gender affect the adoption of agricultural innovations? The case of improved maize technology in Ghana. *Agricultural Economics,* 25, 27-39.

Etwire, P. M., Al-Hassan, R. M., Kuwornu J, K. M., and Osei-Owusu, Y. (2013b) Smallholder farmers' adoption of technologies for adaptation to climate change in Northern Ghana. *Journal of Agricultural Extension and Rural Development,* 5(6), 121-129.

Etwire, P. M., Al-Hassan, R. M., Kuwornu, J. K., and Osei-Owusu, Y. (2013a) Application of the Livelihood Vulnerability Index in Assessing Vulnerability to Climate Change and Variability In Northern Ghana. *Journal of Environment and Earth Science, 3(2),* 157-170.

Fafchamps, M. (1992) Cash crop production, food price volatility, and rural market integration in the third world. *American Journal of Agricultural Economics,* 74(1), 90 - 99.

Feder, G., Just, R. E., and Zilberman, D. (1985) Adoption of agricultural innovation in developing countries: A survey. *Economic Development and Cultural Change,* 33(2), 255-298.

Food Agruculrure and Natural Resources Policy Analysis (FANRPAN) (2011) Measuring Vulnerability - Challenges and Opportunities. *FANRPAN, 2(11).*

Food and Agricultural organisation, FAO (2011) *FAO-Adapt Framework Programme on climate change Adaptation.* Rome, Italy.

Fosu-Mensah, B., MacCarthy, D., Vlek, P., and Safo, E. (2012) Simulating impact of seasonal climatic variation on the response of maize (Zea mays L.) to inorganic fertilizer in semi-humid Ghana. *Nutrient Cycling in Agroecosystems*, 94(2-3), 255-271.

Gbetibouo, G. A. (2009) Understanding Farmers' Perceptions and Adaptations to Climate Change and Variability: The Case of the Limpopo Basin, South Africa. *International Food Policy Research Institute, IFPRI Discussion Paper 00849.*

Greene, W. H. (2003) *Econometric Analysis.* Pearson Education, India.

Ghana Statistical Service, GSS. (2012) *2010 Population and Housing Census.* Analytical Report, Northern Region.

Ghana Statistical Service, GSS. (2014) *Ghana Living Standards Survey Round 6 (GLSS 6).* Poverty Profile in Ghana (2005-2013), Accra, Ghana.

Harvey, C. A., Rakotobe, Z. L., Rao, N. S., Dave, R., Razafimahatratra, H., Rabarijohn, H. R., et al. (2014) Extreme vulnerability of smallholder farmers to agricultural risks and climate change in Madagascar. *Philosophical Transactions of the Roral Society B, 369,* 20130089.

Hassan, R., and Nhemachena, C. (2008) Determinants of African Farmers' Strategies for Adapting to Climate Change: Multinomial Choice Analysis. *Journal of Agricultural and Resource Economics,* 2, 83-104.

Inter-Governmental Panel on Climate Change, IPCC (2007) *Climate Change 2007: Sythesis Report.Contribution of Working Groups I, II and III to the Fourth Assessment Report of the Intergovernmental Panel on Climate Change.* In: IPCC, 104.

IPCC (2014) *Climate Change: Impact, Adaptation and Vulnerability. Contributions of Working Groups I, II nad III to the Fourth Assessment Report.* Cambridge University Press, Cambridge, UK.

Kankam-Yeboah, K., Adjei, K. A., Ren, L., Appiah-Adjei, E. K., and Agyapong, A. A. (2011) Validation of TRMM Data in the Black Volta Basin of Ghana. *Journal of Hydrologic Engineering,* 17(5), 647-654.

Koru, B. and Holden, S. (2008) Differences in maize productivity between male- and female-headed households in Uganda. *Ethopiopian Development research Institute, Email: betye97@yahoo.com.*

Maddison, W. P. (2006) Confounding asymmetries in evolutionary diversification and character change. *Climate Change and Indigenous People in the United States,* 93-106.

Mandleni, B. K., and Anim, F. D. (2011) Climate Change Awareness and Decision on Adaptation Measures by Livestock Farmers. *85th Annual Conference of the Agricultural Economics Society, Warwick University,* (p. 26).

Mazvimavi, K., Nyamangara, J., Mashingaidze, N., Masvaya, E. N., Nyengerai, K., Kunzekweguta, M., et al. (2014) Weed growth and labor demand under hand-hoe based reduced tillage in smallholder farmer's field in Zimbabwe. *Agriculture, Ecosystems and Environment,* 187, 146 - 154.

Mbakahya, G. M. and Ndiema, A. C. (2015) Farming households' vulnerability and resilience to climate change in Nambale Sub-Country of Kenya. *International Journal of Science, environment and technology,* 4(6), 1608 - 1617.

Mensah-Bonsu, A., Sarpong, D. B., Al-Hassan, R., Asuming-Brempong, S., Egyir, I., Kuwornu, J. K. M., et al. (2011) Technology Adopyion and Land and Water Management Practices among Maize Farmers in Ghana.

Minia, Z. (2004) Climate scenarios developed for climate change impacts assessment in Ghana. *The Netherlands climate Change Studies Assistance Programme (NCCSAP)*, Phase 2 - Part I.

Ministry of Food and Agriculture (MoFA). (2009) *National Rice Development Strategy.* Accra: Ministry of Food and Agriculture.

MoFA. (2011) *Agriculture in Ghana: Facts and Figures.* Accra, Ghana: MoFA.

Ministry of Food and Agriculture, MoFA. (2014) *2014 Agric Sector Annual Progress Report.* Accra, Ghana: MoFA.

Ministry of Food and Agriculture, MoFA. (2015) Statistics, Research and Information Directorate, Accra, Ghana.

Morris, M. L., Tripp, R., and Dankyi, A. A. (2012) Adoption and Impacts of Improved Maize Production Technology: A Case Study of the Ghana grains Development Project. *Economics Program Paper 99-01.* Mexico, DF: CIMMYT.

Obayelu, O. A., Adepoju, O. A., and Idowu, T. (2014) Factors influencing farmers' choices of adaptation to climate change in Ekiti State, Nigeria. *Journal of Agriculture and Environment for International Development,* 108(1), 3-16.

Peterman, A., Quisumbing, A., Behrman, J., and Nkonya, E. (2010) Uderstanding gender differences in agricultural productivity in Uganda and Nigeria: Citeseer.

Quisumbing, A. R. (2007) Investment, bequest, and public policy: International transfers and the escape from poverty. *Chronic Poverty Research Center Working Papers (98).*

Uddin, M. N., Bokelmann, W., and Entsminger, J. S. (2014) Factors Affeccting Farmers' Adaptation Strategies to Environmental Degradation and Climate Effects: A Farm Level Study in Bangladesh. *Climate,* 2, 223-241.

World Bank (2008) *Biodiversity, Climate Change and Adaptation. Nature-Based Solution fromthe World Bank Portfolio.* . 1818 H Street NW Washington DC, USA. P102.

World Bank (2010) *Economics of Adaptation to Climate Change: Social synthesis Report.* 1818 H Street NW, Washington DC, USA.

CHAPTER 10

Perceptions and determinants of the adoption of indigenous strategies for adaptation to climate change: Evidence from smallholder livestock farmers in North-West Ghana

Mohammed Tiyumtaba Shaibu[1], Suhiyini Issah Alhassan[2], Franklin Kodzo Avornyo[1], Elaine Tweneboah Lawson[3], Adelina Mensah[3], Christopher Gordon[3]

[1]Council for Scientific and Industrial Research – Animal Research Institute, Tamale, Ghana

[2]Department of Agricultural Economics and Agribusiness, University of Ghana, Legon, Accra

[3]Institute for Environment and Sanitation Studies, University of Ghana, Legon, Accra

Abstract

Given the challenges that climate variability and change brings to bear, farmers adopt certain measures to adapt to their environment. Some of these measures include planting certain tree species for livestock feeding and the creation of fencing material to kraal livestock by planting shady trees. This also provides a micro-climate for the wellbeing of insect fauna, which helps to build soil properties such as through the decomposition of organic materials and retention of soil moisture. This paper used both qualitative data derived from focus group discussions in eight communities in north-west Ghana to understand livestock farmers' Indigenous perceptions of climate change, and quantitative methods (i.e., a multinomial logit model) to analyze the factors that influence their adoption of Indigenous climate change adaptation strategies. A semi-structured questionnaire was administered to 200 randomly selected livestock farmers in two districts of the Upper West Region. Four categories of Indigenous adaptation strategy were identified. These are feed-, health-, house- and breed-related strategies. The health-related strategy was used as a base category. The results from the focus group discussions revealed that farmers have used the occurrence of thunder and lightning, fruiting time of certain tree species, germination of certain grass species during certain periods of the year, the appearance of certain insect and bird species, the nature of the moon and the appearance of rainbows to estimate the local changes in climate. Further empirical results showed that the factors that statistically and significantly influenced the decision by smallholder livestock farmers to adopt Indigenous adaptation strategies were climate variability, number of extension contacts, time to market, access to credit, flock size, household size and number of livestock types. Based on these findings, policies that will enhance farmers' knowledge of climate variability should be encouraged. Also, farmers should be encouraged to intensify the diversification of the livestock species they rear.

Keywords: Adoption; climate change; Indigenous adaptation strategies; multinomial logit model; north-west Ghana

10.1 Introduction

Climate change has a heavy impact on resources that are highly sensitive to environmental conditions, and agriculture is one such sector (Barnett et al., 2007). There is global consensus about the adverse effects of climate change on agriculture and food production (Gatiso, 2015). The contribution of livestock to the economy of Ghana cannot be over emphasized. In 2012, the contribution of livestock to Ghana's agricultural GDP was 7.5% (MOFA, 2013). Findings like those of Rotter and van de Geijn (1999), Houghton et al. (2001), Seo and Mendelsohn (2006), Nardone et al. (2010) and Upadhyay et al. (2016) have reported negative impacts of climate change on livestock production, especially in semi-arid areas like north-west Ghana.

Due to the impacts of climate change, there is growing interest in farmer perceptions and local knowledge about climate change. Farmers adopt conservation practices that preserve the environment and the species and breeds of livestock they rear. After perceiving changing climate, farmers are often unable to afford the costs of introduced adaptation strategies and therefore rely on traditional measures as ways of adapting to the adverse effects of climate change and variability.

The traditional or Indigenous measures adopted cut across the livestock production sector to include feeding, health, breeding and housing strategies. An Indigenous adaptation strategy is a type of strategy developed and used by local farmers to adapt to perturbations of climate change (Kihila, 2018), while an introduced strategy is the improvement being made to farmers' Indigenous strategies by development partners who then encourage the farmers to adopt the improved strategy (Tabbo and Amadou, 2017). Adoptions of these adaptation strategies are therefore influenced by local perceptions about the existence of climate change and a host of other factors, and Nhemachena and Hassan (2008), Deressa et al. (2010), Baffoe-Asare et al. (2013), Ngombe et al. (2014) and Ghimire et al. (2015) have elaborated several factors that affect adoption decisions. However, most of these factors have been skewed towards crop farmers, without considering their relationship with livestock farmers. Motivated by this research gap, this study, having identified the Indigenous livestock strategies, examines the Indigenous perceptions of climate change and the factors that are responsible for the adoption of the identified Indigenous adaptation strategies.

10.2 Materials and Methods

10.2.1 Study area, sampling and data

Data used in this study was collected through primary and secondary sources. Primary data was collected through focus group discussions (FGDs) and individual farmer interviews. Prior to this, a reconnaissance survey was conducted in some selected communities within Ghana's Upper West Region. This provided a platform where farmers were engaged to provide information about their farm activities. The visited communities were Metor, Naburnye, Orbilli and Kasalgre in the Lawra District, and Zedung, Wallanteng, Zimoupare and Segru in the Nandom District. These communities were selected because they are locations in which Climate Change, Agriculture and Food Security (CCAFS) projects are being undertaken.

A multi-stage sampling technique was applied to sample respondents for this study. The two districts (i.e., Lawra and Nandom) in the Upper West Region were purposively selected because an intervention known as the 'Adaptation at Scale in Semi-Arid Regions' (ASSAR) project was being implemented in these districts. Each district was then stratified into four strata. The basis of using stratified sampling was to group the CCAFS communities, from which the sample communities would be chosen so that one community would be randomly selected from each group. A simple random sampling technique was used to select one community from each stratum, making a total of eight (8) communities. Again, a simple random sampling technique was used to disproportionately select 25 smallholder livestock farmers from within each community, giving a total of two hundred (200) livestock farmers. A semi-structured questionnaire was administered to each of these 200 smallholder livestock farmers.

The purpose of the FGDs was to obtain information on local perceptions and knowledge on climate change and identify livestock-related climate change and variability adaptation strategies being adopted by farmers. Prior to the start of the FGDs, a team comprising a facilitator, a note-taker and an observer was trained on how to obtain information during the FGDs. There was also a meeting with key people in the communities to inform them of the purpose of the research. Community mobilisation was conducted through which the team invited 10 male and 10 female livestock farmers to engage in the FGDs. The FGDs were then conducted for each group of male and female participants in the eight (8) communities.

10.2.2 Theoretical framework

The theoretical framework underpinning this study is the utility theory. This is important because it provides an accurate understanding of the probabilities. It also motivates and makes distinctions among the alternative model specifications. Finally, it provides the theoretical basis for calculation of changes in consumer surplus from changes in the attributes of the alternatives. This theoretical framework, which is based on the random utility model as specified by Greene (2003), is specified as follows:

$$U = X\beta + \varepsilon \qquad (10.1)$$

Where U denotes a farmer's decision to adopt an adaptive strategy;
X = the explanatory variables;
β = the parameter to be estimated; and ε = the error term.

It is assumed that Y_a and Y_b are smallholder livestock farmers' utility of two choices, which can be represented as $U^a$and U^b. For example, U^a could be the utility derived from Feed-Related Strategies, while U^b could represent the utility derived from Health-Related Strategies. The choice by the farmer between the two strategies indicates which one provides a higher utility; the farmer's utility is, however, latent. Hence, the observed indicator is equal to one (10.1) if $U^a > U^b$ and zero (0) if $U^a \leq U^b$.

A common formulation is the linear random utility model, specified as:

$$U^a = X'\beta_a + \ell_a \quad (10.2)$$

$$U^b = X'\beta_b + \ell_b \quad (10.3)$$

Factors involving choices are obtained by using multinomial models. There are several types of such models, but for ease of computation, this study adopted the multinomial logit model.

Following Greene (2003), a smallholder livestock farmer choosing, for example, an Indigenous Feed-Related strategy, is modeled to be a function of three main characteristics: socioeconomic, institutional and environmental characteristics, as shown in equation (10.1).

That is, the probability that an ith smallholder livestock farmer chooses a jth climate change adaptation strategy is given as:

$$\Pr ob(Y_i = 1) = \frac{e^{\beta_j X_i}}{\Sigma_{k=1}^{3} e^{\beta_k X_i}}, \quad j=0,1,\dots\dots\dots3 \quad (10.4)$$

Where β_j is a vector of coefficients; X_i's are the exogenous variables,

Assuming $\beta_0 = 0$, equation (10.4) will be normalized to remove indeterminacy, and the probability estimated as follows:

$$\Pr ob(Y_i = 1|X_1) = \frac{e^{\beta_j X_i}}{1 + \sum_{k=0}^{3} e^{\beta_k X_i}} \quad j= 0,2,\dots\dots J, \beta_0 = 0. \quad (10.5)$$

Estimating Equation (10.5) using the maximum likelihood method yields

the log-odds ratio presented in Equation (10.6):

$$In\left(\frac{P_{ij}}{P_{ik}}\right) = X_i(\beta_j - \beta_k) = X_i\beta_j, \text{ if } k = 0. \quad (10.6)$$

Where P_{ij}= Maximum utility that the ith smallholder livestock farmer gains in choosing the jth adaptation strategy over the kth adaptation strategy.

The choice of any climate change adaptation strategy is, therefore, the log-odds in relation to the base alternative, non-adopters. According to Greene (2003), the coefficients of the multinomial logit model are difficult to interpret, and so associating β_j with the jth outcome is tempting but misleading. Instead, the marginal effects are usually derived to explain the effects of the exogenous variables on the endogenous variable in terms of probabilities, as presented in Equation (10.7):

$$\frac{\partial P_j}{\partial X_i} = P_j\left[\beta_j - \sum_{K=0}^{J} P_k \beta_j\right] = P_j\left(\beta_j - \overline{\beta}\right) \tag{10.7}$$

The marginal effects measure the expected change in the likelihood of choice of a particular climate-related strategy with respect to a unit change in an exogenous variable (Greene, 2003).

10.2.3 The empirical model

Probit and logit models are the most common models used by researchers to identify and quantify the effects of factors that influence the adoption of agricultural technologies (Hausman and Wise, 1978; Wu and Babcock, 1998); this is due to the nature of the dependent variable. The dependent variable can take on two or more variables. If it takes two variables, then binary logit or probit models are appropriate. The multinomial logit (MNL) or multinomial probit (MNP) is the preferred model to use over the binary logit or probit models if the dependent variable takes on more than two variables.

We considered the MNL model because there is a high probability of farmers adopting more than two strategies in the study area. The MNL model is chosen for this study over the MNP model because it is widely used in studies involving multiple choices, and it is also easier to compute (Hassan and Nhemachena, 2008; Karki and Bauer, 2004).

The multivariate analysis involved the use of the MNL model to analyze the predictors of climate change perceptions (Teye et al., 2015). The MNL model was used to determine the factors that influenced the adoption of climate-related technologies introduced by research (Etwire et al., 2013). Deressa et al. (2010) adopted the MNL model to analyze the factors affecting the choice of coping strategies in response to extreme climate events in the Nile Basin of Ethiopia. According to these authors, the advantage of the MNL model is that it permits the analysis of decisions across more than two categories, allowing the determination of choice probabilities for different categories.

Kurukulasuriya and Mendelsohn (2006) employed the MNL model to determine if crop choice by farmers is climate sensitive. Similarly, Seo and Mendelsohn (2006) used the MNL model to analyze how livestock species choice is climate sensitive.

Therefore, in this study, the dependent variables are the adaptation strategies identified, which are unordered to justify the use of ordered logit and probit models, and hence the use of the MNL model is justified. The strategies adopted by smallholder livestock farmers in adapting to climate change and variability are broadly categorized into five (5) main groups, namely: Feed-Related Strategies (FRS), Health-Related Strategies (HRS), Housing-Related Strategies (HRS) and Breed-Related Strategies (BRS). The fifth category worth considering is the category of livestock farmers who did not adopt any adaptation strategy, and who are usually referred to as Non-adopters (NoA). Each of these strategies was identified by reviewing the literature and conducting the FGDs and key informant interviews at the community level. Each farmer was then asked to list the adaptation strategies they use in their farming activities to ensure effective livestock improvement in the face of climate change. A farmer may be using more than one strategy at a time. Descriptive statistics (in the form of a pie chart) were used to show the percentage of farmers using each strategy.

The empirical model specification is therefore specified as:

$$Y_{ij} = \beta_0 + \beta_1 CV + \beta_2 ExtCts + \beta_3 TMkt$$
$$+ \beta_4 Crdt + \beta_5 FrmSZ + \beta_6 LvskD$$
$$+ \beta_7 HSZ + \beta_8 EDUC + \varepsilon \tag{10.8}$$

Where i = smallholder livestock farmer and j= adaptation strategies.

10.3 Results and Discussions

The four main strategies that were identified during the research include feed-, health-, breeding- and housing-related strategies. Under the Indigenous feeding strategies, because of climate change, feed has become very scarce, so farmers have adapted to the changing climate by initially giving only small quantities of feed to their animals. They then start to add additional small quantities of the feed as the livestock finish eating. This method is adopted to prevent feed wastage.

To store feed, sheds are usually constructed in front of the farmers' house so that lots of feed can be collected during the end part of the wet season and stored on the top of sheds for dry season feeding.

Regarding Indigenous health-related strategies, farmers treat wounds on animals by applying a solution of sooth on the affected area (Table 10.1). Some farmers even use the sooth via drenching to prevent diarrhoea among livestock.

Regarding the breeding-related strategies, the farmers find it very easy to adopt and manage the Indigenous livestock species (i.e., West African Dwarf goat/sheep and West African Short-Horn cattle) because of their adaptability to the West African climate. As a result of feed scarcity in recent times, one of the Indigenous breeding strategies applied by farmers to adapt to climate change is reducing the numbers of their flock to levels that can be easily managed. They do this by selling off some of their herd. Table 10.1 provides details of the broad strategies and their sub-strategies.

Table 10. 1: Identified Indigenous climate change adaptation strategies

Type of Strategy	Description of Strategy/sub-strategies
Indigenous Feeding	➢ Adjustment of the quantity of feed given to animals ➢ Use of agricultural by-products such as yam and cassava peels, groundnut haulms, soya bean haulms, leaves of fig tree as livestock feed (Konlan *et al.*, 2014). ➢ Use of fruits from *Faidherbia albida* (goat biscuit tree) ➢ Storing of feed on top of locally-constructed sheds for dry season feeding
Indigenous Health	The use and application of traditional medicines for the treatment and vaccination of animals: ➢ Wounds of livestock can be treated by using barks of certain tree species (Khan *et al.*, 2013). ➢ Wounds are also treated by applying a solution of "duulang". Duulang is a local name given to accumulated soot (sooth). The sooth is usually harvested from the walls of the kitchen. Since the source of energy for cooking is usually from firewood, soot can easily gather on the kitchen walls. ➢ Certain types of tree leaves (locally called bagna) when pounded into a paste and mixed with water can be used to drench livestock to control diarrhoea. ➢ Leaves of "gorgor", a type of tree, is use as a de-wormer to de-worm livestock.
Indigenous Breeding	➢ Indigenous livestock breeds were identified as local breeds and the introduced breeds as improved (i.e., Fulani type) breeds ➢ Keeping a herd size that is easy to manage
Indigenous Housing	An Indigenous house/pen has a: ➢ A pen without a yard ➢ with no window or opening to allow for ventilation ➢ Small entrance

Other efforts being made to adapt to climate change include the prevention of bushfire and the performance of traditional prayer for the rains to come. Traditional prayer is in the form of consulting the gods and asking the gods for good rains and bumper harvests.

10.3.1. Indigenous perceptions of climate change

10.3.1.1 Local knowledge on climate change

Participants in the FGDs were asked about their knowledge of climate change and variability. Their confirmations that climate change and variability does exist were based on shifts in the onset and duration of rainfall, change in timings and irregular fruiting of certain tree species, stronger winds in recent years compared to some 20 years ago, drought and sunshine intensity. The participants postulated that, about some 20-30 years ago, the rains began in February/March and ended in October/November, but now they start in May/June and end in September/October.

Seasonal variability of rainfall over recent years shows that the rains now stop earlier than previously. Though located in a semi-arid environment, the study area used to observe substantial rainfall in November, but for more than fifteen (15) years now, it has not rained in the month of November. Also, there are lots of stronger winds now due to rampant deforestation and fewer afforestation activities. This exposes the environment to stronger winds. The strong winds which used to occur at the beginning of the rainy season (i.e., in March) now occur two to three months into the rainy season.

In addition, average drought conditions used to last for at the most two (2) weeks, but now it is very common for droughts to last as long as two (2) months. The participants also revealed that sunshine intensity and temperature are both on the increase.

Irregular fruiting of trees nowadays is also a sign of the changes in climate. Common among the irregularly-fruiting trees are mangoes and ackee apple *(Blighia sapida)*. In the study area, the ackee apple is locally known as "Achiapoo" (Figure 10.1). This tree used to fruit in March but now it fruits in October, which is a striking change in fruiting time.

Figure 10. 1: Fruits of Achiapo Tree

Source: Field Survey (2016)

Depletion of soil fertility and loss of biodiversity were other confirmations that the climate has changed in the study area. Traditionally, the local farmers did not apply fertilizer to their crops, but over the past 30 years, it has become increasingly common to have a low harvest if farmers do not apply fertilizer.

Another important observation of climate change by farmers in the study area is the onset of harmattan. Harmattan is the hot, dry and dusty wind that blows from north-eastern parts of the West African Sahara, causing dry, cracked skin on the human body. It coincides with usually very cold early mornings (i.e., 9°C) before the temperature steadily rises to 30°C during the day. The study revealed that some 20-30 years ago, the harmattan period used to be from January to February, but it now sets in from late October or November and persists until the end of February or early March.

10.3.1.2 Local knowledge or perceptions that indicate or provide a forecast of future climate

This analysis was based on six indicators identified during the FGDs. These include thunder and lightning, germination of certain grass species, the appearance of certain insect and bird species, the nature of the moon and the appearance of rainbows.

The first local knowledge indicator of the existence of climate change is the occurrence of thunder and lightning. In the past, thunder and lightning were used to indicate that millet was about to mature. This observation is still being used by crop-livestock farmers in the study area. Also, when thunder and lightning occur, it is spiritually aligned to harming/striking someone having done something bad, and which is culturally unacceptable in the society. It was also revealed that thunder and lightning could be used to determine whether the rains will cut short or persist longer.

Secondly, the germination of "kerekere grass" indicates that there will be a good harvest that year. In the past, the germination of Makumpagsule, Gemiea, Miea and bolo grasses showed there would be high rainfall. The grass germinates with the onset of the first rainfall. On the other hand, the appearance of the grass species "duurong duurong" shows that the dry season is imminent.

Furthermore, the appearance of grasshoppers at the beginning of the rains is an indication of good rainfall. Also, types of birds known locally as Chipele, Zier and Pempele appearing prior to the onset of rains indicates that the rains will start early and there will be more rains during the season. White birds that follow cattle signify that the rainy season is imminent. If a type of bird called "Dachir" is seen at the beginning of the rainy season, this indicates there will be bumper harvest of maize.

In addition, the appearance of an insect called "Zulu" signifies an onset of early rainfall and that there will be a longer rainy season. This method is still being used by the people. The appearance of other insect species that signify rainfall includes Pachi, fiifi and ants from termitarium. However, the appearance of butterfly-like insects called "Walpempele" and "Tantulu" shows that there will be low rainfall. Also, when an insect called "Brugmanyili" is seen, it shows that the dry season is imminent. The appearance of rainbow is also an indication of no rainfall, while also signaling that the wet season is coming to an end.

Though not scientifically proven, it has been observed that in the moon there are two objects that when observed critically, and especially at night, indicate the dry and rainy seasons. One of these objects is to the east and the other to the west. The eastern one (i.e., dry season) pushes the western one (i.e., rainy season) away and sets in as the dry season. The opposite is true when the rainy season sets in.

Finally, if the moon appears sharply and quickly at night, it shows that it is likely to rain and there will be good rainfall., In contrast, if the moon appears slowly in a curvature movement, it shows that there will be low rainfall and that it is most likely will not rain.

10.3.2 Factors influencing adoption of Indigenous strategies for climate change adaptation

There were four dependent and eight independent variables that were considered in the MNL model. The independent variables were considered for the model after a multicollinearity test was conducted using the Variance Inflation Factor (VIF) approach. This allowed us to drop some variables such as gender of the farmer and noticed a decrease in rainfall amounts. Thus, variables that did not fall below the tolerance value of less than 0.4 and were more than the VIF value of 2.5, as proposed by Allison (1999), were retained. Therefore, multicollinearity did not pose any problem to the model. The wald chi square ratio was 70.35, with the overall model being significant at 1%. The Hausman test of Independence of Irrelevant Alternatives (IIA) was conducted, and the result showed that the IIA was not violated, justifying the suitability of the MNL model for the dataset. Since the coefficients only provide the magnitude of the change in the dependent variable due to a unit change in the independent variable, the marginal effects which were considered for discussion, provides both the magnitude and direction of change. The marginal effects of the model are shown in Table 10.2. The health-related strategies were used as the base outcome. That is, any other adaptation strategy was used to compare with the health-related strategy.

Noticed climate variability, number of extension contacts, time to market, and access to credit by smallholder livestock farmers were the factors that influence the decision to adopt feed-related strategies. Also, only flock size and household size positively and negatively influenced the decision to adopt breed-related strategies, respectively. Finally, noticed climate variability, time to market and number of livestock types diversified were the factors that influenced the use of house-related strategies.

Table 10. 2: Marginal effects of factors influencing the adoption of Indigenous adaptation strategies

Independent Variables	Feed-Related Strategies	P-value	Breed-Related Strategies	P-value	House-Related Strategies	P-value
Noticed climate variability	0.0744** (0.9811)	0.007	0.0221 (0.4703)	0.122	0.0987** (1.1354)	0.008
Number of extension contacts	-0.0402*** (-0.4334)	0.000	-0.0075 (-0.1112)	0.637	0.0062 (-0.0161)	0.742
Time to market	-0.0346*** (-0.3990)	0.000	0.0089 (-0.0381)	0.672	-0.0257*** (-0.3098)	0.000
Access to credit	-0.0991** (-1.3196)	0.002	-0.0687 (-0.6179)	0.365	0.0464 (0.1751)	0.742
Farm/Flock size	-0.0039 (0.0136)	0.959	0.0291* (0.2078)	0.062	0.0038 (0.0820)	0.666
Number of livestock type diversified	-0.0055 (-0.0594)	0.659	-0.0331 (0.1711)	0.938	-0.0334* (-0.2998)	0.081
Household size	-0.0083 (-1.0736)	0.271	-0.3787*** (-2.2437)	0.000	-0.0124 (-1.1055)	0.309
Years of education	-0.0068 (-0.0767)	0.221	-0.0014 (-0.0231)	0.689	-0.0007 (-0.0216)	0.920

Base outcome: Health-Related Strategies
Number of observations: 200
Pseudo R^2: 0.138
Log Pseudo Likelihood: -258.71
Wald Chi^2: 70.35
Prob> Chi^2: 0.002

***, ** and * denote statistical significance at 1%, 5% and 10%, respectively.

*Figures in parenthesis are the coefficients of the explanatory variables

The probability of adopting the Indigenous feed-related strategies as against the Indigenous health-related strategies as a climate change adaptation strategy increases by 7.4% if a smallholder livestock farmer with past livestock rearing experience noticed climate variability. One way that the environment can be conserved is by an interplay of climate variability, pasture productivity and livestock feeding. Climate variability can lead to poor pasture production which is used as browse for livestock feeding. The effect of this could have led to farmers adopting adjustment measures when feeding livestock. If there is inadequate or a lack of feed for livestock, productivity would be affected since a significant number of livestock would die of hunger. This is contrary to findings that a smallholder farmer is about 17% less likely to adopt soil and plant health strategies if temperatures are unpredictable (Etwire et al., 2013). Unpredictability of temperatures could influence farmers to notice climate variability.

The number of extension contacts received by a livestock farmer significantly and negatively reduces the decision to adopt feed-related strategies by about 4%. The opportunity cost of spending time with an Agricultural Extension Agent (AEA) to be trained to adopt introduced feed-related strategies could be used by farmers to practice what they already know regarding feeding of their livestock such as the adjustment of the quantities of feed given to animals, and planting tree species for lean season livestock feeding. These findings are different from Etwire et al. (2013) and Hassan and Nhemachena (2008), who opined that if farmers have access to an extension service, their possibility of adopting improved breeds and varieties, mixed mono crop-livestock systems under dry land, and irrigation as an adaptation response to climate change would increase.

The average time to market reduces the probability of adopting the feed- and house-related strategies as a climate change adaptation option by 3.5% and 2.6%, respectively. The average time to market influences the decision to adopt introduced strategies. In most cases, inputs such as medications for treating livestock diseases, prepared rations or concentrated feeds are found at market centers. If the market centers are beyond the farmers' reach, they would not have access to these inputs that will help them to increase their productivity. This finding is in line with Hassan and Nhemachena (2008), who found the distance to market to have a significantly negative influence on the decision to adopt mixed-mono and multiple crop-livestock systems under irrigation as an adaptation strategy.

Access to credit reduces the probability of adopting the feed-related strategy as a climate change adaptation option by 9.9%. This can be attributed to the fact that a minority of the farmers (i.e., 20%) have access to credit. This is contrary to Hassan and Nhemachena (2008), as they opined that access to credit has a significantly positive influence on farmers' decisions to adopt mixed-mono crop-livestock systems under dryland conditions as a climate change adaptation option.

The results of this study also show that smallholder livestock farmers in the region with an average farm/flock size of eight (8) goats, for example, have a 2.9% chance of choosing the Indigenous breed-related strategies as against the Indigenous health-related strategies. A majority of the farmers (i.e., 92%) adopted the local livestock breeds and applied Indigenous practices such as the "duulang"- an accumulated soot (sooth) - in the treatment of diarrhea and wounds. This may be part of the reason for the farmers' low average livestock numbers, as compared to the national average of 12 (GSS, 2014). This is not a good practice as far as conservation and sustainable livestock production is concerned. The finding is in line with Etwire et al. (2013), and Hassan and Nhemachena (2008), who found that farm size has a significant and positive influence on the decisions by smallholder farmers in northern Ghana to adopt a recommended agricultural practice to adapt to climate change and variability such as planting different fodder trees, integrating crops with livestock production, and a mixed-mono crop-livestock system under dryland conditions, respectively.

The probability of choosing the Indigenous breed-related strategy as against the Indigenous health-related strategy for climate change adaptation reduces by about 38% as household size of the farmer increases. An increase in household size means that the needs of household members would increase, especially as school fees and hospital bills would have to be paid. Therefore, it makes it very difficult for livestock owners who are mostly the household heads to expand their production by adopting more breed-related strategies such as the introduced ones. This finding agrees with Hassan and Nhemachena (2008), who found household size to be significant, although it reduces the probability of adoption of a mixed-mono crop-livestock system cultivated/reared under dryland conditions.

The likelihood of choosing the Indigenous house-related strategy against the Indigenous health-related strategy increases by 9.9% if a smallholder livestock farmer in the region, with past livestock keeping experience, noticed climate variability. Climate variability can lead to erratic rainfall patterns, sometimes with high rainfall within a short period of weeks. This could lead to cold-stress and humidity-related diseases among livestock when they are left to sleep in the open. When this happens, farmers would prefer to provide shelter for their animals rather than providing treatment, hence the high (i.e., 88%) level of adoption of the Indigenous housing-related strategies.

The number of livestock types diversified reduces the probability of choosing the Indigenous house-related strategy as an adaptation option by 3.3% as against the Indigenous health-related strategies. In the study area, a majority of the livestock owners (i.e., 92%) rear more than one species (including cattle, sheep, goats, pigs, local fowls and guinea fowls) of livestock. However, there is usually only one room/pen available to accommodate these species, and this does not help to improve and expand production. A decrease in production is a measure of hampering the sustainability of livestock types. The semi-extensive system of livestock production in the study area lowers the adoption of improved housing strategies.

In summary, examination of the environmental and socio-economic factors resulted in the level/degree of adoption of the climate change adaptation strategies that are illustrated in Figure 10.2.

Figure 10. 2: The degree of farmers' adoption of the Indigenous climate change adaptation strategies

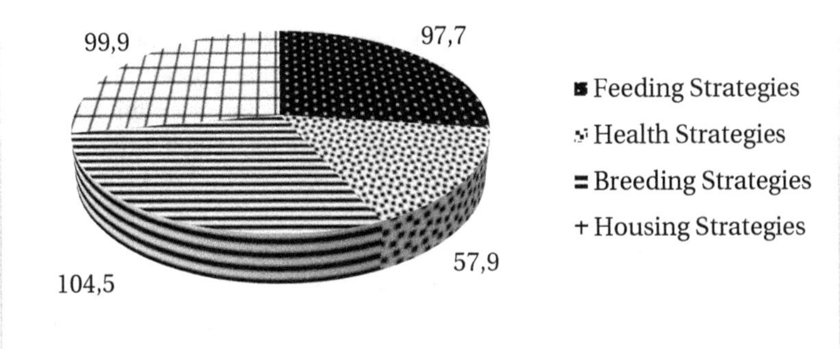

10.4 Conclusions

This study identified the perceptions of climate change and the Indigenous adaptation strategies of smallholder livestock farmers in rural farming districts of north-west Ghana. Farm households in the study area have noticed that the climate is changing due to decreasing rainfall and increasing temperatures. They have noticed these changes locally, based on the occurrence of thunder and lightning, the fruiting time of certain tree species, the germination of certain grass species during particular periods of the year, the appearance of certain insect and bird species, the nature of the moon and the appearance of rainbows. Due to their perceptions of a changing climate, adaptation measures of an Indigenous nature are being adopted. These include planting of certain tree species for livestock feeding, the use of agricultural by-products as feed for animals, the use of herbs in the treatment and vaccination of animals, the use of accumulated soot to treat diarrhoea in livestock, the use of Indigenous livestock breeds which are perceived to be more resilient to the local climate, and the use of Indigenous housing which is usually small and not spacious for the free movement of animals.

The study has further revealed that the degrees of adoption of the climate change adaptation strategies are determined by certain household, institutional and environmental factors. The MNL regression results showed that noticing climate variability, the number of extension contacts, the time to market, access to credit, flock/farm size, the number of livestock types diversified, and household size are the determinants of adopting Indigenous strategies for adaptation to climate change.

Intervention projects must be tailored to match the local Indigenous farming knowledge and associated farmer innovations in the study area. In this way, the local people will have an increased sense of pride in themselves as their innovations are being formally recognized by extension agents. This will encourage them to continue to improve upon their current farming practices. Also, farmers should be encouraged to adopt rainwater harvesting techniques both at their farms and homesteads so they can have year-round access to water for livestock watering. And last but not the least, more extension agents with backgrounds in livestock production should be recruited to help educate farmers on improved farm management practices.

Acknowledgements

The authors of this paper wish to acknowledge the funding support from two bodies. First, the "Adaptation at Scale in Semi-Arid Regions (ASSAR)" project, being implemented by the Institute for Environment and Sanitation Studies, University of Ghana, and second from the A. G. Leventis Foundation scholarship, who both partly supported this research. Also, two support (veterinary) staff of the Ministry of Food and Agriculture in the study districts deserve our acknowledgement, having contributed immensely during the data collection phase.

References

Allison, P. D. (1999) Multiple regression: A primer. Pine Forge Press.

Barnett, J., Dessai, S., and Jones, R. N. (2007) Vulnerability to Climate Variability and Change in East Timor. *Current Opinion in Environmental Sustainability*, 36(5), 372–378.

Baffoe-Asare, R., Daquah, J.A., and Annor- Frempong, F. (2013) Socioeconomic Factors Influencing Adoption of Codapec and Cocoa High-tech Technologies among Small Holder Farmers in Central Region of Ghana. *American Journal of Experimental Agriculture*, 3(2), 277–292.

Deressa, T. T., Ringler, C., and Hassan R. M. (2010) Factors Affecting the Choices of Coping Strategies for Climate Extremes: The Case of Farmers in the Nile Basin of Ethiopia, Environment and Production Technology Division, IFPRI Discussion Paper, 01032.

Etwire, P. M., Al-hassan, R. M., Kuwornu, J. K. M., and Osei-owusu, Y. (2013) Smallholder farmers ' adoption of technologies for adaptation to climate change in Northern Ghana. *Journal of Agricultural Extension and Rural Development*, 5, 121–129.

Gatiso, T. G. (2015) Effectivesness of Climate Change Adaptation Strategies: Household-Level Evidence from Ethiopia. Journal of Economics and Sustainable Development, 6(19), 99-110.

Ghimire, R., Huang W., and Poudel, M. P. (2015) Adoption Intensity of Agricultural Technology : Technology Empirical Evidence from Smallholder Maize Famers in. *International Journal of Agriculture Innnovations and Research*, 4(1), 139–146.

Greene, W. H. (2003) Econometric Analysis, Pearson Education, India.

Ghana Statistical Service, GSS. (2014) *Ghana Living Standards Survey Round 6 (GLSS 6).* Poverty Profile in Ghana (2005-2013), Accra, Ghana.

Hassan, R., and Nhemachena, C. (2008) Determinants of African Farmers' Strategies for Adapting to Climate Change: Multinomial Choice Analysis. *African Journal of Agricultural and Resource Economics, 2*(1), 83–104.

Hausman, J. A., and Wise, D. A. (1978) A Conditional Probit Model for Qualitative Choice : Discrete Decisions Recognizing Interdependence and Heterogeneous Preferences. *The Econometric Society, 46*(2), 403–426.

Houghton H. J., Ding Y., Griggs D. J., Noguer M., van der Linden P. J., Dai X., Maskell K., and Johnson, C. A. (2001) *Climate Change 2001 : The Scientific Basis.*

Karki, L. B., and Bauer, S. (2004) Technology, Adoption and Household Food Security. Analyzing Factors Determining Technology Adoption and Impact of Project Intervention: A case of smallholder peasants in Nepal. A paper presented in the Deutscher Tropentag, Humboldt University, Berlin.

Khan, R. U., Mehmood, S., Khan, S. U., Khan, A., Shah, I. A., and Bokhari, T. Z. (2013) Medicinal Value of Indigenous Flora in the Vicinity of District Bannu. *Advances in Pharmaceutical and Ethnomedicines.* 1(1), 7–14.

Kihila, J. M. (2018) Indigenous Coping and Adaptation Strategies to Climate Change of Local Communities in Tanzania: a Review. Climate and Development, 10(5), 406 - 416.

Konlan, S. P., Ayantunde, A. A., Dei, H. K., and Avornyo, F. K. (2014) Evaluation of existing d potential feed resources for ruminant production in northern Ghana.

Kurukulasuriya, P., and Mendelsohn, R. (2006) Crop Selection: Adapting to Climate Change in Africa. CEEPA Discussion Paper No. 26. Centre for Environmental Economics and Policy in Africa. Pretoria, South Africa: University of Pretoria.

MOFA. (2013) Agriculture in Ghana, Facts and Figures. Statistics, Research and Information Directorate (SRID).

Nardone, A., Ronchi, B., Lacetera, N., Ranieri, M. S., and Bernabucci, U. (2010) Effects of climate changes on animal production and sustainability of livestock systems. *Livestock Science*, 130(1-3), 57–69.

Ngombe, J., Kalinda, T., Tembo, G., and Kuntashula, E. (2014) Econometric Analysis of the Factors that Affect Adoption of Conservation Farming Practices by Smallholder Farmers in Zambia. *Journal of Sustainable Development*, 7(4), 124–138.

Nhemachena, C., and Hassan, R. M. (2008) Micro-Level Analysis of Farmers ' Adaptation to Climate Change in Southern Africa, IFPRI Discussion Paper No. 714, Washington, DC: International Food Policy Research Institute.

Rotter , R., and van de Geijn S.C. (1999) Climate change effects on plant growth, crop yield and livestock. *Climate Change*, 43,651-681.

Seo, N., and Mendelsohn, R. (2006) Climate Change Adaptation in Africa: A Microeconomic Analysis of Livestock Choice.

Tabbo, A. M., and Amadou, Z. (2017) Assessing Newly Introduced Climate Change Adaptation Strategy Packages among Rural Households: Evidence from Kaou Local Government Area, Tahoua State, Niger Republic. *Journal of Disaster Risk Studies*, 9(1), a383. https://doi.org/10.4102/jamba.v9i1.383

Teye, J. K., Yaro, J. A., and Bawakyillenuo, S. (2015) Local farmers' experiences and perceptions of climate change in the Northern Savannah zone of Ghana. *International Journal of Climate Change Strategies and Management*, 7(3), 327–347.

Upadhyay, R. C., Singh, S. V., Kumar, A., and Gupta, S. K. (2016) Impact of Climate change on Milk production of Murrah buffaloes Impact of Climate change on Milk production of Murrah buffaloes. *Italian Journal of Animal Science*, 6(sup2), 1329–1332.

Wu, J. J., and Babcock, B. A. (1998) The Choice of Tillage, Rotation and Soil Testing Practices: Economic and Environmental Implifications. *American Journal of Agricultural Economics*, 80, 494–511.

CHAPTER 11

Gender-based climate change impacts and adaptation strategies among smallholder farmers in Northern Ghana –What lessons for nationally-appropriate mitigation actions exist?

Joseph Amikuzuno, John K. M. Kuwornu, Damba, T. Osman

Abstract

If the Food and Agriculture Organization's (FAO's) evidence that women produce over 70% of sub-Saharan Africa's food is true, then targeting women during agriculture-related research, technology development and diffusion is highly appropriate. To achieve this, it is important to recognize that compared to men, women farmers have unequal access to productive resources. But, under climate change when productive resources such as fertile soils and water may become increasingly scarce, the adoption of resilient, yield-enhancing technologies for farming productivity improvements is crucial. This study analyses the gender differences in climate change impacts and adaptation by farmers in the Northern, Upper East and Upper West Regions of Ghana; and the implications of these differences for nationally-appropriate mitigation actions. Using cross-sectional data based on 800 farms in these three districts of northern Ghana, we estimate the gender-disaggregated impacts of climate change on smallholder farms using the Trade-Off Analysis for Multi-Dimensional Impact Assessment (TOA-MD) model. The immediate impacts include a reduction in crop yields by 62.7% for men and 41.8% for women farmers, and disruption in the annual cropping calendar for 25.4% of men and 30.9% of women farmers. Considering the impacts of climate change on future agricultural systems with adaptation, the results show that aggregate percentage impacts constitute about 35%, while aggregate gains, losses and net losses of average net per farm revenues are 17%, 35% and 15%, respectively. We conclude that identifying and effectively addressing the needs of women farmers requires delineating climate change adaptation policies in a gender-disaggregated way. The findings illustrate the need to prioritise women in designing adaptation strategies, and thus contribute to the gender policy discourse.

Key words: Climate change; gender; impacts; northern Ghana; smallholder farmers

11.1 Introduction

The world is faced with the problem of global climate change and its deleterious effects on the availability of already scarce natural resources. In response, various adaptation and mitigation strategies are being developed by governments, farmer-based/civil society organizations, communities and individuals to help solve and/or mitigate the adverse consequences of the climate change problem.

Women increasingly play a key role in agricultural production and sustainable food security (Carvajal-Escobar et al., 2008). It is estimated that women's equal access to productive agricultural lands could increase yields by 20-30% and result in the reduction in world hunger by 12-17% (FAO, 2011). The changing climate may have different impacts on male and female farmers. In the face of climate change, the important role of women is likely to be hindered for many reasons if the right steps are not taken to address the resource constraints already facing them.

Smallholder women farmers are already limited in terms of access to land and other yield-enhancing agricultural inputs such as fertilizers, pesticides and extension services. Climate change is likely to further aggravate this limited access due to its impacts on the availability of productive but scarce resources. Secondly, climate change adaptation and mitigation strategies are largely driven by new, advanced and often labour-intensive technologies. More women than men may be exposed to the adverse impacts of climate change due to the gender-biased nature and high labour demands of these technologies (Jost et al., 2015). Moreover, women are reportedly facing higher risks of the negative impacts of climate change due to the burden of their productive and reproductive responsibilities (Goh, 2012; Kakota et al., 2011). According to the Food and Agriculture Organization (FAO), gender and social norms as well as the responsibilities of women and their lack of access to productive resources and services such as land, general yield-enhancing inputs and extension services that are essential to adapting to climate variability and change, render them more climate-sensitive than men (FAO, 2011).

Different causal mechanisms have been proposed to explain the gender dimension of climate change adaptation globally. However, understanding of the gender dimension of climate change in agriculture in developing countries such as Ghana is limited. This is because the gender roles of women in developing country agriculture such as in Ghana are complicated and often punctuated by a mix of productive and reproductive roles whose contribution to agriculture are difficult to isolate. Predictions of the outcomes of gender dimensions of climate change on agriculture in northern Ghana are not expected to be different from the global case, as Ghanaian women's roles in agricultural production are limited by cultural and social norms which may influence their capacity to effectively access and adopt adaptation technologies and interventions.

The current literature on the gender dimensions of climate change and adaptation is supported by findings of Abdul-Razak and Kruse (2017), who revealed that compared to men, women in Ghana have significantly lower adaptive capacities in terms of economic resources, technology, and knowledge and awareness. Ravera et al. (2016) assessed the diversity of gendered adaptation strategies to climate change of Indian farmers using a feminist intersectional approach and found the presence of geographically-dependent, gendered preferences and adaptation strategies which are influenced by socio-economic and institutional factors such as age and gender. Decision-making influence, power dynamics and negotiation skills, both at the household and community levels, have also been identified as constraints to adoption among groups (Ravera et al., 2016).

Although the above studies and earlier ones by Kuwornu et al. (2013), Nantui et al. (2012), Etwire et al. (2013a) and Armah et al. (2011) have analyzed farm-level adaptation strategies and the adaptive capacities of smallholder farmers, the gender dimension of adaptation was not captured. In addition, various climate change interventions by governments are specific and often do not address the low adaptive capacities of vulnerable smallholder farmers such as women. Thus, the social and institutional dimensions of gendered climate change adaptation strategies have been neglected by policy institutions due to a lack of interventional specificity and targeted designs.

The conceptual literature is well-aware of the need to address the gender dimension of climate change adaptation and the possibility of creating gender-sensitive adaptation strategies. These concerns, however, hardly gain a foothold in current adaptation programs which tend to focus on improving production practices, crop choice and varieties, access to inputs, output markets, credits, and information and extension services for farms or farming communities without regard for the gender differences that determine the capacity of these factors in the production improvement. In Ghana, attempts have been made to bring to the fore the lessons learnt on gender-based climate change impacts and adaptation, with the aim of inculcating these lessons into nationally-appropriate mitigation actions (NAMAs). However, these attempts have not been well documented, hence the need for this study.

The main objective of this paper is to examine the gender-based climate change impacts and adaptation strategies among smallholder farmers in northern Ghana. The focus on northern Ghana is due to the region's vulnerability to drought, constant bush burning and the dependence on ever-dwindling levels of rainfall for agricultural production.

Knowledge of the gender-based vulnerability of smallholder farmers to climate change and the determinants of their adaptive capacities is critical for supporting sustainable agricultural production in northern Ghana. Specific and targeted climate change adaptation strategies applied elsewhere are sources of critical information about their strengths and weaknesses. This information can be used to develop similarly targeted adaptation strategies suited to the northern Ghana context. Such strategies will form the basis for geographically-based climate change adaptation strategies that are mindful of the adaptive capacities of vulnerable groups such as women.

An understanding of the gender dimension of climate change will provide a significant starting point for understanding the impacts of climate change in a gender-disaggregated way and shaping the outcomes of external interventions. Specifically, the results will inform Ghana's NAMAs. This will fundamentally minimise the risks of policy failure resulting from the adoption of often erroneous but popular assumptions about the different roles that women and men in sub-Saharan Africa play and must continue to play, to achieve food security in the face of climate change (Okali and Naess, 2013).

11.2 Materials and Methods

11.2.1 Literature review

Under a changing climate, gender roles and dimensions are of importance to researchers and national governments. Fundamentally, adaptation involves core gender responsibilities, which in northern Ghana determine access to land and other productive agricultural inputs. Aside from the traditional role of women in northern Ghana, some of the climate change adaptation strategies are technology-based and relegate women to the background due to their lack of access to tools and other equipment to support their adoption of preferred strategies.

Adger et al. (2007) assessed climate change adaptation practices, options, constraints and capacity, and concluded that adaptive capacity is uneven across and within societies, especially among women engaged in subsistence farming. In southern Africa, such women are unevenly burdened with the costs of recovery and coping with drought.

Damptey and Essel (2012) conducted a study on gender perspectives of climate change coping and adaptation strategies in the Kwanyaku and Keta Districts of Ghana. Their findings revealed that more men than women were aware of climate change. Also, due to gender divisions of labour in communities, they discovered differences in climate change impacts between men and women, while the access to and control of natural resources also varied by gender in communities. In Keta District, more men were landowners and were also less vulnerable to climate change impacts than women. Due to cultural norms, women lost out in activities in which they had equal access to and control over resources. Men also had more access to and control over money compared to women, and hence had more options for drought-coping strategies than women. Generally, a gender gap in access to financial resources favoring men was identified (Damptey and Essel, 2012).

Mangheni et al. (2010) tried to understand the gender dimensions of the impacts of climate change on agriculture and the adaptation among smallholder farmers in Eastern Uganda using the Minimum-Data Tradeoff Analysis (TOA-MD) model. The study generated gender-targeted disaggregated data on climate change adaptation based on agricultural, economic and environmental assessments of adaptation strategies, farmers' understanding of climate change, and their experiences with adaptation.

Jost et al. (2015) adopted the climate analogue approach that identified areas with similar future climatic conditions to identify potential adaptations in Uganda, Ghana and Bangladesh. This approach increases farmer-to-farmer learning in locations with a high degree of climate variability. Findings revealed that women are less adaptive to climate change strategies due to their financial and physical resource constraints, that men have more access to information and extension services than women, and that there was an inability of women to meet the high labour demands of existing adaptation strategies.

Reyes (2009) also studied the gender responses to El Nino conditions in rural Peru and realized that gender inequalities in agricultural activities within households were prevalent, including periods where households were food insufficient. Increased migration of men to coastal regions and cities for employment increased the number of temporarily female-headed households who faced challenges in attempting to survive. Women-headed households were not given recognition by a majority of the study region's rural community organizations. Also, the increased burden of agricultural work placed on women in the absence of men was a limitation on their ability to seek paid employment.

Nath and Behera (2011) critically reviewed the impact of adaptation to climate change in developed and developing economies, concluding that adaptation is essential to enhance resilience against increasing climate variability, especially among vulnerable groups. These authors concluded that adaptation enhances resilience against rising climate variability, and hence a systematic and critical review of the impacts of climate change and choices of adaptations within countries are required to gain insights for a comprehensive policy framework, particularly for developing countries. They highlighted that government and civil society organizations have crucial roles to play in developing and promoting efficient adaptation methods.

Perez et al. (2015) compared the agricultural and livelihood systems of male and female smallholder farmers with their productive resources, organization and access to services in nine East and West African countries. Findings revealed that women have less access than men to common property resources, including cash to obtain goods and services. Also, women control less land than men, and this is often limited to poor fertile lands with an unsecured land tenure system. Hence, women were more exposed to the high costs of climate change adaptation strategies. Perez et al. (2015) suggested a need to formulate policies and strategies that will eliminate these differences so that men and women can equally increase their resilience and better manage their changing environments.

Goh (2012) reviewed the literature on gender-differentiated impacts of climate change on women's and men's assets and their wellbeing in developing countries. The review tested two hypotheses on gender-differentiated impacts of climate variability, revealing that climate-related shocks affect women more negatively than men. The review further showed that women's and men's assets and wellbeing are affected in six areas of impact: (i) agricultural production, (ii) food security, (iii) health, (iv) water and energy resources, (v) climate-induced migration and conflict, and (vi) climate-related natural disasters. Also, women suffered more in terms of their assets and wellbeing due to the social and cultural norms towards gender roles and their general lack of access to and control of assets.

Kianjun et al. (2015) applied a randomized sampling technique to assess the gender differences in farmers' responses to climate change adaptation in Yonqiao District in China. The assessment targeted 220 households and a binary logit model was used to analyze the data. It was revealed that gender does not influence knowledge and perceptions of climate change, but gender differences in climate change adaptation measures were significant. Male-headed households were more likely than female-headed households to adopt new water conservation technology involving an increased investment in irrigation facilities. The findings further revealed that adaptation decisions of male- and female-headed households were influenced by different factors. These findings were considered to provide lessons for increasing adoption of adaptation strategies in similar geographical areas.

Huyer (2016) concluded that technologies to support smallholder farmers' resilience and adaptation to climate change is a way to promote women's empowerment and transformation of the gender relations coupled with sustainable agricultural production. But this would only be possible if the farmers' capacities were strengthened and the technologies were implemented within a framework of equal access to resources, control of assets, and equitable decision-making among women and men.

Holvoet and Inberg (2013) assessed 31 Sub-Saharan African National Adaptation Programs of Action (NAPAs) that integrated gender dimensions into the different phases of program implementation. They realized that ignoring the mediating influence of gender relations will impact on the effectiveness of climate change adaptation and mitigation policies. These authors further analyzed the extent to which women participate in diagnosis and decision-making including gender sensitivity, and found a decline in gender sensitivity throughout the intervention cycle.

Nakuja et al. (2012) applied an Ordered Logit model to assess water storage for dry season vegetable farming as an adaptation to climate change in the Upper East Region of Ghana. They found gender to be one of the factors influencing farmers' decision to participate in dry season vegetable farming because dry season vegetable farming is typically the preserve of men in this region. This further supports arguments that women are often sidelined in the design and implementation of climate change adaptation strategies (Rodenberg, 2009).

Recent climate change effects on the vulnerability of smallholder farmers in Ghana are attributed to their dependence on climate-regulated water resources. Based on this, Adu et al. (2018) assessed the vulnerability of smallholder maize farmers to climate change by employing the Livelihood Vulnerability Index (LVI), with an emphasis on access and utilization of water resources in the Brong Ahafo Region. Results revealed that the region's farming households are vulnerable to food, water and health impacts due to increasing weather variability. The adoption of adaptive capacities was related to the socio-demographic characteristics, social networks, and livelihoods of the farming households. The authors, therefore, concluded that there was a need for a broad stakeholder implementation of climate change resilience intervention projects in the region. But Saxena et al. (2016) earlier argued that resilience has since moved from a peripheral ecological concept to a central development agenda among intervening agencies, and a hindering factor is operationalizing resilience. They concluded that the next goal of climate change stakeholders will largely depend on handling the dynamic features of complex adaptive socio-ecological systems.

Kumar et al. (2016) concluded that: (i) adaptive farmers obtained higher profits than non-adaptive farmers based on differences in their model; (ii) non-adaptive farmers with <4 acres require either altering existing agricultural practices for minimum management costs and increased profit or incurring more costs for adaptation; (iii) large-scale farmers require rationalizing investments for higher profits; (iv) profits do not directly depend on the costs of adaptation; (v) income from agriculture alone is not sustainable among smallholder farmers; (vi) appropriate adaptation will likely reduce the costs of farm operations, and increase profits; and (vii) additional cost is at all times not required for adaptation but requires rationalizing the agricultural expenditure which is required for adaptation. However, these various points failed to address the gender dimensions of climate change adaptions, especially in developing countries.

Armah et al. (2013) questioned the factors that influence farmers' choice of Indigenous adaptation strategies to address agro-biodiversity loss in northern Ghana, and the findings revealed that among the identified factors, gender of the household head was a contributory factor. This was found through an MNL model analysis of solicited datasets from the farming households. Other identified factors included household size, access to credit, and farm size. To enhance adaptation, these researchers recommended effective government policies to enhance the off-farm income generating activities that women are largely engaged in.

Al-Hassan et al. (2013) applied an MNL regression model to assess the determinants of choice of Indigenous climate-related strategies by smallholder farmers in northern Ghana. Results showed that among other factors such as farmer location, farmer-to-farmer extension and a decrease in rainfall, gender roles and dimensions also indirectly come into play during farmer-to-farmer extension services. Culturally, farms owned by women were least serviced during extension visits, especially in the savanna ecological zone.

Etwire et al. (2013b) further applied the LVI in assessing vulnerability to climate change and variability in northern Ghana to 320 farm households, augmented with rainfall and temperature data. Findings identified the Northern Region as the most vulnerable in terms of the overall LVI. Lack of climate information services, illiteracy, family size and lack of access to medical care were the drivers of vulnerabilities in farming communities in the Northern Region, but these results lacked a gender dimension. In the Northern Region, smallholder women farmers are the most affected in terms of vulnerability rates since illiteracy and access to various forms of information are still major challenges.

From this review of past studies, it is evident that there is a gap in documenting gender-based lessons for efficient climate change adaptation in sub-Saharan Africa, and especially in Ghana where the savanna ecological zone is faced with extreme drought, bushfires and lack of irrigation facilities. The literature further supports the view that women are most vulnerable to climate change due to the high labour demands of adaptation strategies and their inability to access fertile lands for sustainable agricultural activities and resilience- and yield-enhancing technologies, and the relatively lower social position of women farmers.

11.2.2 Study area

The study was conducted in three selected sites in the three regions of northern Ghana. The selected sites include the Sisili-Kulpawn Basin in the Northern Region, the White Volta Basin in the Upper East Region and Lawra District in the Upper West Region. Agriculture at these sites is rain-fed and dominated by smallholders (i.e., farm sizes ranging from 1 to 2ha) practicing mixed farming systems. Farmers in the study area rear livestock and cultivate cereals such as millet, maize, rice and sorghum, and legumes including groundnut, soybean and cowpea.

The rainfall in northern Ghana is seasonal in nature. It typically commences in May and ends in October. The dry, off-farm season lasts for about seven months (i.e., November to April). The average temperature in northern Ghana is higher than in the rest of the country. In the case of Navrongo in the Upper East Region, minimum temperatures range between 15.6 and 20.6°C, with an average of 19.2°C; and maximum temperatures range from 38.0 to 42.8°C, with an average of 40.4°C. Annual rainfall ranges between 688 and 1365 mm, with an average of 969 mm. Changes in the observed trends are not significant for maximum temperature and annual rainfall, but minimum temperatures have shown a slight increase (Adiku et al., 2015).

11.2.3 Data collection and modelling

The study used a Trade-Off Analysis for Multi-Dimensional Impact Assessment (TOA-MD) model to estimate the gender-disaggregated impacts of climate change on smallholder farms. The TOA-MD model is a unique simulation tool for multi-dimensional impact assessment that uses a statistical description of a heterogeneous farm population to simulate the adoption and impacts of new technology, a change in environmental conditions, and ecosystem services supply on current farm systems.

The farm-level socio-economic data for calibrating the TOA-MD model was obtained from a sample of 800 farmers using a semi-structured questionnaire in the 2014 and 2015 production seasons. The data included forms of land-use, household size, farm size, herd (livestock) size, input levels and prices, costs of production, yields, output levels and prices, non-agricultural income levels, and net per farm returns on different crop and livestock enterprises. A summary of some of the key variables used to estimate the TOA-MD model parameters is presented in Table 11.1.

Table 11. 1: Summary of the key variables used to estimate the TOA-MD model

Parameter/ Strata	Female			Male		
Farm characteristics	Mean	Stand. Dev.	CV (%)	Mean	Stand. Dev.	CV (%)
Household size	9.95	4.14	41.57	11.54	7.11	61.60
Farm size	2.68	1.44	53.71	3.59	2.94	81.96
Herd size	18.17	18.53	102.01	27.66	28.21	102.00
Off-farm income	939.87	506.94	53.94	2588.40	3514.23	135.77
Maize						
Yield/farm (kg)	417.67	265.38	63.54	722.28	1047.73	145.06
Var. cost/farm (GHS)	539.22	488.84	90.66	907.77	1147.44	126.40
Net rev./farm (GHS)	458.36	716.50	156.32	275.97	987.29	357.75
Cowpea						
Yield/farm (kg)	265.71	163.71	61.61	628.73	903.20	143.65
Var. cost/farm (GHS)	191.38	201.75	105.42	600.94	796.47	132.54
Net rev./farm (GHS)	588.03	516.81	87.89	1642.11	2007.38	122.24
Livestock						
Var. cost (GHS)	112.64	170.17	151.08	323.97	550.91	170.05
Net. rev (GHS)	137.60	294.10	213.74	403.46	1465.27	363.18

11.2.4 Economic analysis of gender-disaggregated practices

In an environment where men and women play equally significant roles in agricultural production, designing agricultural policies needs to consider gender roles to provide male and female farmers alike with incentives for the adoption of sustainable practices that are capable of addressing the current effects of climate change on agriculture. Gender differences are known to impact on farm practices, productivity and output.

Following Antle and Valdivia (2010), we consider a TOA-MD model of a farmer's choice between two production systems - the current or "base" production system (i.e., hereafter called System 1), and a new production system (i.e., hereafter called System 2). Following Adiku et al. (2015), we use farm survey data with yield projections under five global circulation models under intensive and extended (I & E) irrigation adaptation. We simulate and compare gender-disaggregated economic outcomes in terms of gains and losses, percent gainers, per capita income, net returns per farm, and poverty rates of farmers under Systems 1 and 2.

The TOA-MD model, as used in this study, is designed to simulate what would be observed if it were possible to conduct a controlled experiment in which a population of farms is offered the choice of continuing to use the current or "base" production system (i.e., System 1), or choosing to adopt a new production system (i.e., System 2) under changed climatic conditions. According to Ilukor et al. (2014), the TOA-MD model is useful for an ex-ante assessment of the adoption of a practice to estimate an adoption rate of alternative practices based on their economic feasibility. This further involves differences in returns of alternative practices applied at various scales (i.e., land and farm levels) and biophysical data to estimate profitability. With this, adoption capabilities and investment towards adoption rates of technologies varies based on the biophysical and economic factors (Khanna et al., 1999, 2000).

System profiles are summarised as follows:

Suppose that a farmer at site (s) is using a production system (h) with inputs prices (p) earns returns/ha equivalent to $V_t = V_t(s, h)$ each season. When the production system changes, because of a change in technology or climate or both, expected returns at each site also change. The effect on a farm's returns of changing from system j to system k is $\omega(p, s, h) = V(p, s,) - V(p, s, h)$. Thus, if $\omega(p, s, h)$ is positive, it represents the loss associated with switching from one system to another.

The proportion of farms using System 2, referred to as the adoption rate of System 2, is given by the cumulative distribution function $\mathrm{r(p, 2, a)} = \int_{-\infty}^{a} \emptyset(\omega|\mathrm{p})d\omega$, and the share of farms using System 1 is r(p, 1, a) ≡ 1 – r(p,2,a). In addition to economic outcomes $\mathrm{V(a, h)}$, Antle et al. (2014) considered other environmental or social outcomes *z*(a, h). In this case, the distribution for the sub-population using each system is the joint outcome distribution between ω and $\mathrm{k = v, z}$, truncated according to $\omega > \mathrm{a}$ for System 1 and $\omega < a$ for System 2 (Antle et al. 2014):

$$\emptyset(\omega, k|a, h) = \emptyset(\omega, k \mid h)/r(a, h) = \emptyset(k \mid \omega, h)\varphi(\omega)/r(a, h)$$

Thus, the outcomes (economic and social/environmental) are indexed by k = v, z, while systems are indexed by h = 1,2.

Antle (2011a) and Antle et al. (2014) note that $\emptyset(\omega, k \mid h)/r(a, h)$ indicates a distribution that is truncated from below by ***a*** for System 1, and truncated from above by ***a*** for System 2, whereas $= \emptyset(k \mid \omega, h)$ shows a distribution defined over the entire population. The joint distribution of ω and $k = v, z$ in a population using both systems is a mixture of the distributions defined in equation (a), with mixing proportions $r(a, h)$.

Climate Projections: The study is based on earlier simulated yields of selected crops under five global circulation models (GCMs). The GCMs are CCSM4 (GCM E), GFDL-ESM2M (GCM I), HadGEM-ES (GCM K), MIROC5 (GCM O) and MPI-ESM-MR (GCM R) for mid-century (i.e. 2050).

Firstly, with regards to temperature scenarios, all the GCM's predict an increase in temperature ranging from 1.1°C to 2.4°C. The greatest warming is predicted by GCM K and the least warming is predicted by GCM E. Secondly, regarding precipitation, the GCM I scenario predicts a dramatic decrease in rainfall (i.e., 15.7% reduction), while the other GCM's predicted positive changes ranging from 3.5% to 9%. (Adiku et al., 2015). Table 11.2 shows the temperature and precipitation projections.

Table 11. 2: Temperature and Precipitation Projections of the Selected GCMs

GCM's	Temperature	Precipitation
GCM E	1.1°C	9%
GCM I	1.7 °C	-15.7%
GCM K	2.4 °C	8%
GCM O	2.1 °C	8%
GCM R	2.3 °C	3.5%

Source: Adiku *et al.* (2015)

11.2.5 Crop Models (DSSAT and APSIM)

The study adapted the crop models of Adiku et al. (2015), which are the Decision Support System for Agro-Technology Transfer (DSSAT) and the Agriculture Production Systems Simulator (APSIM). The crop models were used to project the mean yield changes for the crops in 2050. Table 11.3 shows the projections for the various GCM's under the crop models.

However, the crop model's projections for groundnut was used as a proxy for cowpea in this study.

Table 11. 3: Yield projections for maize and cowpea by DSSAT and APSIM across 5 GCMs

Maize	**Yield Change (No Adaptation) DSSAT/(APSIM)**	**Yield Change (Adaptation) DSSAT/(APSIM)**
GCM E	-32% (-15%)	26% (44%)
GCM I	-34% (-10%)	36% (47%)
GCM K	-62% (-18%)	30% (38%)
GCM O	-41% (-9%)	30% (47%)
GCM R	-57% (-23%)	31% (48%)
COWPEA		
GCM E	14% (16%)	19% (12%)
GCM I	1% (12%)	18% (12%)
GCM K	-14% (10%)	18% (13%)
GCM O	-1% (8%)	19% (12%)
GCM R	-10% (6%)	18% (13%)

Source: Adiku *et al.* (2015).

11.2.6 Representative Agricultural Pathways (RAP)

Representative Agriculture Pathways (RAPs) were used for System 2 projections in this study. According to Adiku et al. (2015), crop production will be farmer-led with the adoption of intensive and expanded (I & E) irrigation technology using hand-dug, shallow wells. The RAP assumes that farms with access to irrigation will be better-off following reduced precipitation and increased temperature under a future climate (i.e., in 2050). This is because irrigation mutes the negative impacts of climate change on farm systems. Details of parameters used in the RAP analysis are indicated in Table 11.4.

Table 11. 4: Representative Agricultural Pathways (RAPs) Analysis for northern Ghana

Variables	Percent Change	Direction of Change	Rational for Change
Household Size	50	Decrease	Internal migration of young people to the cities, split in households and education
Farm Size	100	Increase	People migrate and the few that stay have more available land to mechanize and increase farm sizes.
Non-Agriculture Income	30	Increase	Service sector is developed, causing people to find alternatives sources of income outside farming.
Production Cost	10	Increase	Increased input cost due to high fuel prices and greater concentration of input providers.
Livestock Number	55	Increase	Increased demand, increasing incomes and purchasing power.
Livestock Yield	30	Increase	Animal breeding programs, improved animal health programs.

Source: Adiku et al. (2015).

11.3 Results and Discussion

In this section, we present and discuss the results of the study. We then examine what these results could imply for nationally-appropriate climate change mitigation actions in Ghana. Since our primary objective is to determine gender-disaggregated patterns in climate change impacts and adaptation, the results are presented and discussed in a gender-disaggregated way. All results presented here are based on yield projections from the DSSAT crop model across the five GCMs - CCSM4 (GCM E), GFDL-ESM2M (GCM I), HadGEM-ES (GCM K), MIROC5 (GCM O) and MPI-ESM-MR (GCM R), as described in the last section.

Fundamentally, the results combine the strengths of simulations based on projections of integrated climate-crop production models to estimate the differences in the economic livelihood outcomes for male and female farmers who adapt (or fail to adapt) to climate change. Some of the results are presented graphically to synthesize a large body of evidence in a relatively convenient format. All results are simulated for mid-century (i.e., in 2050) from a 30-year baseline period (i.e., from 1980-2009). Note that simulations are made on a per farm rather than a per acre or hectare basis, and farms herein defined include all crop and livestock enterprises of the farmer.

11.3.1 Climate change impacts without adaptation

Without adaptation to the adverse effects of climate change, one of the primary consequences will be a decrease in some if not all livelihood outcome indicators of the affected farmers. For smallholder farmers, one such indicator that may be adversely affected by climate change is revenue. As illustrated in Figures 11.1 and 11.2, revenue drastically falls for male and female farmers alike if there is no adaptation to climate change.

Figure 11. 1: Climate change impact on Mean Net Returns Per farm (Without Adaptation) under GCMO

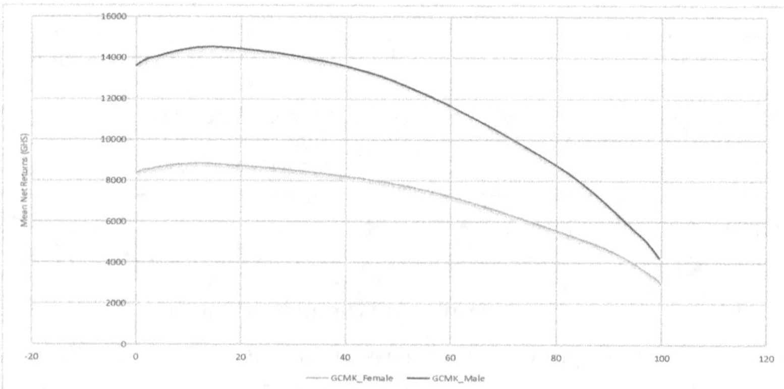

Source: Plots from TOA-MD Simulation Results.

Under the GCMO (i.e., Figure 11.1), the mean net revenue per farm for female farmers falls by GHS5840.00, from GHS8346.00 to GHS3162.00, representing a reduction in mean returns following climate change by 62%. Similarly, the mean net returns of male farmers decrease from GHS13612.00 to GHS4418.00, representing a fall of 67% by mid-century.

Under the GCMK (i.e., Figure 11.2), female farmers experience an income decline from GHS8346.00 to GHS3018.92, thus a reduction of GHS5327.00 (or 64%). Male farmers, on the other hand, incur a reduction in their mean net returns by GHS9360.82 (or 69%), from GHS13613.00 to GHS4252.00.

Figure 11. 2: Climate change impact on mean net returns per farm (without adaptation) under GCMK

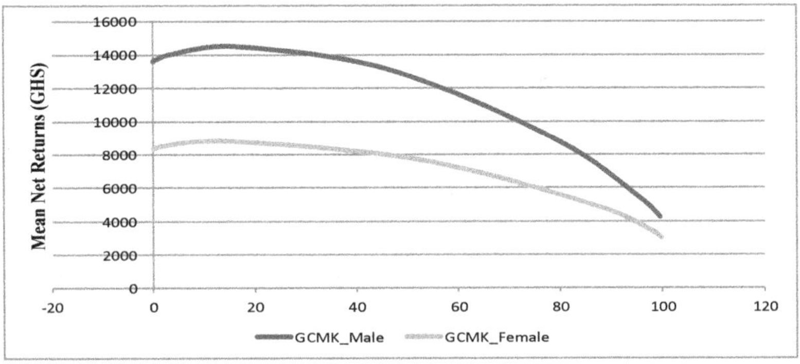

Source: Plots from TOA-MD Simulation Results

Under both climate change scenarios, the mean net returns of female farmers is consistently lower than that of male farmers, but the decline in the mean net returns of male farmers, which is 67% and 69% under GCMO and GCMK, respectively, is more drastic than that of the female farmers, which is 62% and 64% under the GCMO and GCMK scenarios, respectively.

Overall, the results mean that without adaptation, lower future yields are simulated by the DSSAT crop yield model for northern Ghana. Thus, the mean net returns for both male- and female-owned farms will decline, with the decline for male-owned farms being higher on average than female-owned farms. This also implies that adaptation by female-owned farms may lead to higher net revenue gains than that of male-owned farms.

11.3.2 Percentage gains, losses and the net impact of climate change

Gains refer to the incremental income benefits under climate change, whereas losses mean the additional costs incurred in terms of a reduction in farm incomes following climate change. Both gains and losses, computed as a percentage of net returns per farm, occur even under adaptation with different causal mechanisms, including adaptation itself.

In Figures 11.3 – 11.5, we illustrate the gains, losses and net impacts defined as the differences between gains and losses for three of the GCMs

– GCME, GCMK and GCMR. The results reveal that adaptation to climate change by male and female smallholders leads to substantial economic gains, some losses and overall positive net impacts. For instance, under GCME, adaptation by male farmers results in 57.5% gains and 34.5% losses, giving a net positive impact of 23%. Adaptation by female farmers leads to 78.4% gains and 46.2%, thus with a net positive impact of 32.2%. Under GCMK, the gains, losses and net impact for males are 49.9%, 33.4% and 16.5% (positive), respectively, while those for female farmers are 71.1%, 47.1 and 24% (positive), respectively. Finally, under GCMR, the gains, losses and resulting net impacts for male farmers are 51.7%, 34.4% and 17.3% (positive), respectively, as against 72.3% gains, 46.9% losses and 25.4% net positive impact for female farmers.

In aggregate, adaptation to climate change reduces losses. It results in net positive impacts of 32%, 23.9% and 25.3% under GCME, GCMK and GCMO, respectively, for all farms (i.e., male- and female-owned farmers combined). Nevertheless, adaptation by female farmers leads to higher gains than adaptation by male farmers, and likewise for the losses by female farmers compared to male farmers. The net impacts of adaptation are also consistently higher for female farmers than the net impacts of male farmers. This means investing in adaptation strategies that benefit women will be a better policy choice.

11.3.3 Climate change impact on farmers' revenue with and without adaptation

In this section, we empirically examine the impact of climate change on the mean net monetary returns of farmers with and without adaptation. Fundamentally, the question is how much profit (in GHS) are male and female farmers likely to make with and without adaptation to climate change by 2050?

Again, the results demonstrate strong evidence of the benefits of adaptation (Table 11.5). It can be seen that the mean net returns of male farmers are higher than that of female farmers. This is perhaps due to male farmers' better access to productive farm resources. However, adaptation largely benefits female farmers more than male farmers. Across all the GCMs, the percentage in net returns for female farmers is higher than that of male farmers.

Figure 11. 3: Percent Gains, Losses and Net Impact under GCME

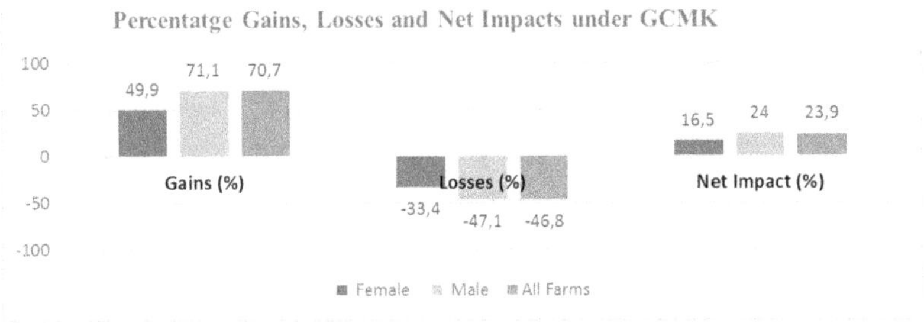

Figure 11. 4: Percent Gains, Losses and Net Impact under GCME

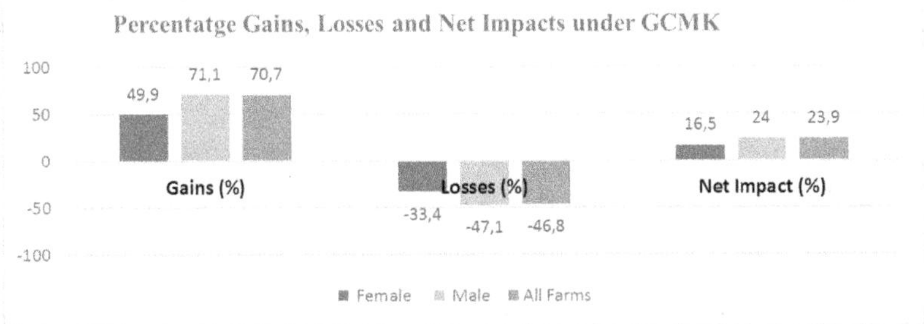

Figure 11. 5: Percent Gains, Losses and Net Impacts of GCMR

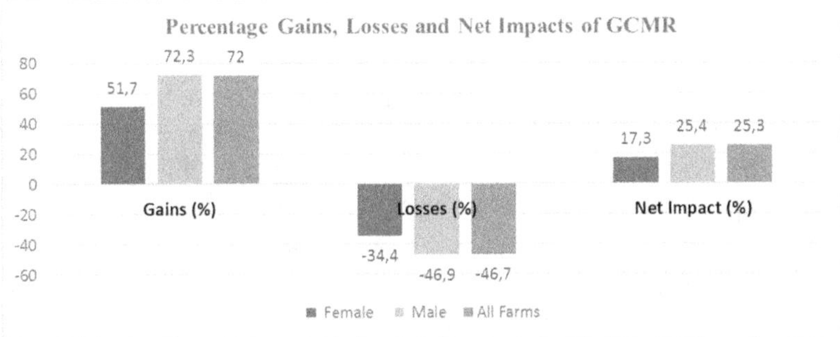

Table 11. 5: Impacts of climate change on mean net returns per farm with and without adaptation (DSSAT Model)

GCM	DSSAT Crop Model			
	Stratum	**Without Adaptation (GHS)**	**With Adaptation (GHS)**	**Change in Mean Net Returns (%)**
GCME	Female	896.43	2347.42	162
	Male	2085.49	4978.68	139
GCMI	Female	977.32	3833.90	292
	Male	1860.46	5655.35	204
GCMK	Female	705.03	1607.15	128
	Male	1784.99	3614.66	103
GCMO	Female	790.93	2110.30	167
	Male	1891.57	4390.51	132
GCMR	Female	725.91	1740.78	140
	Male	1823.76	3821.37	110

Source: TOAMD Model Simulations

The above results show that across all five GCMs, the increase in mean net returns of female farmers is higher than the change in mean net returns of male farmers. For instance, under GCME, whereas mean net returns for male farmers increase by 139% due to adaptation, that of female farmers is 162%. The same increments are 292% for female farmers and 204% for male farmers under GCMI, and 128% and 103% under GCMK, 167% and 132% under GCMO, and 140% and 110% under GCMR for female and male farmers, respectively.

The contribution of adaptation to yield is expected to contribute to the simulated increments in the mean net returns of farmers. However, autonomous increases in price trends of food commodities, especially of livestock products, and reductions in production costs due to the availability of improved technology, are absolutely expected to act as conditioning variables in increasing the net returns of northern Ghana's smallholder farmers by 2050.

Like the results on the net impacts of climate change, the above results inspire the policy discourse for the inclusion or even prioritizing of women farmers in the development of adaptation strategies. The results are confirmed by some of the literature on gender roles in climate change adaptation that portray women as having a particularly special ability to adapt to climate change compared to men (Okali and Naess, 2013).

11.3.4 Impact on farmers' per capita income with and without adaptation

Without adaptation, climate change impacts on the per capita income of both male and female smallholder farmers in northern Ghana will be lower than with adaptation. With the per capita income of male farmers being greater than that for female farmers across all the GCMs, the results show that the contribution of adaptation to the per capita incomes of female farmers is consistently higher than that of male farmers.

We note from the results that across all climate scenarios, the value of per capita income for male farmers is positive and overall higher than that for female farmers with and without adaptation (Table 11.6). This may again be attributed to the higher access of men to productive resources and alternative sources of wage income than women. However, the percentage changes in per capita income for female farmers are higher than that for male farmers.

Table 11. 6: Impacts of climate change on per capita income with and without adaptation (DSSAT Model)

GCM	Stratum/Impacts			
	Stratum	Without Adaptation (GHS)	With Adaptation (GHS)	Change in per capita income (%)
GCME	Female	224.38	431.04	92
	Male	558.84	997.62	79
GCMI	Female	233.92	606.28	159
	Male	531.94	1078.53	103
GCMK	Female	201.82	343.78	79
	Male	522.91	834.53	60
GCMO	Female	211.95	403.09	90
	Male	535.66	927.29	73
GCMR	Female	204.28	359.53	76
	Male	527.55	859.24	63

Source: TOAMD Model Simulations

For instance, under GCME, the per capita income of female farmers increases from GHS 224.38 without adaptation to GHS 431.02 with adaptation, representing a 92% change. In contrast, the increase for male farmers is from GHS 558.84 to GHS 997.62, representing a 79% change. The respective change in per capita income for female and male farmers, respectively, across the other GCMs are: 159% and 103% under GCMI, 79% and 60% under GCMK, 90% and 73% under GCMO, and 76% and 63% under GCMR.

The broad patterns that are observable demonstrate that female farmers display higher impacts of climate change adaptation than male farmers, despite the real income inequalities against women. The inequality in opportunities, especially due to gender gaps in education and economic participation, appears to give male farmers an advantage over female farmers in terms of net incomes. Nevertheless, helping female farmers in northern Ghana to adapt, have equal access to productive resources or engage in alternative sources of employment is more beneficial as it boosts their per capita income by a higher rate than it does or male farmers.

11.3.5 Impact on poverty rates with and without adaptation

Climate change adaptation can make an important contribution to poverty reduction, just as the failure to adapt to climate change can increase poverty, especially for smallholder farmers whose livelihoods are more sensitive to changes in the weather.

As shown in Table 11.7, although projected poverty rates without adaptation among female smallholder farmers in northern Ghana is higher across all GCMs than the poverty rates among male farmers, adaptation contributes to a greater reduction in poverty rates for female farmers than it does for male farmers. This means that the impact of adaptation on the poverty rates of female smallholder farmers in northern Ghana will be higher than the impact on the poverty rates among male farmers, based on crop yield simulations from the DSSAT model.

Table 11. 7: Impacts of climate change on poverty rates with and without adaptation (DSSAT model)

GCM	Stratum/Impacts			
	Stratum	**Without Adaptation (GHS)**	**With Adaptation (GHS)**	**Change in Poverty Rates (%)**
GCME	Female	78.76	31.49	-47.3
	Male	32.29	16.78	-15.5
GCMI	Female	72.79	18.71	-54.1
	Male	34.30	15.38	-18.9
GCMK	Female	87.90	46.55	-41.4
	Male	33.47	20.26	-13.2
GCMO	Female	83.46	35.27	-48.2
	Male	33.19	18.03	-15.2
GCMR	Female	86.82	43.06	-43.8
	Male	33.31	19.62	-13.7

Source: TOAMD Model Simulations

These results are consistent with and linked to those presented in the previous sections. The declines in poverty rates are attributed to the higher impacts of adaptation on the mean net returns per farm and per capita income of female farmers presented earlier. Since the projected mean net returns per farm and per capita incomes of female farmers are higher under all scenarios than for male farmers, the reduction in poverty rates for female farmers is likewise expected to be higher. In addition, this means that although women are considered more vulnerable than men to climate change impacts, adaptation has the potential to reverse the vulnerability of women faster than it does for that of men in northern Ghana. More rapidly decreasing poverty levels among female farmers implies an increased ability for them to adapt to the climate change problem in the long-run, since the more prosperous one is, the easier it becomes to adapt (Edenhofer and Lotze-Campen, 2008).

11.4 Conclusions

What conclusions may be drawn from the findings of this study? The projected economic impacts based on gender disaggregation under the DSSAT model and across all of the five GCMs showed that all the economic outcome indicators - net impacts, mean net returns per farm, per capita income and poverty rates - among female smallholder farmers are higher than similar indicators for male smallholder farmers in the Northern Region of Ghana. This means female smallholder farmers stand to gain more from future adaptation strategies than male smallholder farmers in northern Ghana.

This revelation contributes to the policy discourse on the development of gender-friendly climate change adaptation and resilience strategies and the involvement of women in the associated discussions. Thus, current technologies and adaptation strategies should be designed to empower female farmers in agriculture. This will be a sure way of reducing poverty and increasing food security among farm households in northern Ghana.

Reference

Abdul-Razak, M., and Kruse, S. (2017) the Adaptive Capacity of Smallholder Farmers to Climate Change in the Northern Region of Ghana. *Climate Risk Management*, 17, 104-122. doi:10.1016/j.crm.2017.06.001.

Adger, W. N., et al. (2007) Assessment of adaptation practices, options, constraints and capacity. Climate Change 2007: Impacts, Adaptation and Vulnerability. Contribution of Working Group II to the Fourth Assessment Report of the Intergovernmental Panel on Climate Change

Adiku, S. G. K., Dilys, S.D., MacCarthy, S., Hathie, I., Diancoumba, M., Freduah, B.S., Amikuzuno, J., Traore, P.C.S., Traore, S., Koomson, E., Agali, A., Lizaso, J.I., Fatondji, D., Adams, M., Tigana, L., Diarra, D.Z., N'diaye, O., and Valdivia, R.O. (2015). Climate Change Impacts on West African Agriculture: An Integrated Regional Assessment (CIWARA). *World Scientific Books.* Imperial College Press. 25-73.

Adu, D. T., Kuwornu, J. K. M., Anim-Somua, H., and Sasaki, N. (2018) Application of livelihood vulnerability index in assessing smallholder maize farming households' vulnerability to climate change in Brong-Ahafo region of Ghana. *Kasetsart Journal of Social Sciences*, 39(1), 22-32.

Al-Hassan, R. M., Kuwornu, J. K. M., Etwire, P. M., and Osei-Owusu, Y. (2013) Determinants of choice of indigenous climate related strategies in Northern Ghana. *British Journal of Environment and Climate Change* 3(2), 172-187.

Antle, J.M. (2011) Parsimonious multi-dimensional impact assessment. *American Journal of Agricultural Economics*, 93(5), 1292-1311.

Antle, J. M., and Valdivia, R. O. (2010) TOA-MD Version 4: Minimum-Data Tradeoff Analysis Model. www.tradeoffs.oregonstate.edu.

Antle, J. M., Stoorvogel, J., and Valdiva, R. (2014). New parsimonious simulation methods and tools to access future food and environmental security of farm population. Philosophical transactions of the Royal Society B 369:20120280

Armah, F. A., Odoi, J. O., Yengoh, G. T., Obiri, S., Yawson, D. O., and Afrifa, E. K. (2011) Food security and climate change in drought-sensitive savanna zones of Ghana. *Mitigation and adaptation strategies for global change* 16 (3), 291-306.

Armah, R. N. A., Al-Hassan, R. M., Kuwornu, J. K. M., and Osei-Owusu, Y. (2013) What influences farmers' choice of indigenous adaptation strategies for agrobiodiversity loss in Northern Ghana? *British Journal of Applied Science and Technology* 3(4), 1162-1176.

Carvajal-Escobar, Quintero-Angel, and Garcia-Vargas (2008) Women's role in adapting to climate change and variability, *Advances in Geo Sciences*, Issue14, 277-280.

Damptey, P. T. M., and Essesl, A. K. (2012) Gender perspective of climate change coping and adaptive strategies in Ghana. http://wedo.org/wpcontent/uploads/2015/06/ghanagenderesearch.compressed.pdf

Edenhofer, O., and Lotze-Campen, H. (2008) "Emissions must have a price" In: Climate Change and Poverty: A challenge for a fair world policy. welt-sichten 5, 2008.

Etwire, P. M., Al-Hassan, R. M., Kuwornu, J. K. M., and Osei-Owusu, Y. (2013a). Smallholder farmers' adoption of technologies for adaptation to climate change in Northern Ghana. *Journal of Agricultural Extension and rural Development* 5(6), 121-129.

Etwire, P. M., Al-Hassan, R. M., Kuwornu, J. K. M., and Osei-Owusu, Y. (2013b) Application of livelihood vulnerability index in assessing vulnerability to climate change and variability in Northern Ghana. *Journal of Environment and earth Science.* 3(2), 157 – 170.

FAO. (2011) Climate change, water and food security; water report; Food and Agriculture Organization of the United Nations. Rome.

Goh, A. H. X. (2012) A Literature review of the gender-differentiated impacts of climate change on women's and men's assets and well-being in developing countries. A Collection Action and Property Rights (CAPRi) Working Paper no. 106. Washington, D.C.: International Food Policy Research Institute. http://dx.doi.org/10.2499/CAPRiWP106.

Holvoet, N., and Inberg, L. (2013) Gender sensitivity of Sub-Saharan Africa National Adaptation Programmes of Action: findings from a desk review of 31 countries. *Climate and Development* 6(3), 266-276.

Huyer, S. (2016) Closing the gender gap in agriculture. *Gender, Technology and Development* 20(2), 105-116.

Ilukor, J., Bagamba, F., and Bashaasha, B. (2014) Application of the TOA-MD model to assess adoption potential of improved sweet potato technologies by rural poor farm households under climate change: the case of Kabale district in Uganda. Food Security 6:359-368.

Jost, C., Kyazze, F., Naab, J., Neelormi, S., Kinyangi, J., Zoughmore, R., Aggarwal, P., Bhatta, G., Chudhury, M., Tapio-Biostrom, M., Nelson, S., and Kristjanson, P. (2015) Understanding gender dimension of agriculture and climate change in smallholder farming communities. *Climate and Development.* http://dx.doi.org/10.1080/17565529.2015.1050978.

Kakota, T., Nyariki, D., Mkwambisi, D., and Kogi-Makau, W. (2011) Gender vulnerability to climate variability and household food insecurity. *Climate and Development,* 3(4), 298-309.

Khanna, M., Epouhe, O. F., and Hornbaker, R. (1999) Site-specific crop management: adoption patterns and incentives. *Applied Economic Perspectives and Policy,* 21(2), 455–472. doi:10.2307/1349891.

Khanna, M., Isik, M., and Winter-Nelson, A. (2000) Investment in sitespecific crop management under uncertainty: implications for nitrogen pollution control and environmental policy. *Agricultural Economics,* 24(1), 9–12. doi:10.1111/j.1574-0862.2000.tb00089.x.

Kianjun, J., Xiaomin, W., and Yiwei, G. (2015) Gender differences in farmers' responses to climate change adaptation in Yongqiao District, China. *Science of the Total Environment* 538, 942-948.

Kumar, S. N., Anuja, Rashid, M., Banyopadyay, S. K., Padaria, R., and Khanna, M. (2016) Adaptaion of farming community to climate risk: does adaptation cost for sustaining agricultural profitability? *Current Science; Special Section: Climate Change Impacts and Adaptation,* 110(07), 1216 -1224.

Kuwornu, J. K. M., Al-Hassan, R.M., Etwire, P.M., and Osei-Owusu, Y. (2013) Adaptation strategies of smallholder farmers to climate change and variability: evidence from Northern Ghana. *Information Management and Business. Review,* 5 (5), 233 - 239.

Mangheni, M., N., Bashasha, B., and Majaliwa, J. G. M. (2010) Understanding the gender dimensions of the impact of climate change on agriculture and adaption among smallholder Farmers in eastern Uganda. A paper presented at the Second RUFORUM Biennial Meeting 20-24 September 10, Entebbe, Uganda.

Nakuja, T., Sarpong, D.B., Kuwornu, J.K.M., and Felix, A.A. (2012) Water storage for dry season vegetable farming as an adaptation to climate change in the Upper East region of Gha-na. *African Journal of Agriculture Research,* 7(2), 298-306.

Nantui, M.F., Bruce, S.D., and Yaw, O.A. (2012) Adaptive capacities of farmers to climate change adaptation strategies and their effects on rice production in the northern region of Ghana. *Russian Journal of Agricultural and Socio-Economic Sciences* 1 (11), 9–17.

Nath, P. K., and Behera, B. (2011) A critical review of impact of and adaptation to climate change in developed and developing economies. *Environment, Development and Sustainability* 13(1), 141-162.

Okali, C., and Naess, L. O. (2013) Making Sense of Gender, Climate Change and Agriculture in sub-S aharan Africa: Creating Gender-Responsive ClimateAdaptation Policy. Future Agriculture Working Paper Working Paper 057. Sourced from: www.future-agricultures.org

Perez, C., Jones, E. M., Krisjanson, P., Cramer, L., Thornton, P. K., Forch, C., and Barahona, C. (2015) How resilient are farming households and communities to a changing climate in Africa? A gender-based perspective. *Global Environmental Change* 34, 95-107.

Ravera, F., Martin-Lopez, B., Pascula U., and Drucker, A. (2016) The diversity of gendered adaption strategies to climate change of Indian farmers: A feminist intersectional approach. *Ambio*, 45(S3), 335-351.

Reyes, R. R. (2009) *Gender responses to El Nini in rural Peru,* Climate Change and Gender Justice. Action in Gender and Development. Practical Action Publishing Ltd, Warwickshire, UK.

Rodenberg, B. (2009) Climate Change Adaptation from a Gender Perspective: a cross-cutting analysis of development-policy instruments / Birte Rodenberg. DIE Research Project "Climate Change and Development" – Bonn : DIE, 2009. – (Discussion Paper / Deutsches Institut für Entwicklungspolitik ; 24) ISBN 978-3-88985-474-2

Saxena, A., Guneralp, B., Bailis, R., and Oliver, C. (2016) Evaluating the resilience of forest dependent communities in Central India by combining the sustainable livelihoods framework and the cross scale resilience analysis. *Current Science; Special Section: Climate Change Impacts and Adaptation,* 110(07), 1195 – 1207.

Appendix 11.1: Adaptation (DSSAT)

IMPACT OF CLIMATE IN FUTURE WITH ADAPTATION	GCME			GCMI			GCMK			GCMO			GCMR		
	Female	Male	All Farms	Female	Male	All Farms	Female	Male	All Farms	Female	Male	All Farms	Female	Male	All Farms
Gains (% mean net returns)	61.627	74.593	74.187	66.029	77.345	77.117	61.953	236.972	230.226	66.578	77.637	77.411	63.792	75.807	75.571
Losses (% mean net returns)	-46.208	-43.635	-43.716	-52.864	-47.102	-47.218	-44.125	-25.816	-26.521	-53.714	-47.193	-47.326	-51.366	-47.450	-47.527
Net impact (% mean net returns)	15.419	30.958	30.471	13.165	30.243	29.898	17.828	211.157	203.704	12.864	30.444	30.085	12.426	28.357	28.044
Projected net returns without adaptation (GHS)	870.259	2240.316	2165.099	850.250	2120.534	2076.187	861.120	337.561	355.839	834.908	2058.855	2016.125	797.155	2023.240	1980.436
Projected net returns with adaptation (GHS)	2951.523	5062.264	4946.381	3102.833	5103.964	5034.102	2726.321	4256.604	4203.180	3096.570	5012.801	4945.903	2782.815	4763.269	4694.129
Change in mean net returns	2.392	1.260	1.285	2.649	1.407	1.425	2.166	11.610	10.812	2.709	1.435	1.453	2.491	1.354	1.370
Projected per-capita income without adaptation (GHS)	221.299	577.355	417.130	218.940	563.033	408.191	220.221	349.848	291.516	217.131	555.658	403.321	212.681	551.400	398.976
Projected per-capita income with adaptation (GHS)	1004.517	2015.226	1560.407	520.096	1012.599	790.972	475.711	911.282	715.275	519.357	1001.699	784.645	482.371	971.863	751.591
Percent change in per capita income	3.539	2.490	2.741	1.376	0.798	0.938	1.160	1.605	1.454	1.392	0.803	0.945	1.268	0.763	0.884
Projected poverty rate without adaptation (%)	78.725	31.376	52.683	79.315	32.170	53.385	79.412	48.269	62.283	79.933	32.501	53.846	82.028	32.558	54.819
Projected poverty rate with adaptation (%)	8.774	11.492	10.269	23.124	16.552	19.509	26.656	19.703	22.832	23.178	16.701	19.616	26.004	17.226	21.176
Change in poverty rate	-69.952	-19.884	-42.415	-56.191	-15.619	-33.876	-52.756	-28.565	-39.451	-56.755	-15.800	-34.230	-56.025	-15.331	-33.643

Appendix 11.2: Adaptation (APSIM)

IMPACT OF CLIMATE IN FUTURE WITH ADAPTATION	GCME			GCMI			GCMK			GCMO			GCMR		
	Female	Male	All Farms	Female	Male	All Farms	Female	Male	All Farms	Female	Male	All Farms	Female	Male	All Farms
Gains (% mean net returns)	61.627	74.593	74.187	66.029	77.345	77.117	61.953	236.972	230.226	66.578	77.637	77.411	63.792	75.807	75.571
Losses (% mean net returns)	-46.208	-43.635	-43.716	-52.864	-47.102	-47.218	-44.125	-25.816	-26.521	-53.714	-47.193	-47.326	-51.366	-47.450	-47.527
Net impact (% mean net returns)	15.419	30.958	30.471	13.165	30.243	29.898	17.828	211.157	203.704	12.864	30.444	30.085	12.426	28.357	28.044
Projected net returns without adaptation (GHS)	870.259	2240.316	2165.099	850.250	2120.534	2076.187	861.120	337.561	355.839	834.908	2058.855	2016.125	797.155	2023.240	1980.436
Projected net returns with adaptation (GHS)	2951.523	5062.264	4946.381	3102.833	5103.964	5034.102	2726.321	4256.604	4203.180	3096.570	5012.801	4945.903	2782.815	4763.269	4694.129
Change in mean net returns	2.392	1.260	1.285	2.649	1.407	1.425	2.166	11.610	10.812	2.709	1.435	1.453	2.491	1.354	1.370
Projected per-capita income without adaptation (GHS)	221.299	577.355	417.130	218.940	563.033	408.191	220.221	349.848	291.516	217.131	555.658	403.321	212.681	551.400	398.976
Projected per-capita income with adaptation (GHS)	1004.517	2015.226	1560.407	520.096	1012.599	790.972	475.711	911.282	715.275	519.357	1001.699	784.645	482.371	971.863	751.591
Percent change in per capita income	3.539	2.490	2.741	1.376	0.798	0.938	1.160	1.605	1.454	1.392	0.803	0.945	1.268	0.763	0.884
Projected poverty rate without adaptation (%)	78.725	31.376	52.683	79.315	32.170	53.385	79.412	48.269	62.283	79.933	32.501	53.846	82.028	32.558	54.819
Projected poverty rate with adaptation (%)	8.774	11.492	10.269	23.124	16.552	19.509	26.656	19.703	22.832	23.178	16.701	19.616	26.004	17.226	21.176
Change in poverty rate	-69.952	-19.884	-42.415	-56.191	-15.619	-33.876	-52.756	-28.565	-39.451	-56.755	-15.800	-34.230	-56.025	-15.331	-33.643

CHAPTER 12

Smallholder maize farmers' constraints to climate change adaptation strategies in the Brong-Ahafo Region of Ghana

Derick T. Adu, John K. M. Kuwornu, Avishek Datta

Abstract

Climate change is believed to have adverse effects on smallholder farmers in Ghana and all of Africa because most of their activities rely on weather- and climate-dependent water availability. This study examined smallholder maize farmers' constraints to the adoption of climate change adaptation strategies. The study used primary data gathered from 150 smallholder maize farmers in the Brong-Ahafo Region of Ghana. A comparative analysis was conducted for smallholder maize farmers in Wenchi and Techiman Municipalities in this region. The empirical results revealed that the three main constraints to the adoption of climate change adaptation strategies were the farmers' lack of access to credit, the lack of a ready market for produce, and poor access to extension services. To the best of our knowledge, this is the first study that has examined the constraints on Ghanaian smallholder maize farmers' adaptation to climate change. The results have implications for government policies and strategic actions to assist smallholder farmers' adaptation to climate change.

Keywords: Brong-Ahafo Region; climate change; constraints; Ghana; maize farmers

12.1 Introduction

Climate change can be a barrier to food security and development agendas in most global economies including Ghana (Scholze et al., 2006). The impacts of climate change and variability on agricultural production and food security is becoming severe in many African countries (Boko et al., 2007). The negative impacts already being felt in Africa include direct effects on climate-dependent activities such as agriculture and indirect effects on socioeconomic indicators such as health, education, conflict and livelihoods (Orindi and Murray, 2005). According to the Intergovernmental Panel on Climate Change (IPCC), Africa has already experienced worsening food production (IPCC, 2007). This has made it difficult in achieving the United Nations' Millennium Development Goals (MDGs) of reducing hunger by half by 2015. Maize (*Zea mays* L.) is the major cereal crop cultivated in Ghana where the domestic demand is rapidly increasing. Under optimum rain-fed conditions, and with a 1°C increase in temperature, the regions in Africa which are known for growing maize could be hit by yield losses as high as 65% (Lobell et al., 2011). There would be substantial crop yield reductions if drought conditions prevail in these regions. Reports show variation in the rainfall pattern in Ghana, with a trend towards a longer dry season and disappearing short dry spells (Owusu et al., 2008). The result of high temperatures in Ghana is evident in the form of low maize yields due to a reduction in the optimal growing period and an increase in evapotranspiration (Dazé, 2007). Ghana's domestic demand for maize is projected to grow at 2.6% annually between 2010 and 2015 (MiDA, 2009). Nevertheless, there has been a shortage in the per capita global food production by 7% in the 21st century. Climate change and variability, and low soil productivity are believed to be the causes of the shortage (Rosenzweig and Parry, 1994). Adger et al. (2013) asserted that climate change is believed to be the paramount challenge of the 21st century and that its effect is being felt in every aspect of human life such and on the biodiversity of the forest, wetland, coral reef, fisheries and agricultural ecosystems.

The negative effects of climate change and variability are now widely perceived by the rural poor in developing countries. Changes in climatic variables such as rainfall, temperature, wind speed, relative humidity and sunshine duration are critical in the determination of crop yields, especially maize (Chang, 2002). The number of smallholder farmers who are already vulnerable would increase as a result of variations in climate (Schipper and Pelling, 2006). This could be attributed to poverty, sensitivity of the geographical locations, farmers' high dependence on natural resources, and their inability to adopt new livelihood strategies.

Although, smallholder farmers do have some level of experience and local knowledge in adapting to climate change that helps them cope during difficult periods. However, the intense and sustained levels of variability associated with long-term climate change cannot be managed by traditional coping strategies (Pettengell, 2010). Poor people in the poorest countries are the most vulnerable to the impacts of climate change (Stern, 2007). Extension officers have real challenges in providing smallholder farmers with knowledge that is contextually- and ecologically-specific. This is because models and information about climate change and variability are only available at global, national and continental levels. Therefore, smallholder farmers will find it extremely difficult to undertake sustainable crop production and animal rearing in environments with erratic climatic conditions, and particularly if suitable policies or adaptive measures are not put in place (Nelson et al., 2007).

Previous studies have examined farming household vulnerability to climate change (e.g., Etwire et al., 2013), but there is still limited information on household vulnerability to extreme climatic events in Ghana. This is because the majority of past studies on vulnerability in Ghana have mostly been based on poverty using national aggregates without household-level data (Norton et al., 1995; Klein and Nicholls, 1999; Novignon et al., 2012). Therefore, the objective of this study was to examine smallholder maize farmers' constraints to the adoption of climate change adaptation strategies in the Brong-Ahafo Region of Ghana.

12.2 Materials and Methods

12.2.1 Climate change

The global scientific community has not been able to agree on a common definition of climate change. This has led to institution-specific definitions of the term, although some similarities exist among various institutions. Climate change can, therefore, be defined as 'any change in climate over time irrespective of the cause of the change' (FAO, 2006). The Food and Agriculture Organization (FAO) added that the change is continuous in the global climatic system. The change is said to emanate from global warming, which is mostly caused by anthropogenic activities leading to the emission of greenhouse gases and the loss of vegetation cover. The IPCC (2007) has defined climate change as 'the mean weather conditions that are found in a place over a long period of time, usually three decades'. According to the United Nations Environment Program (UNEP), climate change may be caused by natural events (internal processes or external forces) or through the activities of humans, which modify the composition of the atmosphere or land-use (UNEP, 2009).

12.2.2 Constraints to climate change adaptation

Financial constraints

Numerous studies have revealed that financial constraints are one of the key barriers to climate change adaptation by smallholder farming households in Sub-Saharan Africa (Bryan et al., 2009; Antwi-Agyei et al., 2012; Peterson et al., 2013). Poor farming households are believed to have limited capital assets that may be needed to adopt a climate change adaptation strategy to reduce the adverse effects on their livelihoods (Dasgupta and Baschieri, 2010).

Institutional constraints

Lamb and Davis (2003) have defined an institution as a social bonding agent that interconnects stakeholders, including to access various kinds of capital and to facilitate their exercising of power. This defines the channels through which institutions disseminate climate change adaptation strategies. Agrawal and Perrin (2009) reported that institutions play a significant role in society. They can help local communities withstand the adverse effects of climate change and variability. They can also provide the mechanisms to help shape the social and individual interactions within a society. According to Biesbroek et al. (2013), institutions that are established by the government play a key role in assisting the adoption of climate change adaptation strategies. These authors also concluded that the climate change adaptation policies of many Sub-Saharan African countries tend to be informed by few people in authority rather than by the smallholder households who are most affected. In Sub-Saharan Africa, the supposed duties of extension personnel are to provide the essential link between research institutions and farming households (Antwi-Agyei et al., 2013). This facilitates the smooth flow of scientific technologies from the researchers to the farming households and a greater flow of knowledge of the household problems to the researchers through the extension officer for the development of required solutions. However, in most countries, the ratio of agricultural extension officers to farmers is far too small, making it extremely difficult for the extension personnel to respond to the needs of all farming households. Antwi-Agyei et al. (2013) revealed that institutional constraints in one way or another have some relationship with political constraints regarding adaptation to climate change and variability. This was also supported by Ford et al. (2011), who reported that the lack of political will at the local level is a major barrier to climate change adaptation.

Lack of climate change information

Availability of and access to information on climate change is important for planning and implementation of climate change adaptation strategies as well as adoption by smallholder farming households. According to the IPCC (2007), smallholder farming households represent the majority of the agricultural sector in Sub-Saharan Africa, and these households depend on climatic stability for all their livelihood activities. However, there are few reliable climate projections for this region due to a lack of appropriate climate data. This is consistent with the assertions of Adger et al. (2009), who noted that a lack of information and awareness about climate change characteristics could potentially serve as a barrier to successful implementation of adaptation strategies. Previous studies have revealed that data on the features of climate change is periodic, so even when it is forecasted, it might not be beneficial for the long-term planning of agricultural activities (Ziervogel et al., 2010). The usefulness of periodic climate projections may be limited by issues related to credibility, measurement, intellectual capacity, legality, and procedural and institutional mechanisms (Patt and Gwata, 2002). Antwi-Agyei et al. (2013) also argued that the lack of adequate state-of-the-art equipment at the meteorological agencies across the regions of Sub-Saharan Africa hinders timely predictions and forecasts of rainfall and temperature patterns. This makes it extremely difficult for smallholder farming households to make concrete decisions about their activities that are climate-dependent. This challenge means the smallholder farming households in most villages in Ghana will resort to the use of local agro-ecological knowledge. They also forecast the weather based on their past farming experiences, which fails in most cases (Antwi-Agyei et al., 2013).

Technological constraints

Technology is considered as one of the principal hindrances to climate change adaptation (Kithiia, 2011). Indeed, Smit and Skinner (2002) have argued that improvements in technology, such as the development of improved crop varieties, early warning systems and establishing irrigation techniques, are crucial to climate change adaptation. Farming households can only use the adaptation strategies and options that have been already developed (Kolikow et al., 2012) and which can be applied in their areas (regions) of operation. The choice of technology for climate change adaptation is also made within appropriate socio-economic, legal and institutional structures (Frich et al., 2002). However, countries in Africa tend to lack these structures. According to the Africa Climate Report commissioned by the UK Government, low technical know-how in climate modelling is one of the key challenges to climate change research in Africa (Washington et al., 2006). Enete and Amusa (2010) also reported that the technological constraint has been a major challenge in Africa. This is because governments in Africa do not sufficiently budget for agriculture and climate change-related research.

Lack of infrastructure development

It is believed that the lack of access to a ready market is linked with the poor physical infrastructure development such as road networks in the majority of farming communities across Africa. The lack of markets for farm produce is interconnected with the lack of appropriate storage facilities in most Sub-Saharan African countries (Antwi-Agyei et al., 2013). Vermeulen et al. (2012) posited that the lack of storage facilities weakens the bargaining power of the smallholder farming households compared with other agents within the agricultural marketing channel. This is because most of the smallholder farming households cannot store their produce and, therefore, accept whatever price the other agents offer them for the fear that their produce would lose its value. In situations where farming households do not receive good prices for their produce, it will be difficult for them to fulfil their credit obligations with financial institutions. This might have adverse effects on their ability to access future credits to adopt a climate change adaptation strategy.

Sociocultural constraints

The way a group of people lives and their belief systems can be a barrier to the adoption of climate change adaptation strategies (Jones and Boyd, 2011; Smith et al., 2011; Antwi-Agyei et al., 2013). Thus, the deep value systems, beliefs, cultural practices and perceptions of a group of people greatly affect their attitudes toward climate change and adaptation strategies. Culture is one of the principal factors influencing the decision to adopt. It also helps in identifying risks and implementation of appropriate climate change adaptation strategies. People who live in the same locality with dissimilar cultural backgrounds may respond differently to risks such as the negative impacts of climate change (Moser and Ekstrom, 2010; Adger et al., 2013).

12.2.3 Sampling and data

The primary data were collected from 150 smallholder maize farmers in the Brong-Ahafo Region of Ghana. A questionnaire was designed, pre-tested, revised and then administered at the household level to obtain the primary data. A multi-stage sampling procedure was employed in this study. First, the Brong-Ahafo Region was purposively selected due to the predominance of smallholder maize farmers in this region compared with other regions in Ghana. Second, Techiman and Wenchi Municipalities were purposively selected for the study based on their ecological, population and land area characteristics. Third, simple random sampling was employed to select two communities from within each of the municipalities. Aworowa and Oforikurom were the two communities selected from Techiman Municipality, and Wurompo and Awisa were selected from Wenchi Municipality. Fourth, 75 smallholder maize farming households were randomly selected from each municipality. Thus, the questionnaire was administered to a total of 150 maize-growing households from within the four communities in the two municipalities.

12.2.4 Data analysis

First, based on a literature review, we identified the following general factors that may constrain the adoption of climate change adaptation: access to credit, ready market for produce, extension service, access to meteorological information, access to improved crop varieties, education on climate change, and climate change adaptation policy. Second, a questionnaire was developed, pre-tested and then revised to remove all ambiguities in the questions. Third, the questionnaire was presented to the respondents for constraints ranking. The ranking was based on the most pressing constraint to the least pressing constraint using a numeric scale of 1–7, respectively. The total rank score of each constraint was calculated and the constraint with the lowest score was ranked as the most pressing one, whereas the constraint with the highest score was ranked as the least pressing one. The Kendall's coefficient of concordance (W) was used to measure the degree of agreement among the rankings of the constraints by the respondents using their rank scores. This coefficient (W) has a positive value ranging between 0 and 1. It is 1 in situations where the ranks assigned by each household are the same as those assigned by other households, and 0 when there is a maximum disagreement among the households.

Given that T represents the sum of ranks of each factor being ranked, the variance of the sum is given by:

$$\text{var}_T = \frac{\sum T^2 - (\sum T^2)/n}{n} \tag{12.1}$$

The maximum variance of **T** is then given by:

$$\frac{m^2(n^2-1)}{12} \tag{12.2}$$

Where, m = the number of sets of ranking by the farmers, and n = the number of specific constraints being ranked.

The Kendall's coefficient concordance (W) is, therefore, given as:

$$W = \frac{\left[\sum T^2 - (\sum T^2)/n\right]/n}{m^2(n^2-1)/12} \tag{12.3}$$

Equation 12.3 is then further simplified as:

$$W = \frac{12\left[\sum T^2 - (\sum T^2)/n\right]/n}{nm^2(n^2-1)} \tag{12.4}$$

The hypothesis and the significance of the rankings were further assessed using the chi-square (X^2) test, as follows:

The null (H_o) and the alternative (H_a) hypotheses were stated as follows:

H_o: There is no agreement among the rankings of the constraints by the households

H_a: There is agreement among the rankings of the constraints by the households

The Kendall's coefficient of concordance (W) was tested for significance using the chi-square (X^2) distribution.

12.3 Results and Discussion

12.3.1 Socioeconomic characteristics of respondents

The majority (i.e., about 84%) of the smallholder maize farmers in Wenchi Municipality were male and 16% were female (Table 12.1). There were 72% male farmers and 28% female farmers in Techiman Municipality. The pooled sample also showed a similar result, with 78% male and 22% female farmers. This result is consistent with the Regional Analytical Report from the Ghana Statistical Services' (GSS) Population and Housing Census (PHC), which reported that in terms of gender, the male population is larger than the female population by 65.3% in the Brong-

Ahafo Region of Ghana (GSS, 2010). The GSS further reported that 52.7% of the smallholder farmers involved in agriculture were male, with a female smallholder farmer population of 47.3% (GSS, 2010). This is attributed to the fact that agriculture is labour intensive, hence males are the majority.

Table 12. 1: Socioeconomic characteristics of the smallholder maize farmers in the Brong-Ahafo Region of Ghana (n = 150).

Characteristics		Municipality				Pooled	
		Wenchi		Techiman			
		Frequency of farmers	Percentage of farmers	Frequency of farmers	Percentage of farmers	Frequency of farmers	Percentage of farmers
Gender	Male	63	84	54	72	117	78.00
	Female	12	16	21	28	33	22.00
Age	20–34	13	17.33	6	8.00	19	12.67
	35–49	20	26.67	18	24.00	38	25.33
	50–64	21	28.00	29	38.67	50	33.33
	Above 65	21	28.00	22	29.33	43	28.67
Marital status							
	Single	11	14.67	17	22.67	28	18.67
	Married	54	72.00	44	58.67	98	65.33
	Divorced	6	8.00	11	14.66	17	11.33
	Widowed	4	5.33	3	4.00	7	4.67

The data regarding the age of the farmers indicated that 17.33% of smallholder maize farmers in Wenchi Municipality were within the age category of 20–34 years, 26.67% were within 35–49 years and 28% were within both the 50–64 and above 65 years categories. In Techiman Municipality, 8% were within 20–34 years, 24% were within 35–49 years and 38.67% were within 50–64 years, while 29.33% were above 65 years. The pooled sample indicated that 12.67% were within 20–34 years, 25.33% were within 35–49 years and 33.33% were within 50–64 years, while 28.67% were above 65 years. The PHC report indicated that 29.9% of smallholder farmers in the Brong-Ahafo Region were within the age category of 20–34 years, 34.1% were within the range of 35–49 years, 20.7% were within 50–64 years, and 13.4% were 65 years and above (GSS, 2010).

Nearly 15% of the smallholder maize farmers in Wenchi Municipality were single, 72% were married, 8% were divorced and 5.33% were widowed. In Techiman Municipality, 22.67% were single, 58.67% were married and 14.66% were divorced, while 4% were widowed. The pooled sample showed that 18.67% of the household respondents were single, 65.33% were married, 11.33% were divorced and 4.67% were widowed (Table 12.1). This result is consistent with GSS (2010), which reported that 45.4% of the population in the rural locality were married, 38.9% were single, 3.6% were divorced and 4.2% were widowed.

The data regarding access to credit indicated that in Wenchi Municipality, only 6.67% of the smallholder maize farmers had access to formal credit and 93.33% had no access to formal credit. In Techiman Municipality, only 24% of the households had access to formal credit and 76% had no credit access. From the pooled sample, only 15.33% of the households had access to formal credit while 84.67% of them had no access to formal credit (Figure 12.1).

In Wenchi Municipality, 61.33% of the smallholder maize farmers had a basic education, 6.67% had secondary education and 6.67% had tertiary education, while 25.33% of the household respondents had no formal education. In Techiman Municipality, 44% had a basic education, 1.33% had secondary education and 5.33% had tertiary education, while 49.33% had no formal education. The pooled sample showed that 52.67% of the respondents had a basic education, 4% had secondary education and 6% had tertiary education, while 37.33% had no formal education. This result is somewhat consistent with the Ghana Statistical Service (GSS) (2010) report, which revealed that 51.7% of the population in the region had a basic education (Figure 12.2). Furthermore, the data indicated that 34.67% of the respondents in Wenchi Municipality belonged to a local social group, while 65.33% of the respondents did not belong to any local social group. In Techiman Municipality, 30.67% of the respondents belonged to a local social group, while 69.33% of the respondents did not belong to any local social group. The pooled sample showed that 32.67% of the respondents belonged to a social group, while 67.33% of the respondents did not belong to a social group. This implies that the majority (i.e., 67.33%) of the respondents in the Brong-Ahafo Region did not belong to any local social group (Figure 12.3). Regarding extension services available to the farmers, 49.33% of the respondents in Wenchi Municipality had access to an extension service, while 50.67% had no access. In Techiman Municipality, only 24% of the households had access to an extension service, while 76% of the households had no access. For the pooled sample, 36.67% of the respondents had access to an extension service, while 63.33% had no access (Figure 12.4).

Figure 12. 1: Smallholder Maize Farmers' Access to Formal Credit

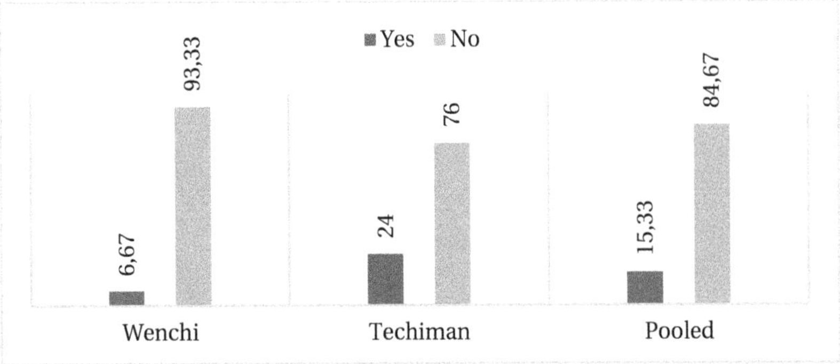

Figure 12. 2: Educational Level of Smallholder Maize Farmers

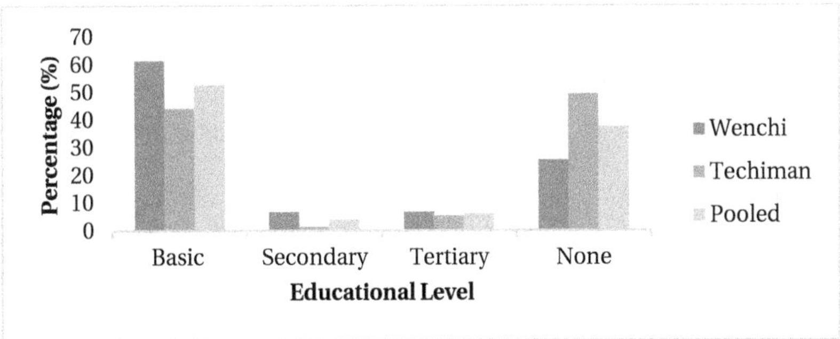

Figure 12. 3: Smallholder Maize Farmers' Membership of a Local Social Group

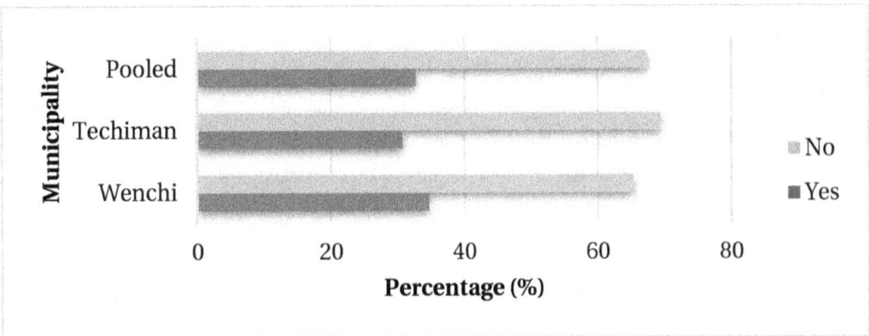

Figure 12. 4: Smallholder Maize Farmers' Access to Extension Services

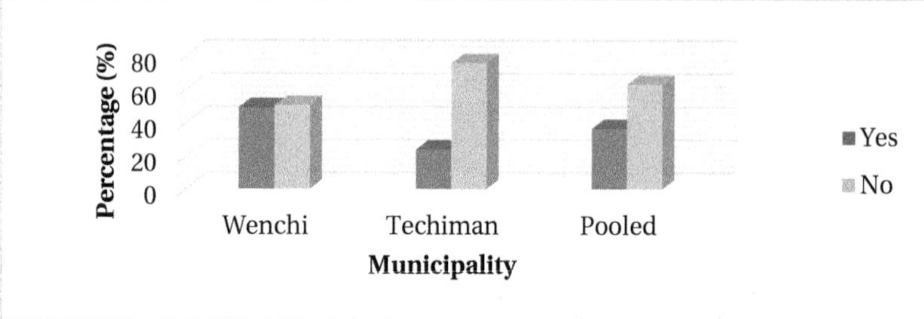

12.3.2 Constraints to the adoption of climate change adaptation strategies

The results presented in Table 12.2 indicate that for the pooled sample, the Kendall's rank technique showed that the lack of access to credit, the lack of a ready market for produce, and poor access to extension services were the three most pressing constraints identified by the respondents. These three constraints had mean scores of 2.01, 2.85 and 3.48, respectively. The result regarding the lack of access to credit is consistent with previous studies (Peterson et al., 2013; Antwi-Agyei et al., 2013), which reported that a financial constraint is one of the key barriers to the adoption of climate change adaptation strategies by smallholder farming households. The result regarding the lack of a ready market for produce agrees with Vermeulen et al. (2012), who found this to be one of the main constraints for smallholder farming households in their quest to adopt climate change adaptation strategies. Vermeulen et al. (2012) also found that due to a lack of storage facilities, smallholder farming households accept whatever price they are offered for their produce. The result regarding poor extension service is also consistent with Antwi-Agyei et al. (2013), who reported that poor access to extension services is one of the main constraints to adopting climate change adaptation strategies in Sub-Saharan Africa. The Kendall's *W* coefficient of concordance was 0.374, indicating that there was 37.4% agreement among the rankers of the constraints ($p < 0.001$).

In Techiman Municipality, the result of the Kendall's rank technique similarly revealed that the lack of access to credit, poor access to extension services and the lack of a ready market for produce were the three most pressing constraints identified by the respondents. These three constraints had mean scores of 2.25, 2.61 and 3.45, respectively. The Kendall's *W* coefficient of concordance was 0.414, indicating that there was 41.4 % agreement among the rankers of the constraints ($p < 0.001$).

In Wenchi Municipality, the Kendall's rank technique showed that lack of access to credit, the lack of a ready market for produce, and the lack of access to meteorological information were the three most pressing constraints identified by the respondents. These three constraints had mean scores of 1.77, 2.24 and 4.28, respectively. This result corroborates with Adger et al. (2009), who reported that information and awareness on climate change is one of the principal constraints to smallholder farming households' adoption of adaptation strategies. The Kendall's *W* coefficient of concordance was 0.438, indicating that there was 43.8 % agreement among the rankers of the constraints ($p < 0.001$) (Table 12.2).

Table 12. 2: Constraints to the adoption of climate change adaptation strategies of the smallholder maize farmers in the Brong-Ahafo Region of Ghana (n = 150).

Constraints	Municipality				Pooled	
	Techiman		Wenchi			
	Mean score	Rank	Mean score	Rank	Mean score	Rank
Lack of access to credit	2.25	1st	1.77	1st	2.01	1st
Lack of ready market for produce	3.45	3rd	2.24	2nd	2.85	2nd
Poor access to extension services	2.61	2nd	4.35	4th	3.48	3rd
Lack of access to meteorological information	3.49	4th	4.28	3rd	3.88	4th
Lack of access to improved crop varieties	5.06	5th	4.83	5th	4.94	5th
Lack of education on climate change	5.67	6th	5.02	6th	5.35	6th
Lack of climate change adaptation policy	5.46	7th	5.51	7th	5.48	7th
Diagnostic statistics						
Number of observations	75		75		150	
Kendall's *W* coefficient of concordance	0.414		0.438		0.374	
Chi-square calculated	186.377		197.012		336.414	
Chi-square critical	12.5916		12.5916		12.5916	
Degrees of freedom	6		6		6	
Asymptotic significance	0.000		0.000		0.000	

12.4 Conclusions

The present study examined smallholder maize farmers' constraints to adopting climate change adaptation strategies in the Brong-Ahafo Region of Ghana using the Kendall's coefficient concordance ranking technique. The study used primary data collected from 150 maize farmers in Techiman and Wenchi Municipalities. The study examined the pooled data for Techiman and Wenchi Municipalities as well as the datasets for the individual municipalities. The empirical results for the pooled sample indicated that the lack of access to credit, the lack of a ready market for produce and poor access to extension services were the three most pressing constraints on the farmers' adaptation to climate change, with mean scores of 2.01, 2.85 and 3.48, respectively. Similarly, the three most pressing constraints for the farming households in Techiman Municipality were the lack of access to credit, poor access to extension services and the lack of a ready market for produce, whereas in Wenchi Municipality they were the lack of access to credit, the lack of a ready market for produce, and the lack of access to meteorological information.

There is a need to formulate financial policies that will facilitate access to credit by smallholder maize farming households in Techiman and Wenchi Municipalities. This would encourage the households to adopt climate change adaptation strategies related to innovative production practices, thereby enabling these farmers to become resilient to climatic shocks. The recommendation is based on the empirical results showing that the lack of access to credit is an important constraint to climate change adaptation by the smallholder maize farming households in both municipalities. The Municipal Assemblies and Meteorological Agencies as well as other non-governmental organizations should inform smallholder maize farmers in Techiman and Wenchi Municipalities about impending natural disasters such as flood, drought and pest outbreaks. This is based on the empirical results that the majority of the respondents in both municipalities did not receive early warnings about impending disasters. Finally, the government should collaborate with agricultural extension agents and other development agencies to play their respective roles in assisting smallholder maize farmers to better cope with the negative impacts of climate change.

References

Adger, W. N., Barnett, J., Brown, K., Marshall, N., and O'Brien, K. (2013) Cultural dimensions of climate change impacts and adaptation. *Nature Climate Change*, 3(2), 112–117.

Adger, W. N., Dessai, S., Goulden, M., Hulme, M., Lorenzoni, I., Nelson, D. R., and Wreford, . (2009) Are there social limits to adaptation to climate change?. *Climatic change,* 93(3-4), 335–354.

Agrawal, A., and Perrin, N. (2009) Climate adaptation, local institutions, and rural livelihoods. In Adger, W. N., Lorenzoni, I., and O'Brien, K. L. (eds), *Adapting to climate change: Thresholds, values, governance,* Cambridge, UK: Cambridge University Press, 350–367.

Antwi-Agyei, P., Dougill, A. J., Fraser, E. D., and Stringer, L. C. (2013) Characterizing the nature of household vulnerability to climate variability: empirical evidence from two regions of Ghana. *Environment, Development and Sustainability,* 15(4), 903–926.

Antwi-Agyei, P., Fraser, E. D., Dougill, A. J., Stringer, L. C., and Simelton, E. (2012) Mapping the vulnerability of crop production to drought in Ghana using rainfall, yield and socioeconomic data. *Applied Geography,* 32(2), 324–334.

Biesbroek, G. R., Klostermann, J. E., Termeer, C. J., and Kabat, P. (2013) On the nature of barriers to climate change adaptation. *Regional Environmental Change,* 13(5), 1119–1129.

Boko, M., Niang, A., Nyong, A., Vogel, C., Githeko, M., Mednay, M., and Yanda, P. (2007) In Climate Change 2007: Impacts, Adaptation and Vulnerability', Contribution of Working Group II to the Fourth Assessment Report of the Intergovernmental Panel on Climate Change, ed. Parry, M. L., Canziani, J. P., Palutikof, J. P., van der Linden, P. J., & Hanson, C. E. Cambridge: Cambridge University Press.

Bryan, E., Deressa, T. T., Gbetibouo, G. A., and Ringler, C. (2009) Adaptation to climate change in Ethiopia and South Africa: options and constraints. *Environmental Science and Policy,* 12(4), 413–426.

Chang, C. (2002) The potential impact of climate change on Taiwan's Agriculture. *Agricultural Economics,* 27(1), 51–64.

Dasgupta, A., and Baschieri, A. (2010) Vulnerability to climate change in rural Ghana: Mainstreaming climate change in poverty-reduction strategies. *Journal of International Development,* 22(6), 803–820.

Dazé, A. (2007) Climate change and poverty in Ghana. *Unpublished report by CARE International, Ghana.*

Enete, A. A., and Amusa, T. A. (2010) Challenges of agricultural adaptation to climate change in Nigeria: a synthesis from the literature. *Field Actions Science Reports,* 4, 1–10.

Etwire, P. M., Al-Hassan, R. M., Kuwornu, J. K. M., and Osei-Owusu, Y. (2013) Application of livelihood vulnerability index in assessing vulnerability to climate change and variability in northern Ghana. *Journal of Environment and Earth Science,* 3(2), 157–170.

FAO. (2006) Global forest resources assessment 2005–progress towards sustainable forest management. FAO Forestry Paper No. 147, Rome, Italy.

Ford, J. D., Berrang-Ford, L., and Paterson, J. (2011) A systematic review of observed climate change adaptation in developed nations. *Climatic change,* 106 (2), 327–336.

Frich, P., Alexander, L., Della-Marta, P., Gleason, B., Haylock, M., Tank, A. K., and Peterson, T. (2002) Observed coherent changes in climatic extremes during the second half of the twentieth century. *Climate Research,* 19 (3), 193–212.

Ghana Statistical Service (2010) Population and Housing Census, Analytical Report, Brong-Ahafo Region, Ghana.

IPCC. (2007) 'Climate Change 2007: Synthesis Report', Contribution of Working Groups I. *II and III to the Fourth Assessment Report of the Intergovernmental Panel on Climate Change.* In: IPCC, 104.

Jones, L., and Boyd, E. (2011) Exploring social barriers to adaptation: Insights from Western Nepal. *Global Environmental Change,* 21(4), 1262–1274.

Kithiia, J. (2011) Climate change risk responses in East African cities: need, barriers and opportunities. *Current Opinion in Environmental Sustainability,* 3(3), 176–180.

Klein, R. J., and Nicholls, R. J. (1999) Assessment of coastal vulnerability to climate change. *Ambio,* 28(2), 182–187.

Kolikow, S., Kragt, M.E., and Mugera, A. W. (2012) An interdisciplinary framework of limits and barriers to climate change adaptation in agriculture, Working Paper 1202, School of Agricultural and Resource Economics, University of Western Australia, Crawley, Australia.

Lamb, S., and Davis, P. (2003) Cenozoic climate change as a possible cause for the rise of the Andes. *Nature,* 425(6960), 792–797.

Lobell, D.B., Banzinger, M., Magorokosho, C., and Bindiganavile, V. (2011) Non-linear heat effects on African maize as evidenced by historical yield trials. *Nature Climatic Change,* 1, 42–45.

MiDA. (2009) Investment Opportunity in Ghana; Maize, Soya and Rice, Millennium Development Authority (MiDA) and United States Millennium Challenge Corporation.

Moser, S. C., and Ekstrom, J. A. (2010) A framework to diagnose barriers to climate change adaptation. *Proceedings of the National Academy of Sciences,* 107(51), 22026–22031.

Nelson, D. R., Adger, W. N., and Brown, K. (2007) Adaptation to environmental change: Contributions of a resilience framework. *Annual review of Environment and Resources,* 32(1), 395–419.

Norton A., Aryeetey, E. B., Korboe, D., and Dogbe, D.K.T. (1995) Poverty Assessment in Ghana Using Qualitative and Participatory Research Methods, PSP Discussion Paper Series No. 83, Washington D.C., World Bank.

Novignon J., Mussa, R., and Chiwaula, L. (2012) Health and vulnerability to poverty in Ghana: Evidence from the Ghana living standards survey round 5. *Health Economics Review,* 2(11), 1–9.

Orindi, V. A., and Murray, L. A. (2005) *Adapting to climate change in East Africa: a strategic approach.* IIED, London, UK.

Owusu, K., Waylen, P., and Qiu, Y. (2008) Changing rainfall inputs in the Volta basin: Implications for water sharing in Ghana. *GeoJournal,* 71(4), 201–210.

Patt, A., and Gwata, C. (2002) Effective seasonal climate forecast applications: examining constraints for subsistence farmers in Zimbabwe. *Global Environmental Change,* 12(3), 185–195.

Peterson, L. C., Deplazes, G., Lückge, A., Timmermann, A., Hamann, Y., Hughen, K. A., and Haug, G. H. (2013) Links between tropical rainfall and North Atlantic climate during the last glacial period. *Nature Geoscience,* 6(3), 213–217.

Pettengell, C. (2010) Climate Change Adaptation: Enabling people living in poverty to adapt. *Oxfam Policy and Practice: Climate Change and Resilience,* 6(2), 1–48.

Rosenzweig, C., and Parry, M. L. (1994) Potential impact of climate change on world food supply. *Nature,* 367, 133–138.

Schipper, L., and Pelling, M. (2006) Disaster risk, climate change and international development: Scope for, and challenges to, integration. *Disasters* 30(1), 19–38.

Scholze, M., Knorr, W., Arnell, N. W., and Prentice, I. C. (2006) A climate-change risk analysis for world ecosystems. *Proceedings of the National Academy of Sciences,* 103(35), 13116–13120.

Smit, B., and Skinner, M. W. (2002) Adaptation options in agriculture to climate change: A typology, *Mitigation and Adaptation Strategies for Global Change,* 7(1), 85-114.

Smith, M. S., Horrocks, L., Harvey, A., and Hamilton, C. (2011) Rethinking adaptation for a 4 C world, *Philosophical Transactions of the Royal Society A: Mathematical, Physical and Engineering Sciences,* 369(1934), 196–216.

Stern, N. (2007) *The economics of climate change: the Stern review,* Cambridge University Press.

UNEP. (2009) *The defining Challenge of our age: United Nations climate Change Conference.* United Nations Environment programme, Copenhagen, Denmark.

Vermeulen, S. J., Aggarwal, P.K., Ainslie, A., Angelone, C., Campbell, B.M., Challinor, A.J., Hansen, J.W., Ingram, J.S.I., Jarvis, A., Kristjanson, P., Lau, C., Nelson, G.C., Thornton, P.K., and Wollenberg, E. (2012) Options for support to agriculture and food security under climate change. *Environmental Science and Policy,* 15(1), 136-144.

Washington, R., Kay, G., Harrison, M., Conway, D., Black, E., Challinor, A., & Todd, M. (2006) African climate change: taking the shorter route. *Bulletin of the American Meteorological Society,* 87(10), 1355–1366.

Ziervogel, G., Johnston, P., Matthew, M., & Mukheibir, P. (2010) Using climate information for supporting climate change adaptation in water resource management in South Africa. *Climatic Change,* 103(3-4), 537-554.

CHAPTER 13

General discussion on the vulnerability and adaptation of food supply chain actors to climate change

John K. M. Kuwornu and Derick T. Adu

13.1 Introduction

In both private and public decision-making over recent years, climate change has been one of the principal topics. This now common perception of climate change is a radical adjustment compared to perceptions only a few years ago. According to Stern (2006), change in climate which emanates from global warming has been recognized globally for adversely impacting on human societies, business operations and the environment itself. In fact, businesses have to undertake their climate change-prone operations in increasingly vigorous and risky environments where institutional, resource-based, supply chain and stakeholder views are all important to characterize and understand corporate strategic responses to sustainability issues (Dasaklis and Pappis, 2013; Kolk and Pinkse, 2007). Drivers of climate change such as environmental treaties, shareholders' value and customers' changing buying patterns pose both opportunities and threats for businesses and their supply chain networks. During the last few years, foremost companies from different sectors have recognized the emerging opportunities and threats caused by a change in climate, and they are now taking actions to carry out a more robust climate change agenda. It is well known that the food sector has been highly compromised by changes in climate.

Figure 13. 1: Implications of climate change drivers for supply chain management

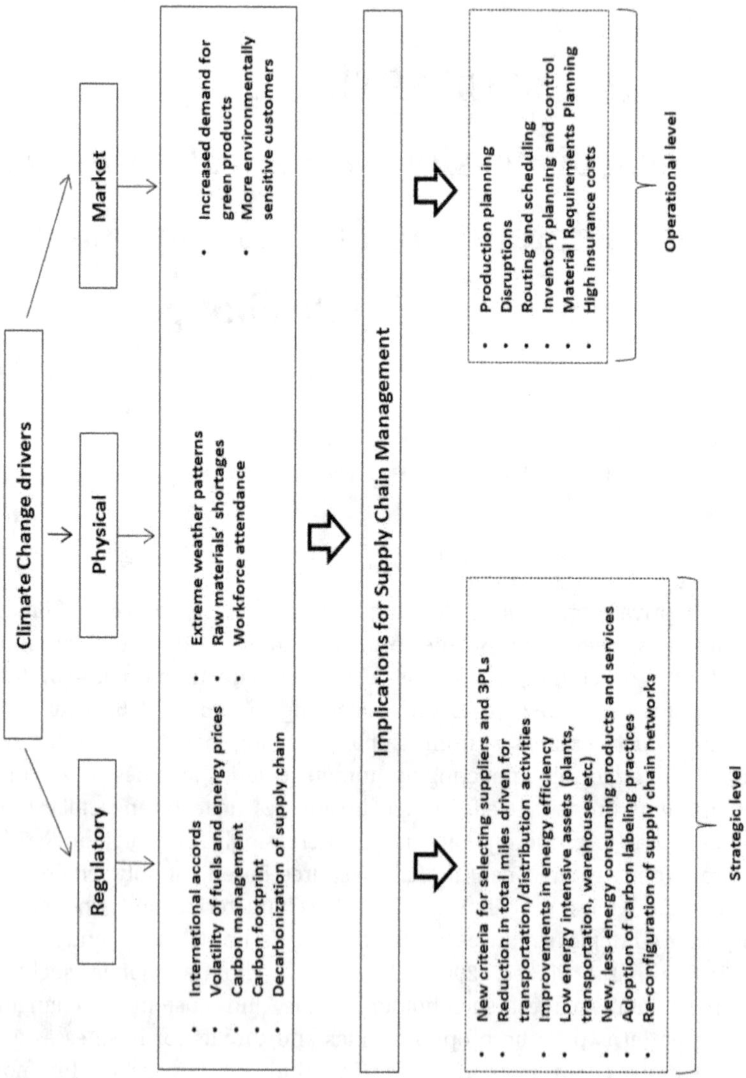

Source: Dasaklis and Pappis (2013)

13.2 Discussion

Climate change and its effects compel enterprises to compete in an increasingly unstable environment. Nowadays, customer demands for products and services include low prices, high quality and growing concerns about environmental impacts such as low carbon footprints and high energy efficiency (Song and Lee, 2010). Businesses declining to account for this message in their own supply chain strategies or that are unwilling to address climate change issues will eventually be "punished" by environmentally-sensitive customers. Also, given it is considered that supply chains and not companies compete (Christopher, 1998), it is apparent that supply chain management will play an important part in meeting new customers' requirements in view of climate change. It is also apparent that appropriate supply chain strategies will enable a firm to take full advantage of the opportunities associated with minimizing the threats of climate change. In the low carbon era we have entered, it seems that low carbon supply chain management will be the driver for enhancing business competence (Yang and Zhang, 2011). Regulatory, physical and market drivers stemming from climate change create both threats and opportunities for supply chains. Consequently, several implications for the strategic and operational management of supply chains must be adequately addressed (Figure 13.1).

One ingredient that is critical for human survival and civilization is food. Apart from providing energy for day-to-day activities, proper nutrition plays an essential part in human health, learning, growth and reproduction. When food is efficiently available on a large scale, excess human labor can be devoted to other occupations, allowing society to advance (Diamond, 2005; FAO, 2013). Food availability, especially in the richer economies of the world, depends on a network of producers, distributors, processors and retailers that move products from farm to plate. Indeed, food surpluses ensure that most of the human population can labor in pursuits other than agriculture (FAO, 2013).

The term 'food system' refers to the complex life-cycle of food products from production through consumption and then disposal (MacRae and Donahue, 2013). Although producers and manufacturers generate enough calories to feed most nations' populations, many products are exported to serve markets abroad, while other products are imported to serve domestic consumers. A supply chain can be defined as the full range of activities required to bring products or services from conception, through the different phases of production and delivery to final consumers, and the final disposal after use (Kaplinsky and Morris, 2001). The food supply chain can, therefore, be defined as the network of organizations involved, through upstream linkages to farms and downstream linkages to consumers, in the different processes and activities that produce valued food products and food services for consumers (Christopher, 1998).

The main difference between a food supply chain and a food system is that the latter includes the interactions within and between the bio-geophysical and human environments, and the societal outcomes of the activities (Ericksen, 2008). Food supply chains are critical components of the global food system, comprising all of the inputs, actors, processes, transactions, and transport between raw inputs and consumption (Hawkes, 2009). Figures 13.2 and 13.3 present the common processes or elements that may be found in a food supply chain, though the form and complexity of an individual chain depends on the product and the actors involved (FAO, 2013).

The food supply chain concept is driven by business management scientists in the fields of logistics, physical distribution, operations management, marketing and supply chain management.

Figure 13. 2: Elements in the food supply chain

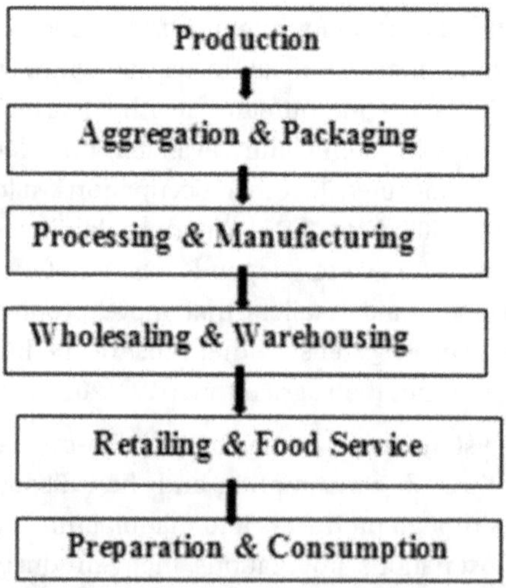

Source: MacRae and Donahue (2013)

Figure 13. 3: A depiction of the food supply chain processes

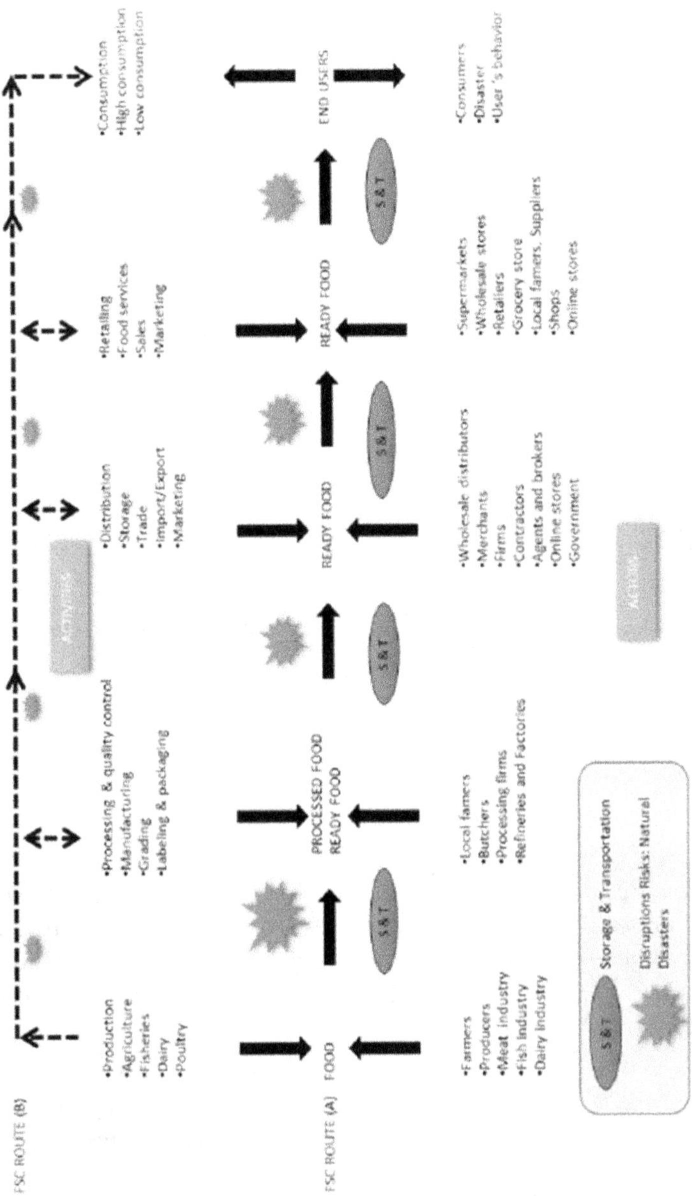

Source: Reddy and Anbumozhi (2017)

Climate change is expected to have significant adverse impacts on the livelihoods of the deprived (Oxfam, 2009; Hepworth and Goulden, 2008; Zake, 2009). Specific adaptation policies, therefore, need to be implemented to decrease these impacts and make available effective options to enhance livelihoods (Allison et al., 2009a). Vulnerability to climate change is expressed as a function of exposure, sensitivity, and adaptive capacity. Vulnerability can be defined as combinations of the extrinsic exposure of individuals, groups or ecological systems to hazards (such as climate change), their intrinsic sensitivity to the hazard, and their lack of capacity to modify exposure to, absorb, and recover from losses stemming from the hazard, and to exploit new opportunities that may arise in the process of adaptation (Allison et al., 2009b, p.175). Following Metzger et al. (2006), vulnerability of food supply chains to climate change can be conceptualized, as depicted in Figure 13.4.

Figure 13. 4: Conceptual model of vulnerability

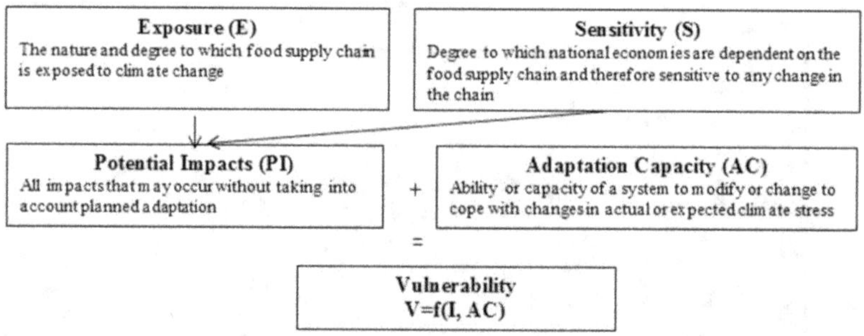

Source: Modified from Allison et al. (2009b)

"Climate change is bad for farmers and for agriculture. Drought, flooding and hotter growing conditions threaten the world's food supply and contribute to food insecurity… Now is the time to meaningfully address the reality of climate change … We are ready to meet the climate challenges that face our businesses"

Open letter in The Washington Post and Financial Times from CEOs of 14 food and beverage companies ahead of the Paris Climate Change Conference

An inclusive climate-adaptive food supply chain initiative would comprise a variety of strategies, to be carried out according to the present and future vulnerability of each link in the chain. Resistance strategies to cushion the food supply chain could include larger climate-controlled storage facilities, improved crop cleaning and drying, better storage to avoid losses to pests, additional processing to reduce spoilage, and more robust packaging to withstand difficult transport conditions. Resilience strategies could include enhancing water storage and the efficiency of irrigation methods, and the selection of seed varieties that maintain yields under current climate variability but decrease the likelihood of crop failure during extremely wet or dry years and whenever credit or fertilizers are in short supply (USAID, 2011).

ADVISEN Ltd. (2013) has identified the following three factors as the most important for determining the climate change vulnerability of different supply chains:

Complexity and dimension of stages and networks: when the supply chain is constituted by a large quantity of suppliers, the possibility of suffering negative impacts generated by disruptions is larger than in the case of small and local supply-production systems.

Concentration of suppliers: the possibility of having different suppliers for the same commodity is an important element to increase the flexibility of the supply chain and reduce the costs and the time of recovery after an unexpected event.

The magnitude of the impact: this is dependent on: i) by how and how much the commodity is susceptible to the effects generated by climate change. Resilience and adaptability or substitution between resources are important elements to determine the magnitude of the impacts; and ii) how capable is the supplying area in coping with unexpected events. This is based on elements such as governance, recovery and risk management strategies.

Accordingly, climate change will affect food supply chains in varying ways, depending on the complexity of the supply chain and on the degree of inter-connectedness of and dependency on other industries (Beermann, 2011). Changing climate can affect the entire chain of value-adding activities for agricultural commodities, from production and processing to marketing and consumption of the final product. Table 13.1 presents an overview of disaster risk at the component level along the food supply chain.

Table 13. 1: Overview of disaster risk at the component level along food supply chains

Supply chain component	Assets	Vulnerabilities
Producers	Standing crops, irrigation systems, livestock shelters, hatcheries, grain reserves, dams, farm access roads, equipment, farm schools and cooperatives, loans, biodiversity, ecosystems, R&D, agriculture technologies.	Increased costs of farm inputs such as fertilizers, seeds, livestock feed, and veterinary care. Reduced demand for inputs, decreased production, limited support of technology, increased ineptness and erosion of livelihoods, depleted savings, sales of vital productive assets.
Processors	Infrastructure, processing facilities, roads, dams, towns, facilities for storage, loans.	Lower revenue, high operational costs, unexpected expenditure, reduced quality, high price index, low revenues.
Distributors/retailers	Buildings, roads, dams, towns, facilities for storage, management, community networks.	Income loss, lower purchasing power, reduced quality, loss of market access, weak social support networks.
Consumers	Commodities price index, ecosystems, infrastructure, roads, dams, towns, community networks.	Food inflation, reduced quantity and quality of food, food insecurity, malnutrition.

Source: Reddy and Anbumozhi (2017)

The International Food Policy Research Institute (IFPRI) has stated that nutritional status, ultimately leading to morbidity and mortality, "can be exacerbated by the effects of climate change at all stages of the food value chain"[1]. The impacts of climate change on the coffee supply chain in Uganda was studied by Dekens and Bagamba (2014) through the application of participatory methods and qualitative analysis. The authors found that every actor along the supply chain is adversely affected by climate change, but in diverse means and to different extents. They added that a good number of the actors have been trying to minimize the adverse impacts of climate change, but nonetheless not all responses are

[1] https://foodtank.com/news/2017/08/report-climate-change-food-value-chain : IFPRI/Milo Mitchell

sustainable. A lack of trust and communication among actors along the supply chain chiefly hinders climate change adaptation. It is well asserted that the development of comprehensive agro-value chains is undermined by changes in climate. Worldwide, the impacts of extreme weather events such as floods, droughts and hailstorms already undermine the performance of supply chains, for example, by leading to losses in agricultural production, the destruction of processing and transport infrastructures, and the deterioration of export earnings. This trend is expected to continue and worsen as climate change brings more frequent and intense extreme events, shifting rainfall patterns and rising temperatures. For example, reduced rainfall may limit hydroelectricity production, restricting agro-processing capacities and leading to higher production costs, processing costs and market prices with distributional impacts not only at the farm level but across a broad array of sectors and industries (Dekens and Bagamba, 2014).

Gregory et al. (2005) noted that climate change might affect the food supply chain in many diverse ways. Extreme events such as floods and droughts, cyclones or windstorm can lead to food safety hazards and increased uncertainty related to agricultural production and costs (IPCC, 2012). Given that the food sector relies on the ability of ecosystems to provide resources, the impacts generated by climate change can be large and costly. Hazards can arise at various stages of the food chain, from primary production, to transport, transformation and consumption, making it difficult to estimate impacts and to plan effective management strategies.

Climate change due to global warming can have severe adverse impacts on transport operations. Extreme weather events may have very costly and even disastrous knock-on effects for the movement of all kinds of materials. For example, extreme weather events may trigger adjustments to transporting speeds and therefore influence travel durations (Maoh et al., 2008). According to Sussman and Freed (2008), a general increase in temperature and a higher frequency of hot summers are likely to result in an increase in buckled railway lines and rutted roads, which results in substantial disruption and repair costs. A comprehensive list of climate change impacts on transport and the corresponding actions required to cope can be found in Humphrey (2008). The health implications of food insecurity such as poor nutrition, premature death or diseases have been considered as one of the main negative impacts generated by climate change, particularly in poor and developing countries. Climate change influencing stability of primary production also affects food manufacturing and trade. In addition, the rising average temperatures can increase the hygiene risks associated with food storage and distribution.

The gradual increase in the world's mean temperature and the increased frequency of extreme weather is expected to have a major impact on food supply chain performance and food security (IPCC, 2014). Every aspect of the food supply chain can be heavily disturbed by climate change. Its effects will endanger fisheries and farms around the globe, possibly causing severe deficits in production (Elver, 2015; IPCC, 2014; Tacoli et al., 2013; World Bank, 2015). For example, in Guatemala, Fresh Del Monte Produce Inc. faced a $9m loss due to heavy rains and flooding which affected banana production. The global commodity trading firm Bunge reported a $56m loss in its sugar and bioenergy business due to drought. Severe droughts in Russia led the country to ban wheat exports, which created a disastrous ripple across global stock markets and prompted a 2.2 percent drop in the share price of General Mills. In 2010, Russia suffered a severe heat wave. The resulting economic losses were estimated to be US$15bn as drought and wildfires destroyed crops, particularly wheat. The resulting effect was export restrictions on wheat in Russia, which contributed to global price increases.

Reductions in productivity, as well as ruined harvests, can bankrupt small producers and upsurge prices, adding to the problem of poverty and growth in population (Tacoli et al., 2013). The susceptibility of infrastructure networks to extreme weather such as severe precipitation, storm surges, intense temperatures, high winds and flooding can adversely affect the efficient functioning of supply chains, including during shipping, manufacturing, wholesaling, retailing, or consumption.

Regarding wholesale and retail trade, both use infrastructure, equipment and processes, which are vulnerable to climate change risks. This implies that wholesalers and retailers use inventory facilities (e.g., buildings, material handling equipment), material movement and transport facilities and infrastructure (e.g., ports, highways, trucks, trains), personnel, and communications. These may be impacted negatively by extreme weather events owing to change in climate. According to the Carbon Disclosure Project (2011), the food retailing sector faces reputational risk as its visibility in the market and the increasing transparency of its operations, combined with NGO power, drives companies to source more sustainable materials. In addition, new regulations for product labelling could increase the costs associated with producing consumer goods. To absorb this increase in costs, the retail sector would have to increase the price of the products, causing reductions in demand for these products leading to product waste with obvious impacts for the environment generally and, more specifically, for global warming (White, 2007). Also, production and packaging are associated with a large share of greenhouse gas emissions and this share is even bigger when products are imported and not domestically produced (Product Policy Institute, 2009).

Tacoli et al. (2013) reported that agricultural production is challenged by changes in climate and long-term shifts in average temperatures and rainfall. Numerous studies have added that transitory events akin to storms and heat waves are difficult to forecast and can produce severe harm to food crops (e.g., Ferreira, 2012; Müller and Elliott, 2015; Wheeler, 2015). The IPCC (2014) asserts that long-term variations in temperature and precipitation patterns will alter the productivity of agricultural lands, thereby benefiting some regions and impairing others.

Viability of commercial fisheries is also threatened by long-term changes in the temperature, acidity, and oxygen content of the oceans which may change species distributions and drive extinctions (Elver, 2015; IPCC, 2014). In the view of authors such as Elver (2015), London Assembly Economy Committee (2015) and Tacoli et al. (2013), a climate-related disaster at one point in a food supply chain at one particular location may result in flow-on impacts for subsequent actors and processes along the chain, wherever they may be located. This indicates that it is not enough for planners to interest themselves just with recognizing and tackling local climate-related risks, but their efforts should be all-inclusive. This is because the distribution of a single food product may involve the function of buildings, infrastructure, market transactions, and ecosystem services in multiple regions, perhaps on multiple continents.

During natural disasters, strong winds and floodwaters may spread hazardous materials that contaminate food products or harm crops, livestock and seafood. Environmental changes can also alter pollutant loads in food products. Changes in climate also pose another challenge for numerous urban regions even now struggling to reinforce and uphold regional food production. Viability of food production will fluctuate with environmental conditions and decreases in production will jeopardize the economic viability of food businesses further down the supply chain. It is believed that interdependency among economies ensures that producers are also vulnerable to climate-related challenges that may influence distant regions and nations (Elver, 2015; Tacoli et al., 2013).

Hurricanes Katrina and Sandy suffered by New York City caused failures in the normal food distribution systems, leading to shortages and the loss of or damage to food products (Bonini, 2014; City of New York, 2013a). Hunts Point, the site of New York's major food wholesaling and distribution hub, is located on a peninsula subject to flooding (City of New York, 2013a; New York City Economic Development Corporation, 2014). Vulnerability, along with variations in distributors' adaptive capacity, means that poor communities may face greater food shortages than wealthier ones. Table 13.2 presents the impacts of climate change on commodity supply chains.

Table 13. 2: Commodity supply chain at glance

Commodity	Global market 2011	Top 5 exporters	Exposure to climate change
Rice	$23.3bn	Clustered in South and South-East Asia, led by Thailand (28% of global exports), followed by Vietnam, India and Pakistan. The US is the 5th largest exporter.	High water intensity and specific climate requirements make rice exposed to changes in temperature and precipitation. These elements, as well as extreme weather conditions such as storms, pose serious threats to the rice supply chain. The key exporters are also vulnerable to rising sea levels.
Wheat	$48.3bn	North America (US, Canada), South America (Argentina), Europe (EU) and Australia.	Changes in temperature can have a positive effect on the wheat yields in North America but are likely to have the opposite effect in Europe, Australia and South America. However, rising sea levels and greater occurrence of heavy rains and storms threaten all top 5 exporters.
Maize	$23.6bn	US, Argentina, Brazil, Ukraine and India.	Increases in temperature are projected to improve harvests in the US as well as other leading maize exporters (the 2012 droughts in the US, however, defied this projection). However, extreme weather conditions and storms pose damage risks which could lead to volatility in supply and therefore global food prices. Maize plays an important role in the food chain, as it is used for feeding livestock. Its availability has a significant effect on the agricultural sector.

Source: Gledhill, Hamza-Goodacre, and Low (2013).

As time goes by, food prices may increase with the rising frequency and intensity of climate-related impacts on food production and transportation. Disasters and their results will also damage business and household budgets and may increase the cost of insurance or make it more difficult to obtain[2]. In addition, climate impacts may result in job and income losses for workers in food-related industries (City of New Orleans OHSEP et al., 2011; Miami-Dade County, NOAA, and NACO, 2011). Impacts on income and wealth could affect residents' ability to afford

[2] City of New Orleans OHSEP et al., 2011; Wolff & Comerford, 2014.

food. If a disaster is significant enough to drive away part of a city's population, there may also be consequences for the recovery period. Climate change impacts such as flooding may reduce the supply of functional retailers, requiring consumers to travel further during a period when transportation systems like subways and streetcars may not be operational (Wolff and Comerford, 2014). The *Fifth Assessment Report* of the IPCC showed that climatic changes over the last 30 years have already reduced global agricultural production in the range of 1–5 percent and that extreme weather events affecting major agricultural producers have helped to drive global food price volatility in recent years (IPCC, 2012, 2014).

Relatively, little has been done to support climate change adaptation beyond the production level, along entire value chains. This is presumably due to a combination of factors such as: a lack of understanding of climate change adaptation (including the distinction between adaptation and mitigation, and between adaptation and sustainable development); the channeling of climate-related finance to mitigation projects, and particularly in emerging economies such as China and Brazil rather than in the least-developed countries (Pauw and Pegels, 2013); the lack of data and information about future impacts of climate change; inherent uncertainties around the information that is available; and the lack of economic and/or costs-benefit analyses on impacts and risk management strategies. As a result, little is known about the impacts of climate change on value chains at the postharvest stage, including the wider macroeconomic implications for a country's competitiveness, especially in the developing world (Stathers et al., 2013). Nonetheless, to secure sustainable investments in value chain development, governments and other actors need to ensure that climate risks are managed not just at the production level, but also throughout the entire value chain from production to marketing. Efforts to adapt food supply chains to threats like climate change must also consider existing shortcomings and potential restraints (Porter et al., 2014). An example of the impact pathways of climate change and variability on food supply chains, in the context of the fisheries value chain in Uganda, is presented in Table 13.3. Based on this, potential adaptation strategies are outlined in Table 13.4. In this analysis, adaptation measures are assigned to the private or public sector appropriately. As recommended by Macfadyen and Allison (2009), while there are many potential adaptation measures that fall on the responsibility of individuals and the private sector, there are two significant areas for the state to foster effective adaptation: creating and supporting an enabling environment for adaptation, and regulating negative informal adaptation.

Studies such as Allison et al. (2009a) and Daw et al. (2009) reported that livelihood diversification support for fishing communities promotes the entire sector's adaptability to climate change and variability. To promote adaptation at the beginning of the fish value chain, support is needed to ensure that physical capital is insured against extreme weather events (Kambewa, 2007).

Table 13. 3: The impact pathways of climate change and variability for Uganda's fisheries sector

Point on value chain	Impact	Potential outcomes for fisheries
Inputs and services	Increased exposure of inputs (e.g., gear, boats, labour) to extreme weather, winds and storms.	Destruction of inputs and gear. Increased danger to boat crew and fishers.
Production	Changes in stream and groundwater temperature. Change in hydrology regimes, a functionof land use, precipitation, soil moisture and evapotranspiration. Hydrologic variability. Eutrophication. Water temperature effects on limnology.	Shifts in primary production. Shifts in secondary production (i.e., volume and distribution). Disease and species invasions. Decreased areas to breed in shallow waters. Less predictable seasonality of lakes.
	Increase in ultra violet (UVB) rays. Water loss from lentic systems; evaporation is expected to be greaterthan precipitation. Higher growth rates. Higher incidence of disease. Changes in water quality.	
Trade and transport	Roads and trade routes become Impassable.	Lack of access to markets.
Processing	Processing areas hit by unpredictablerain patterns.	Postharvest losses. Changes in processing technologies and costs due to abundance of new species.
Marketing	Supply scarcity	Price increases for available supply of fish. Increased number of fishers. Decreased revenues from declines in catch and/or stock abundance.

Adapted from: Ficke et al. (2007); WorldFish Center (2007); Daw et al. (2009)

Table 13. 4: Potential adaptation strategies within the fisheries value chain in Uganda

Point on Value Chain	Impact	Potential Adaptation Measure	Responsibility
Inputs and services	Increased exposure of inputs (e.g., gear, boats, labour) to extreme weather, winds and storms.	Insurance of physical capital equipment.	Private
		Weather warning systems.	Public
		Support for fishers' Advocacy.	Private/Community-based organizations
Production	Changes in stream and groundwater temperature.	Ongoing public research into climate change and variability in freshwater systems.	Public
	Change in hydrology regimes, a function of land-use, precipitation, soil moisture and evapotranspiration.	Support for diversified Livelihoods.	Public/Private
	Hydrologic variability.	Improved monitoring of illegal gear use.	Public
	Eutrophication	Management of wetlands/ lakeshore areas.	Public
Trade and transport	Roads and trade routes become impassable.	Infrastructure provision (e.g., roads).	Public
Processing	Processing areas hit by unpredictable rain patterns.	Improved post-harvest technology (e.g., solar driers).	Private
		Training in postharvest handling suitable for migratory populations.	Private
		Improved market information channels on consumer preferences.	Private
Marketing	Supply scarcity	Promotion and support of cooperative groups.	Public/Private

Climate change adaptation is an example of a societal case for improved food supply chain management. Specifically, it is a social learning process in which the intermediate goal should be to empower all actors in a food system so that they learn how to adapt to climate change (FAO, 2014). Adaptation is also important for more conventional reasons because agriculture and the food sector form an important part of the global economy. For instance, in the EU economy, these sectors provide 15 million jobs (which is 8.3% of the total employment) and 4.4% of its GDP (Moussis, 2013).

Research in the UK has reported that community food enterprises can also play a crucial role in adapting to climate change. The following are adaptation measures suggested in the research:

- Reduced reliance on imports where supply may be more variable and transport more vulnerable to disruption.
- Reduced reliance on inputs that require energy for production.
- Support for methods of food production which are more likely to be able to adapt to changing climatic conditions.
- Creation of a new skills base able to grow more diverse food products and a customer base better equipped to cope with variability in supply[3].

There are barriers to effective climate change adaptation in food supply chains. First, the institutionalization of climate change mitigation is still an ongoing process. The majority of the literature dealing with climate change adaptation in food systems has focused on agricultural activities. This rather obvious, but limited, view of climate change adaptation in the food system has neglected other critical areas beyond the farm gate. Such areas include other activities of the food industry including processing and manufacturing, the supporting logistical system and infrastructure, the wholesale and retail sectors, and consumption and post-consumption activities.

According to Dekens and Bagamba (2014), sustainable supply chain development can only be achieved if all actors along the food supply chain work as a team to deal with climate risks. This indicates that actors must look beyond their own activities in the supply chain to consider how other upstream/downstream actors and activities may be affected by both climate risks and risk management decisions. While such a call for a comprehensive approach to climate change adaptation is not new, there is limited evidence of concrete applications. A value chain analysis is

[3] Local food and climate change-The role of community food enterprises; A publication commissioned by Making Local Food Work.

therefore proposed for a more integrated approach to climate change adaptation.

Three priority actions are recommended for developing a climate-resilient and all-encompassing coffee value chain in Uganda. First, decision-makers should improve networking and partnerships for climate change adaptation along the supply chain by strengthening existing platforms and structures at all levels and exploring the role of incentives (e.g., standards). Second, the government and the financial services industry need to develop new and flexible financial products to support climate-resilient and inclusive agro-value chains through capacity building and innovative public-private partnerships. Finally, investing in climate-resilient infrastructure such as roads, irrigation systems, storage facilities and telecommunications should remain a top priority to support agro-value chain development in a changing climate (Dekens and Bagamba, 2014).

In dealing with the challenges of climate change, governments should expand their export portfolio and promote a combined supply chain development to better link producers to markets and upsurge economic returns to smallholder farmers. Smith (2013) argues that supply chains have become a significant avenue to embed socioeconomic and environmental sustainability. This is due to the interaction among actors from production to consumption, and the necessity for inclusive solutions (i.e., through diverse levels and sectors). However, for supply chain development to be truly sustainable over the long-term, the exposure and vulnerability of supply chain activities to present and anticipated climatic changes must be decreased. In Uganda, an initiative on a climate-resilient coffee supply chain was piloted. The Ministry of Trade Industry and Cooperatives (MoTIC), Makerere University (MAK) and the International Institute for Sustainable Development (IISD) commenced a six-month pilot project to promote the assimilation of climate risks into agro-value chains. In particular, the parties developed a participatory process for discovering connections between the coffee supply chain and climate in Uganda as a means of showcasing the variety of climate-related responses and impacts associated with a particular agro-commodity.

A multi-stakeholder dialogue was employed to analyze climate risk along the coffee supply chain. The application of both qualitative and participatory approaches to climate risk analysis was founded on the knowledge that interpersonal dynamics and perceptions, coupled with trust associations, affect the way supply chain actors adapt (or not) to climate change (Eiser et al., 2012). This means that if some actors decline to sense climate risk as a principal challenge to their activities, it becomes improbable for them to take particular steps to deal with those risks. Although supply chain actors may have similar interests, it is possible there will be a deficiency in trust amongst them that will uphold the current situation and avert innovative developments that would lead to win-win solutions. The method also recognizes that all actors along the supply chain are interwoven, answering questions such as: How do actors at a specific level of the chain (e.g., farmers) influence, or not, other actors at the same levels (e.g., input suppliers) or other levels (e.g., exporters)? How do climate hazards combined with other non-climatic risks affect all value chain actors? Who is affected the most along the value chain? The multi-stakeholder dialogues used in the study mobilized 80 participants representing farm input suppliers, coffee farmers, traders, processors, exporters, and service providers at the production, transformation and marketing stages of the chain. The approach was based on the four-step process described in Table 13.5.

Table 13. 5: Overall four-step approach of the Climate-Resilient Coffee Value Chains (CRCV) initiative

Steps		Purpose	Methods
1.	Engagement (ongoing)	Raise awareness about the project and theme. Secure ownership and support for process.	Meetings
2.	Coffee value chain mapping	Mapping process, functions, actors, geography, resources, governance structure and vulnerability to climate risks.	Literature review Key expert meetings
3.	Climate risk analysis II (vertical integration)	Understanding climate impact and response chains across the different segments of the value chain (focus on risk transmission/distribution as perceived by key actors).	National multi-stakeholder workshop
4.	Reporting and dissemination (ongoing)	Elaborate and validate results, secure key stakeholder ownership of results.	Proposal Final Report and briefs

Source: Dekens and Bagamba (2014)

It has been established that climate risks adversely influence all actors along the supply chain, but in diverse means and to dissimilar extents. The effects of climate risks are experienced along every aspect of the coffee supply chain from production to export. All contributing actors voiced their concerns about the perceived effects of climate risks (principally drought, floods, and changing rainfall patterns) on their activities. Indirectly, climate risks cause a reduction in incomes owing to declines in business activities and services provision, and an upsurge in costs of business and service delivery at three stages: production (e.g. increased human labour), transformation (e.g. increased breakdown of processing equipment and machinery due to high moisture content of beans resulting from heavy rainfall) and distribution (e.g., increased vehicle repair costs due to road deterioration from heavy rainfall). Diminishing coffee quality due to climate risks influences coffee prices and the margins earned by the various actors, with effects on the economy's competitiveness in the global market.

It is not easy to generalize disparities among the supply chain actors regarding the level of exposure and vulnerability to climate risks. This is due to the complexity of the supply chain and the variety of actors in terms of their roles, sizes (in relation to their activities) and locations. Nevertheless, producers and processors usually appear to be more vulnerable to the effects of climate risks compared to traders, middlemen and exporters. The results from the Dekens and Bagamba (2014) study show that farmers and processors have a constrained diversification ability (e.g., the coffee processing plants are unable to process other commodities), feeble organizational capacity (e.g., very limited direct links between farmers and exporters) and they encounter an unfavorable policy environment (e.g., most actors along the supply chain pass on the losses incurred from climate hazards to farmers in the form of lower coffee bean prices). Thus, vulnerability to climate change tends to be more concentrated at the production end of coffee value chains.

Most actors are already making some efforts to minimize the negative impacts of climate hazards on their activities, but not all responses are sustainable. Most responses along the chain are oriented toward loss prevention, sharing or transferring the losses to other actors along the chain, and to a lesser extent on capacity building and awareness creation (see examples, Table 13.6).

Table 13. 6: Examples of responses to climate hazards along the coffee value chain

Main Response types	Examples
Loss prevention	Farmers modify coffee farming techniques; processors advance contract loans to farmers to prevent premature coffee sales.
Sharing or transferring the losses to other actors of the chains	Input suppliers transmit risks to the farmers by selling improper inputs; exporters transfer the risk to actors in the lower stages by offering lower prices for lower quality coffee; insurers increase premiums.
Capacity building and awareness creation	Some exporters support the adoption of good agricultural practices among farmers and the development of drought-resistant coffee varieties in partnership with research institutes.

Source: Dekens and Bagamba (2014)

A lack of communication, exchange of information and trust between and among actors along the value chain hampers climate adaptation.

Figure 13.5 depicts that all actors along the coffee supply chain are extremely inter-reliant, where actions at one level can impact actions at the other levels. Overall, there is a weak linkage between actors in the supply chain due to a lack of trust, which weakens their ability to respond in a sustainable way to the adverse effects of climate risks. This is partially because the supply chain is disjointed with many mediators amongst exporters and farmers. For example, farmers often blame input suppliers for the sale of adulterated chemicals, and input suppliers often complain that the sale of adulterated chemicals is exacerbated by farmers who work with traders who sell at low prices, and exporters perceived the main issue of low-quality coffee as mostly due to poor ethics and lack of discipline on the part of other actors in the value chain (Dekens and Bagamba, 2014).

Figure 13. 5: Simplified drought impact chain along the coffee value chain in Uganda

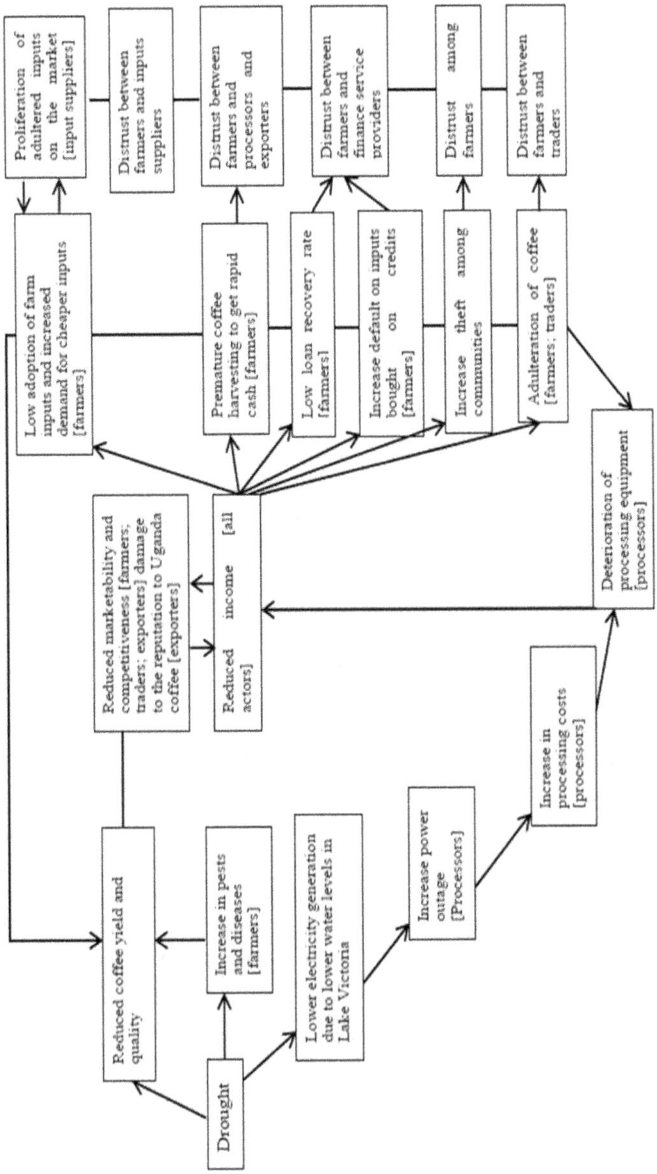

Source: Dekens and Bagamba (2014)

From the perspective of Dekens and Bagamba (2014), the principal obstacle perceived by every actor in the coffee supply chain is access to agricultural credit. This tends to increase the negative effects of climate risks on their activities. A reduction in income resulting from a decrease in yield pushes the farmer to purchase less expensive inputs and fall into the trap of low quality production, low efficiency and future declines in yields and incomes. The credits mostly have excessively high-interest rates (i.e., between 12 to 36 percent). Farmers are usually characterized by low savings and low loan repayments, which heighten the adverse effects of climate risks on their activities. Processors are often limited by the absence of start-up capital and insurance to cover losses and damage to their infrastructure from climate hazards. They also face high-interest rates charged on loans. Few national exporters are involved in coffee marketing due to a lack of initial capital requirements and credit access (due mainly to high-interest rates). In fact, over 60 percent of the production is being traded by five exporting firms. Due to the lack of access to credit, service providers keep their interest rates high and undermine the processes of loan recovery (Dekens and Bagamba, 2014).

It is believed that to enhance climate change adaptation along the supply chain, adaptation partnerships should be built among actors through intensification of prevailing platforms at all stages, and exploring the role of market incentives in reinforcing such activities (e.g., standards) would be a step in the right direction. Decision- and policy-makers should work to shorten the supply chain to promote direct linkage amongst farmers, processors and exporters via methods such as cooperatives, contract farming and strong farmer organizations. Strengthening the cooperatives and farmers organizations would increase the bargaining power of farmers. Efficient use of resources will also be achieved when structures and relationships are reinforced, and the organization and application of regulatory activities are enhanced along the whole supply chain (IISD, 2014). New and flexible financial products should be developed to support climate-resilient and comprehensive agro-value chains through capacity building and innovative public-private partnerships. Easing access to agricultural credit for climate change adaptation requires the development of weather index insurance and the incorporation of climate risk into prevailing financial mechanisms such as the Warehouse Receipts System. This demands climate change adaptation capacity building and awareness raising among financial service providers so that they comprehend the advantages of integrating climate adaptation into their activities and strategies. Investing in climate-resilient infrastructure such as roads, irrigation systems, storage facilities and telecommunications should remain a top priority to support agro-value chain development and build productive capacities in a changing climate. Improving infrastructure to support all coffee value chain actors and especially rural coffee growers has already been identified as a priority action in the 2012-2017 National Export Strategy for the Coffee Sector 2012–2017 (International Trade Center, 2012). For example, the lack of irrigation makes the coffee supply susceptible to drought, and heavy rains affect coffee transportation by destroying roads and bridges and increasing the beans' moisture content during transportation. These impacts can contribute to reductions in the farmers' yields and access to markets and service provisions, increases in the prices of inputs, and decreases in the prices of outputs, which all culminate in reduced incomes for farmers. However, any sustainable investment in physical infrastructure should ensure that the location, the composition and the design of the infrastructure, among other characteristics, account for the expected increase in frequency and intensity of climate hazards or any potential new climate hazards due to climate change.

Table 13.7 presents a roadmap of future research approaches regarding climate change and supply chain management.

Table 13. 7: Future research approaches regarding climate change and supply chain management

Supply chain management attributes	Logistical determinants	Methodological frameworks	Objectives
Supply chain configuration	• defining exposure of infrastructure and assets to climate-related stimuli. • contingency plans for managing disruptions.	• multidisciplinary approaches (mathematical programming bound with Life Cycle Analysis or Risk Management approaches).	• development of tailor-made adaptation frameworks. • enhancement of adaptive capacity. • defining causal linkages in the supply chain leading to maladaptation. • safeguarding continuity.
	• de-localisation of production. • trade-off approaches. • consolidation of nodes due to policy regulation.	• prescriptive approaches (i.e., optimization). • descriptive approaches (i.e., simulation). • exploratory research/case studies.	• accrued benefits from diversification of globalized networks.

	• decoupling point/postponement strategies. • product perspectives.	• exploratory research/case studies.	• efficiency and responsiveness maintained across the supply chain. • reconfiguration of supply chain networks. • achieving robustness and resilience.
Cross-functional drivers (e.g., sourcing, information)	• aspects of suppliers' engagement. • establishment of incentives among supply chain partners for efficiently managing carbon. • climate change-related aspects embodied within contracts. • liability insurance risks.	• game theory. • cost-benefit analysis. • comparative analysis. • surveys/interviews. • exploratory research/case studies.	• tracing of enablers and inhibitors for efficient carbon management. • drafting win-win sourcing agendas. • possible shift of costs. • capturing volatility of markets. • market and economic diversification.
	• sharing of information (e.g., carbon accounting systems). • Information technology and its role in managing carbon or ameliorating the impacts of climate change hazards.	• surveys/interviews. • exploratory research/case studies. • conceptual theory.	• understanding impediments to adapting to climate change. • efficient emissions auditing. • using climate change stimuli for leveraging global supply chains. • theoretical foundation of Supply Chain Management practices.
Alignment, coordination, integration and control mechanisms.	• internal and external functional co-ordination. • collaborative innovation and alignment (when drafting climate change agendas).	• empirical research. • case studies. • exploratory research. • conceptual theory.	• clarification of overarching responsibilities. • defining and prohibiting causal linkages in the supply chain leading to possible misalignment. • assessing coordination and

			integration discrepancies stemming from climate change drivers. • upgrading cooperation to engagement.
	• supply chain integration.	• benchmarking/best practices. • comparative assessments. • conceptual theory. • case studies.	• best adaptation practices (i.e., reactive or proactive). • cross-boundary synergy effects and value creation/sharing (i.e., vertical and horizontal).
	• performance.	• conceptual theory. • case studies.	• long-term economic performance.

Source: Dasaklis and Pappis (2013)

Social issues are adversely affected by climate change to differing degrees. It may be possible for consumers to place a higher value on adapted food products if they fully understand and appreciate how climate change adaptation addresses their broader societal concerns. Consumers can adapt to climate change and take action through their power of choice. However, such choices may not be made solely on the basis of climate change adaptation. Instead, consumers are more likely to adapt if there is an added benefit such as a cost saving, rather than a deliberate act to protect the environment and mitigate climate change risk. The choice of installing insulation in a home, for example, may be a decision based on reducing energy costs, supported by price reductions from policy initiatives, rather than to solely protect the household from extreme temperatures. This situation limits the ability to profit from adaptation, especially in the case where adaptation requires a significant investment (Lim-Camacho et al., 2014).

A step that can be taken by food businesses towards creating value from adaptation is improving their understanding of the beliefs and values of their market segments. For some businesses, some adaptation initiatives may be worth communicating to consumers if that target market is receptive to such messages. For example, businesses engaged in the production and marketing of food products (such as organics) that appeal to eco-friendly and eco-warrior consumers may benefit from exploring specific adaptation initiatives that are valued by their market segments.

On the other hand, businesses operating in highly segmented food markets such as the wine industry may benefit by positioning certain types of products in line with specific environmental values of their consumers. Overall, not all adaptation initiatives will translate to consumer value. For now, most adaptation initiatives will more specifically address internal business risks such as rising input costs, the sustainability of growing regions and reliability of supply. As such, selecting adaptation options on the basis of what consumers prefer is unlikely to be the optimal approach to a climate change adaptation strategy. It can, however, be beneficial to know which initiatives have potential to add value to product offerings and brand equity, and as such it may be worthwhile communicating this to consumers (Lim-Camacho et al., 2014).

References

ADVISEN LTD (2013) The Vulnerability of Global Supply Chain: The Importance of Resiliency in the Face of Systemic Risk. White Papers, 2013.

Allison, E. H., Beveridge, M. C. M., and van Brakel, M. (2009a) Climate change, small-scale fisheries and smallholder aquaculture. In: M. Culberg, ed. *Fish, Trade and Development.* pp. 73–87. Stockholm, Royal Swedish Academy of Agricultur e and Forestry (in press).

Allison, E. H., Perry, A., Badjeck, M-C., Adger, W. N., Brown, K., Conway, D., Halls, A. S., Pilling, G. M., Reynolds, J. D., Andrew, N. L., and Dulvy, N. K. (2009b). Vulnerability of national economies to the impacts of climate change on fisheries. *Fish and Fisheries* 10: 173-196.

Beermann, M. (2011) Linking corporate climate adaptation strategies with resilience thinking *Journal of Cleaner Production 19 836-842.*

Bonini, C. (2014) San Francisco disaster food system report: Analysis and recommendations to advance food resiliency for low-income and vulnerable populations. *Walter and Elise Haas Fund.* Retrieved from http://www.haassr.org/wp-content/uploads/2014/10/wehfDisasterFoodSystems.pdf (Accessed: 10 June 2017).

Carbon Disclosure Project (2011) *CDP Global 500 Report 2011. Accelerating Low Carbon Growth.* Retrieved October 02, 2012 from https://www.cdproject.net/en-us/results/pages/cdp-global-500-report-2011.aspx (accessed: 02 September 2017)

Christopher, M. (1998) *Logistics and supply chain management,* Ritman Publishing London.

City of New Orleans OHSEP (Office of Homeland Security and Emergency Preparedness), GCR and Associates, Dewberry and Davis LLC, Vissering Pardue and Associates, Perez, and Henry Consulting LLC. (2011, March 7) *Orleans Parish 2010 hazard mitigation plan update.* Retrieved from http://www.nola.gov/getattachment/Hazard-Mitigation/Hazards-and-Planning/Orleans-Parish-2010-Hazard-Mitigation-Plan-Final-032311.pdf (Accessed: 04 August 2017).

City of New York. (2013a) *A stronger, more resilient New York.* Retrieved from http://s-media.nyc.gov/agencies/sirr/SIRR_singles_Lo_res.pdf (Accessed: 10 August 2017)

Dasaklis, T. K., and Pappis, C. P. (2013) Supply chain management in view of climate change: An overview of possible impacts and the road ahead. *Journal of Industrial Engineering and Management,* 6(4), 1124-1138.

Daw, T., Adger, W. N., Brown, K., and Badjeck, M. C. (2009) Climate change and capture fisheries: potential impacts, adaptation and mitigation. In: K. Cochrane, C. De Young, D. Soto and T. Bahri (eds). Climate change implications for fisheries and aquacultur e: overview of current scientific knowledge. *FAO Fisheries and Aquaculture Technical Paper* No. 530. Rome, FAO. 151-212.

Dekens, J. and Bagamba, F. (2014) Climate Resilient Value Chains and Food Systems Briefing Note Series February 2014, promoting an integrated approach to climate adaptation: Lessons from the coffee value chain in Uganda, International Institute of Sustainable Development (Climate Change and Development Network).

Diamond, J. (2005) *Collapse: How societies choose to fail or succeed.* New York, NY: Viking.

Eiser, J.R., Bostrom, A., Burton, I., Johnston, D.M., McClure, J. Paton, D. (2012). Risk interpretation and Action: A conceptual framework for responses to natural hazards. *International Journal of Disaster Risk Reduction,* 1, 5–16.

Elver, H. (2015, August 5) *Interim report of the special rapporteur on the right to food.* Retrieved from http://www.un.org/en/ga/search/view_doc.asp?symbol=A/70/287 (Accessed: 02 September 2017).

Ericksen, P.J. (2008) Conceptualizing food systems for global environmental change research. *Global Environmental Change* 18, 234–245.

Eiser, J.R., Bostrom, A., Burton, I., Johnston, D., McClure, J., Paton, D., van der Pligt, J., and White, M. (2012) Risk interpretation and action: A conceptual framework for research in the context of natural hazards, *Int. J. Disaster Risk Reduct.*,doi:10.1016/j.ijdrr. 2012.05.002.

FAO (Food and Agriculture Organization of the United Nations) (2013) *FAO statistical yearbook 2013: World food and agriculture.* Rome, Italy: FAO. Retrieved fromhttp://www.fao.org/docrep/018/i3107e/i3107e.PDF

FAO (2014) LACC Project (http://www.fao.org/climatechange/laccproject/47741/en/) Accessed 10 December 2014.

Ferreira, S. (2012, October 29) Storm damages crops in Haiti, fueling food price woes. *Reuters.* Retrieved from http://www.reuters.com/article/storm-sandy-haiti-idUSL1E8LT5EX 20121029 (Accessed: 10 August 2017).

Ficke, A.D., Myrick, C.A. and Hansen, L.J. (2007) Potential impacts of global climate change on freshwater fisheries. *Rev. Fish Biol.* 17: 581-613.

Gledhill, R., Hamza-Goodacre, D., & Low, L.P. (2013). Business-not-as-usual: Tackling the impact of climate change on supply chain risk. *Resil. Win. Risk PWC*, 2, 15-20.

Gregory, P.J., Ingram, J.S.I., and Brklacich, M. (2005) Climate change and food security. Transactions of the Royal Society B: *Biological Sciences* 360, 2139, 2148.

Hawkes, C. (2009) Identifying innovative interventions to promote healthy eating using consumption-oriented food supply chain analysis. *Journal of Hunger & Environmental Nutrition, 4*, 336-356. doi:10.1080/19320240903321243.

Hepworth, N. and Goulden, M. (2008) Climate Change in Uganda: Un derstanding the Implications and Appraising the Response. LTS International, Edinburgh.

Humphrey, N. (2008) Potential impacts of climate change on U.S. Transportation. *TR News* 256, 21-24.

International Institute for Sustainable Development. (2014) *State of Sustainability Initiatives (SSI) Review, 2014*. Retrieved from: http://www.iisd.org/pdf/2014/ssi_2014.pdf (Accessed: 04 November, 2017).

International Trade Center. (2012) *Uganda national export strategy: Coffee sector export strategy update 2012–2017*. Geneva: ITC.

IPCC (2012) Managing the risk of extreme events and disasters to advance climate change adaptation. Special Report of the Intergovernmental Panel on Climate Change. Cambridge University Press, New York.

IPCC (Intergovernmental Panel on Climate Change). (2014) *Climate Change 2014: Synthesis Report. Contribution of Working Groups I, II and III to the Fifth Assessment Report of the Intergovernmental Panel on Climate Change* [Core Writing Team, R. K. Pachauri and L. A. Meyer (eds.)]. Geneva, Switzerland: IPCC. Available from http://www.ipcc.ch/report/ar5/syr/

Kambewa, E.V. (2007) Contracting for Sustainability, An analysis of the Lake Victoria-EU Nile perch chain.Wageningen: Wageningen Academic Publishers, 176.

Kaplinsky, R. and Morris, M. (2001) A Handbook for Value Chain Research. IDRC.

Kolk, A., and Pinkse, J. (2007) Towards strategic stakeholder management? Integrating perspectives on sustainability challenges such as corporate responses to climate change. *Corporate Governance*, 7(4), 370-378. http://dx.doi.org/10.1108/14720700710820452

Lim-Camacho, L., Ariyawardana, A., Lewis, G., and Crimp, S. (2014) Climate adaptation: What it means for Australian consumers. Consumer Survey - 2014 results. CSIRO, Australia. EP148832ISBN 978-1-4863-0512-4

London Assembly Economy Committee. (2015) *Weathering the storm: The impact of climate change on London's economy*. Retrieved from https://www.london.gov.uk/sites/default/ files/ Economy%20Committee%20Weathering%20the%20Storm_0.pdf (Accessed: 14 September 2017).

MacFadyen, G., and Allison, E. (2009) Climate Change, Fisheries, Trade and Competitiveness: Understanding Impacts and Formulating Responses for Commonwealth Small States . Commonwealth Secretariat,UK.

MacRae, R., and Donahue, K. (2013, June) *Municipal food policy entrepreneurs: A preliminary analysis of how Canadian cities and regional districts are involved in food system change.* Retrieved from http://capi-icpa.ca/pdfs/2013/Municipal_Food_Policy_ Entrepreneurs_ Final_Report.pdf (Accessed: 10 August 2017).

Maoh, H., Kanaroglou, P., and Woudsma, C. (2008) Simulation model for assessing the impact of climate change on transportation and the economy in Canada. *Transportation Research Record: Journal of the Transportation Research Board,* 2067, 84-92.

Metzger, M.J., Rounsevell, M.D.A., Acosta-Michlik, L., Leemans, R., and Schröter, D. (2006) The vulnerability of ecosystem services to land use change. *Agriculture, Ecosystems and Environment* 114: 69-85.

Miami-Dade County, NOAA (National Oceanic and Atmospheric Administration), and NACO (National Association of Counties). (2011) *Miami-Dade County roadmap for adapting to coastal risk: Vulnerability assessment & planning workshop results & recommendations.* Retrieved from https://www.mwcog.org/environment/climate/adaptation/Miami%20 Dade%20Vulnerability%20Assessment.pdf (Accessed: 04 September 2017).

Moussis, N. (2013) *Access to European Union: Law, economics, policies* Intersentia, Rixensart.

Müller, C., and Elliott, J. (2015) The Global Gridded Crop Model Intercomparison: Approaches, insights and caveats for modelling climate change impacts on agriculture at the global scale. In A. Elbehri (Ed.), *Climate change and food systems: Global assessments and implications for food security and trade* (pp. 29-59). Rome, Italy: Food and Agriculture Organization of the United Nations (FAO). Retrieved from http://www.fao.org/3/a-i4332e.pdf (Accessed: 14 November 2017).

New York City Economic Development Corporation. (2014, May 13) *Mayor's Office of Recovery and Resiliency announces request for proposals to study resiliency of city's food supply system.* Retrieved from http://www.nycedc.com/press-release/mayors-office-recovery-and-resiliency-announces-request-proposals-study-resiliency (Accessed: 04 August, 2017).

Oxfam (2009) *Suffering the Science: Climate Change, People, and Poverty,* Oxfam Briefing Paper No. 130, Oxfam International, http://policy-practice.oxfam.org.uk/publications/suffering-the-science-climate-change-people-and-poverty-114606 (Accessed: 04 August 2017).

Pauw, P., and Pegels, A. (2013) Private sector engagement in climate adaptation in least developed countries: An exploration. *Climate and Development, 5*(4), 257–267.

Porter, J.R., Xie, L., Challinor, A.J., Cochrane, K., Howden, S.M., Iqbal, M.M., Lobell, D.B., and Travasso, M.I. (2014) 'Food Security and Food Production Systems' in Climate Change 2014: Impacts, Adaptation, and

Vulnerability, Part A: Global and Sectoral Aspects, Contribution of Working Group II to the Fifth Assessment Report of the Intergovernmental Panel on Climate Change.

Product Policy Institute (2009) Products, Packaging and US Greenhouse Gas Emissions. Product Policy Institute, USA.

Reddy, V.R., and Anbumozhi, V. (2017)Managing Stranded Assets and Protecting Food Value Chains from Natural Disasters. ERIA Discussion Paper Series: ERIA-DP-2017-01.

Smith, T.M. (2013) Climate change: Corporate sustainability in the supply chain. *Bulletin of the Atomic Scientists*, 69, 43–52.

Song, J.S., and Lee, K.M. (2010) Development of a low-carbon product design system based on embedded GHG emissions. *Resources, Conservation and Recycling*, 54(9), 547-556.

Stathers, T., Lamboll, R., and Mvumi, B.M. (2013) Postharvest agriculture in changing climates: Its importance to African smallholder farmers. *Food Security*, 5, 361–392.

Stern Review. (2006) *Stern Review on the Economics of Climate Change. HM Treasury, CabinetOffice.* Retrieved December 3, 2012 from http://webarchive.nationalarchives.gov.uk/+/http://www.hmtreasury.gov.uk/stern_review_report.htm (Accessed: 10 November 2017).

Sussman, F., and Freed, J. (2008). *Adapting to climate change: A business approach. Pew Center on Global Climate Change.* Retrieved October 12, 2012 from http://www.pewclimate.org/docUploads/Business-Adaptation.pdf

Tacoli, C., Bukhari, B., and Fisher, S. (2013, March). *Urban poverty, food security and climate change* (Human settlements working paper no. 37). London, UK: International Institute for Environment and Development (IIED). Retrieved from http://pubs.iied.org/pdfs/ 10623IIED.pdf (Accessed: 04 August 2017).

USAID (2011). Enhancing Food Security in a Changing Climate: Adaptation Options for Smallholder and Microenterprise Development Initiatives. USAID FIELD BRIEF No. 15, Washington D.C, USA.

Wheeler, T. (2015) Climate change impacts on food systems and implications for climate-compatible food policies. In A. Elbehri (Ed.), *Climate change and food systems: Global assessments and implications for food security and trade* (pp. 315-336). Rome, Italy: Food and Agriculture Organization of the United Nations (FAO). Retrieved from http://www.fao.org/3/a-i4332e.pdf (Accessed: 15 December 2015).

White, A. (2007). *A New Wave of SCM Innovation Must Address Climate Change Concerns. Gartner.* ID Number: G00149710. Retrieved October 10, 2012 from http://www.lexian.com.au/upload/Gartner%20New%20Wave%20of%20SCM%20Innovation.pdf

Wolff, M. and Comerford, C. (2014) *San Francisco climate & health profile.* Retrieved from http://www.sfclimatehealth.org/wpcontent/uploads/2015/01/SFDPH_ClimateHealthProfile_FinalDraft.pdf

World Bank. (2015, April 1) *Food security: Overview.* Retrieved from http://www.worldbank. org/en/topic/foodsecurity/overview#1 (Accessed: 04 September 2016).

WorldFish Center. (2007) The Threat to Fisheries and Aquaculture from Climate Change. WorldFish Policy Brief No.1, The WorldFish Center, Penang, Malaysia. Available at: http://www.worldfishcenter.org/resource_center/ClimateChange2.pdf (accessed December 20, 2011).

Yang, H., and Zhang, J. (2011) *The strategies of advancing the cooperation satisfaction among enterprises based on low carbon supply chain management.* Paper presented at the 2010 International Conference on Energy, Environment and Development, Kuala Lumpur.

Zake, J. (2009) A Guide to the Climate Change Effects, Challenges and Options in the Lake Victoria region. Uganda Coalition for Sustainable Development, Kampala.

CHAPTER 14

Conclusions and recommendations

John K. M. Kuwornu

14.1 Conclusions

Natural and man-made systems, infrastructure, food supply chains and entire economies of countries have been adversely affected by climate change. Uncertainties that surround climate change necessitate a variety of comprehensive responses and policy strategies at the local, regional, national and global levels. Mitigation and adaptation have been two basic response strategies emphasized by the United Nations Framework Convention on Climate Change (UNFCCC). This book focuses on the vulnerability of and adaptation to climate change by food supply chain actors, with an emphasis on smallholder farmers.

Farming households' crop and livestock production activities have been vulnerable to the effects of climate change, as revealed in the studies presented in this book in the following ways:

The study of the vulnerability of smallholder women rice farmers to climate variability in the Savelugu/Nanton and West Mamprusi Districts in the Northern Region of Ghana revealed that these farmers are vulnerable to natural disasters and climate variability, socio-demographic factors, health, water, social network, and food. Furthermore, these households are vulnerable in terms of their adaptive capacity, exposure and sensitivity.

The study of the Bui Division of Northwest Cameroon revealed that rainfall has been fluctuating, leading to considerable reductions in the output of the main food crops (i.e., maize, beans and solanum potato) on which smallholder farmers' livelihoods depend. At the household and village levels, adaptation strategies include mitigating the effects of weather through soil and water conservation, micro-irrigation, multiple farms, inter/relay cropping, mixed farming, organic manure usage and drought-resistant crops. However, adoption of these adaptation strategies is constrained by factors such as limited know-how, gender imbalances in land and other resources, cultural constraints, and institutional failures. Again, the low adaptive capacity of most rural households is due to extreme poverty and an increasingly frequent irregular onset of the first rains. The full implications of climate change for food crop production are, however, currently not clearly understood by most rural populations. Many farmers still associate crop failures and food security threats to superstition. At the more local level, the poor often cannot adopt diversification as an adaptive strategy and often have very limited options available to them.

A study using cross-sectional data from 704 farm households in Northern Togo showed that climate change is well recognized, while a lack of capital and limited information and knowledge about adaptation were factors that have constrained farmers to adapt to climate change. Cultural beliefs and modern approaches have been suggested as adaptation measures. The shift in the rainy season, declines in livestock yields and food shortages have led farmers to opt for adaptation to climate change. Farm household socioeconomic characteristics such as the farm household location, gender of household head, number of active household members, land devoted to cereal production, access to credit and climate information, government assistance, off-farm income and extension services were determinants of the farmers' climate change adaptation responses. The interest rate on credit has a negative effect on farmers' decisions regarding adaptation to climate change.

The study regarding the role of social capital on climate change adaptation mechanisms among 225 smallholder farmers in the Central Region of Ghana revealed that these farmers make use of the opportunities of embedded social networks to adapt to the changing climate. The smallholder farmers have networked with people including friends and acquaintances of different socioeconomic status to promote climate change adaptation. Social relationships not only facilitate the flow of information but also influence behavioural change to manage risk. The study found that smallholder farmers have embraced both Indigenous and introduced adaptation strategies. Thus, farmers developed their own Indigenous adaptation strategies due to their experience with the changing climate. The introduced adaptation strategies are those developed through scientific research.

The study using cross-sectional data from 260 households to examine the factors influencing their vulnerability to climate change revealed that most farmers interviewed were highly vulnerable to climate change and variability. It was also revealed that whereas education, age, off-farm income, participation in decision-making, formal extension contacts and farmland ownership have a significant negative influence on women rice farmers' vulnerability to climatic stress; farm size, membership of a farmer-based group, hired labour, access to tractor services and noticing climate variability have a positive influence on farmers' vulnerability to climate variability.

The study that identified 241 smallholder farmers' perceptions about climate change and assessed the impact of participation in climate change-related projects on resilience to climatic shocks, maize yield, farm income, and off-farm income revealed that there was an insignificant difference between participants and non-participants in terms of the outcome variables. This notwithstanding, participation in climate change projects has much potential for positive impacts on the farmers' livelihoods.

The study regarding the vulnerability of 150 smallholder maize farming households to climate variability in Ayensuano District and Yilo Krobo Municipality in the Eastern Region of Ghana revealed that households in Ayensuano District are more vulnerable to the impacts of climate variability than those in Yilo Krobo Municipality.

The study regarding the adaptation strategies of 260 women rice farmers in the Savelugu/Nanton Municipality and West Mamprusi District in the Northern Region of Ghana revealed that soil-related strategies (i.e., loosening of soil, creating bunds or drainage channels, and organic fertilizer) are these farmers' predominant Indigenous adaptation strategy to adapt to climate variability. On the other hand, improved variety strategies (i.e., high yielding varieties, early maturing varieties and drought/flood-resistant varieties) are their most used research-based adaptation strategies. The choice of Indigenous adaptation strategies among women rice farmers is significantly determined by ownership of farmland, farmer-farmer extension contact, distance to market, access to informal credit from friends and relatives, farming experience, farm size, number of economic activities engaged in by a farmer, and family labour. The factors influencing their adoption of research-based adaptation strategies are the age of the farmer, participation in decision-making, farming experience, family labour, access to tractor services, access to formal credit, noticing decreased rainfall and increased temperatures, membership of the farmer-based organization, access to weather information and farmers' location.

The study that explored livestock farmers' Indigenous perceptions of climate change, and analyzed the factors that influenced the adoption of Indigenous climate change adaptation strategies among livestock farmers in the North-Western Region of Ghana revealed that Indigenous adaptation strategies were identified as: feed-, health-, house- and breed-related strategies. Further empirical results showed that noticing climate variability, number of extension contacts, time taken to reach the market, access to credit, flock size, household size and the number of livestock types diversified were factors influencing the decisions by smallholder livestock farmers to adopt Indigenous adaptation strategies.

The study on gender-based climate change impacts and adaptation strategies among smallholder farmers in northern Ghana revealed that immediate impacts include reductions in crop yields by 62.7% for men and 41.8% for women farmers, and disruption in the annual cropping calendar for 25.4% of men and 30.9% of women farmers. Considering the impacts of climate change on future agricultural systems with adaptation, the results show that aggregate percentage impacts as a result constitute about 35%, while aggregate gains, losses and net losses as a percentage of average net per farm revenues are 17%, 35% and 15%, respectively.

The study that explored 241 smallholder farmers' perceptions of climate change and assessed the impact of participation in climate change-related projects on resilience to climatic shocks, maize yield, farm income, and off-farm income in Ghana's Northern Region, and the Afram and Accra Plains, revealed that there is an insignificant difference between participants and non-participants in terms of the outcome variables. Nevertheless, participation in climate change projects has much potential to positively impact farmers' livelihoods. The study also revealed that climate change projects promote adaptation strategies such as irrigation, tree planting, use of inoculum as a soil management practice and capacity building. Therefore, farmers who apply the strategies should be expected to improve their resilience. The lack of impact of these projects probably points to their unsustainability arising from limited geographical coverage, and the duration and depth of the projects. Also, projects that build the capacity of farmers but do not enhance access to inputs are not likely to achieve expected results.

The study that examined 150 smallholder maize farmers' constraints to the adoption of climate change adaptation strategies in the Brong-Ahafo Region of Ghana revealed that the lack of access to credit, lack of a ready market for produce and poor access to extension services were the three most pressing constraints working against the farmers' adaptation to climate change.

Nowadays, consumers demand products and services with low prices, high quality and that are produced under environmentally-friendly conditions and energy-efficient practices (Song and Lee, 2010). Businesses declining to incorporate this message into their own supply chain strategy or that are unwilling to address climate change issues will eventually be "punished" by environmentally-sensitive consumers. Also, given it is considered that supply chains and not companies compete (Christopher, 1998), it is clear that supply chain management will play an important part in meeting new consumers' requirements in view of climate change. It is also clear that supply chain management is the appropriate strategy that will enable a firm to take full advantage of the opportunities of climate change while minimizing threats. In the low carbon era we have entered, it seems that low carbon supply chain management would be the driver for enhancing competitiveness (Yang and Zhang, 2011). Regulatory, physical and market drivers stemming from climate change create both threats and opportunities for supply chains. Consequently, several implications for supply chain strategic and operational management must be adequately addressed for the continuous and efficient functioning of food supply chains in the context of climate change.

14.2 Recommendations

The studies presented in this book provide the following recommendations.

The communities in Savelugu/Nanton Municipality should urgently be provided with health centers, sources of potable water and household latrines by their Local Assembly. Communities in West Mamprusi District should be given priority in local governmental and non-governmental social intervention projects to enable farmers in the district to diversify their livelihoods and reduce youth out-migration as a means of adapting to climate variability.

Climate change adaptation policies should be focused on extension services for better dissemination of knowledge about adaptation strategies, for increased accessibility of weather information, and to provide credit at a special rate. The extension services should also identify and incorporate synergies between traditional and modern beliefs about climate change adaptation.

It is recommended that adult education programs should be vigorously pursued to improve farmers' literacy. Also, regular engagement between farmers and agricultural extension agents is necessary to train farmers on appropriate agronomic practices and other non-farm income generation skills to diversify their livelihoods. Farmers should be encouraged to diversify the livestock species they rear.

Identifying and effectively addressing the needs of women farmers requires delineating climate change adaptation policies in a gender-disaggregated way. There is a need to prioritise women in designing adaptation strategies, and thus contribute to the gender policy discourse. To foster adoption of adaptation strategies to climate variability, it is recommended that governments should assist women farmers to form groups and be trained on modern agronomic practices, given input credits and access to tractor services, to link with the appropriate markets, and be pre-notified of expected seasonal weather patterns for informed farming decision-making.

There is a need for financial policies that will facilitate access to credit by smallholder maize farming households in Techiman and Wenchi Municipalities in Ghana's Brong Ahafo Region. This would encourage the households to adopt climate change adaptation strategies through innovative production practices, thereby enabling these farmers to become resilient to climatic shocks. This recommendation is based on the empirical results that the lack of access to credit is a major constraint to the maize farming households' adoption of climate change adaptation strategies in both municipalities. The Municipal Assemblies, Meteorological Agencies and NGOs should inform smallholder maize farmers in Techiman and Wenchi Municipalities about impending natural disasters such as flood, drought, and pest outbreaks. This is based on the empirical results that a majority of the respondents in both municipalities did not receive early warnings about impending disasters. Finally, the government should collaborate with agricultural extension agents and other development agencies to play their respective roles in assisting smallholder maize farmers to cope with the negative impacts of climate change.

In general, the implementation of climate change adaptation projects for smallholder farmers should be combined with the provision of financial and input credits for these farmers, to enable them to apply the appropriate strategies to enhance their resilience to climate change. Governments, and national and international organizations all have crucial roles to play in this regard.

The vulnerability of food supply chains to climate change depends on the following: the nature and degree to which food supply chains are exposed to climate change (i.e., exposure); the degree to which food supply chains are sensitive to climate change (i.e., sensitivity); and the ability of food supply chains to cope with actual or expected climate stresses (i.e. adaptive capacity). The combined effect of the exposure and sensitivity yields the potential impact of the climate change (i.e., the impact that may occur without taking into account planned climate change adaptation measures).

The vulnerability of food supply chains to climate change has implications for supply chain management at both the strategic and operational levels. At the strategic level, climate change has implications for the selection of suppliers, total mileage for transportation/distribution activities, improvements in energy efficiency, low energy intensive assets (e.g., plants, warehouses), lower energy consuming products and services, adoption of carbon labelling practices, and the re-configuration of supply chain networks. At the operational level, the vulnerability of food supply chains to climate change has implications for production planning, scheduling, inventory planning, control and management, material requirement planning, and insurance costs.

Overall, an inclusive climate-adaptive food supply chain initiative would comprise a variety of strategies, to be carried out according to the present and future vulnerability of each link in the chain. Resistance strategies to cushion the food supply chain could include larger climate-controlled storage facilities, improved crop cleaning and drying, better storage to avoid losses to pests, additional processing to reduce spoilage, and more robust packaging to withstand difficult transport conditions. Resilience strategies could include enhancing water storage and the efficiency of irrigation methods, and the selection of seed varieties that maintain yields under current climate variability but decrease the likelihood of crop failure during extremely wet or dry years and whenever credit or fertilizers are in short supply.

Along entire food supply chains, little has been done to support climate change adaptation beyond the production level. This is presumably due to a combination of factors such as: a lack of understanding of climate change adaptation (including the distinction between adaptation and mitigation, and between adaptation and sustainable development); the channeling of climate-related finance to mitigation initiatives and particularly in emerging economies such as China and Brazil, rather than in the least-developed countries (Pauw and Pegels, 2013); the lack of data and information about future impacts of climate change; inherent uncertainties around the information that is currently available; and the lack of economic and/or costs-benefit analyses on the impacts and associated risk management strategies. As a result, little is known about the impacts of climate change on food supply chains at the post-harvest stage, including the wider macro-economic implications for a country's competitiveness, especially in the developing world (Stathers et al., 2013). Nonetheless, to secure sustainable investments in value chain development, governments and other actors need to ensure that climate risks are managed not just at the production level, but also throughout entire food supply chains from production to marketing. Sustainable supply chain development can only be achieved if all actors along the food supply chains work as a team to deal with climate risks. This indicates that actors must look beyond their own activities in the supply chain to consider how other upstream/downstream actors and activities may be affected by both climate risks and risk management decisions. Investing in climate-resilient infrastructure such as roads, irrigation systems, storage facilities and telecommunications should remain a top priority to support food supply chain development and build productive capacities in a changing climate.

References

Christopher, M. (1998) *Logistics and supply chain management,* Ritman Publishing London.

Pauw, P., and Pegels, A. (2013) Private sector engagement in climate adaptation in least developed countries: An exploration. *Climate and Development, 5*(4), 257–267.

Song, J.S., and Lee, K.M. (2010) Development of a low-carbon product design system based on embedded GHG emissions. Resources, Conservation and Recycling, 54(9), 547-556.

Stathers, T., Lamboll, R., and Mvumi, B.M. (2013) Postharvest agriculture in changing climates: Its importance to African smallholder farmers. *Food Security,* 5, 361–392.

Yang, H., and Zhang, J. (2011) The strategies of advancing the cooperation satisfaction among enterprises based on low carbon supply chain management. Paper presented at the 2010 International Conference on Energy, Environment and Development, Kuala Lumpur.

INDEX

A

C

F

G

H

I

P

T

U

V

W

X

Y

Z

www.ingramcontent.com/pod-product-compliance
Lightning Source LLC
LaVergne TN
LVHW050804220926
848485LV00013B/1428/J
* 9 7 8 1 6 2 2 7 3 2 6 5 4 *